AF801275

New Frontiers in Regional Science: Asian Perspectives

Volume 70

Editor-in-Chief

Yoshiro Higano, University of Tsukuba, Tsukuba, Ibaraki, Japan

This series is a constellation of works by scholars in the field of regional science and in related disciplines specifically focusing on dynamism in Asia.

Asia is the most dynamic part of the world. Japan, Korea, Taiwan, and Singapore experienced rapid and miracle economic growth in the 1970s. Malaysia, Indonesia, and Thailand followed in the 1980s. China, India, and Vietnam are now rising countries in Asia and are even leading the world economy. Due to their rapid economic development and growth, Asian countries continue to face a variety of urgent issues including regional and institutional unbalanced growth, environmental problems, poverty amidst prosperity, an ageing society, the collapse of the bubble economy, and deflation, among others.

Asian countries are diversified as they have their own cultural, historical, and geographical as well as political conditions. Due to this fact, scholars specializing in regional science as an inter- and multi-discipline have taken leading roles in providing mitigating policy proposals based on robust interdisciplinary analysis of multifaceted regional issues and subjects in Asia. This series not only will present unique research results from Asia that are unfamiliar in other parts of the world because of language barriers, but also will publish advanced research results from those regions that have focused on regional and urban issues in Asia from different perspectives.

The series aims to expand the frontiers of regional science through diffusion of intrinsically developed and advanced modern regional science methodologies in Asia and other areas of the world. Readers will be inspired to realize that regional and urban issues in the world are so vast that their established methodologies still have space for development and refinement, and to understand the importance of the interdisciplinary and multidisciplinary approach that is inherent in regional science for analyzing and resolving urgent regional and urban issues in Asia.

Topics under consideration in this series include the theory of social cost and benefit analysis and criteria of public investments, socio-economic vulnerability against disasters, food security and policy, agro-food systems in China, industrial clustering in Asia, comprehensive management of water environment and resources in a river basin, the international trade bloc and food security, migration and labor market in Asia, land policy and local property tax, Information and Communication Technology planning, consumer "shop-around" movements, and regeneration of downtowns, among others.

Researchers who are interested in publishing their books in this Series should obtain a proposal form from Yoshiro Higano (Editor in Chief, higano@jsrsai.jp) and return the completed form to him.

Keiko Nakayama

Eigenvalue Problem and Nonlinear Programming Problem

For Economic Studies

Springer

Keiko Nakayama
Chukyo University Faculty of Economics
Nagoya, Aichi, Japan

School of Economics, Chukyo University
Nagoya, Aichi, Japan

ISSN 2199-5974 ISSN 2199-5982 (electronic)
New Frontiers in Regional Science: Asian Perspectives
ISBN 978-981-99-2942-9 ISBN 978-981-99-2943-6 (eBook)
https://doi.org/10.1007/978-981-99-2943-6

© Springer Nature Singapore Pte Ltd. 2024
This work is subject to copyright. All rights are reserved by the Publisher, whether the whole or part of the material is concerned, specifically the rights of translation, reprinting, reuse of illustrations, recitation, broadcasting, reproduction on microfilms or in any other physical way, and transmission or information storage and retrieval, electronic adaptation, computer software, or by similar or dissimilar methodology now known or hereafter developed.
The use of general descriptive names, registered names, trademarks, service marks, etc. in this publication does not imply, even in the absence of a specific statement, that such names are exempt from the relevant protective laws and regulations and therefore free for general use.
The publisher, the authors, and the editors are safe to assume that the advice and information in this book are believed to be true and accurate at the date of publication. Neither the publisher nor the authors or the editors give a warranty, expressed or implied, with respect to the material contained herein or for any errors or omissions that may have been made. The publisher remains neutral with regard to jurisdictional claims in published maps and institutional affiliations.

This Springer imprint is published by the registered company Springer Nature Singapore Pte Ltd.
The registered company address is: 152 Beach Road, #21-01/04 Gateway East, Singapore 189721, Singapore

If disposing of this product, please recycle the paper.

Preface

Students of economics are well aware that there is a close relationship between nonlinear programming and dynamic theory. Linear programming is insufficient for analyzing the complex dynamics of real economies, and the use of a nonlinear dynamic model is essential. In this publication, therefore, systematically summarizing nonlinear programming problems is the main subject. These have various economic applications. A secondary topic is Frobenius's theorem related to nonlinear transformation. This is grounded in the fact that Frobenius roots in nonlinear transformations can be characterized as upper limits on real numbers that bring about non-negative solutions to nonlinear inequalities and that Frobenius's theorem gives us useful information when trying to solve stability problems.

In this book, while trying to apply them to economic theory as much as possible, I have attempted to describe them systematically and self-contained to the fullest extent possible and summarize my modest research results until now in this area.

In this publication, I have made significant additions and revisions to the book entitled "Nonlinear Programming and Nonlinear Programming Eigenvalue Problems," which was published in September 1995 as a research work in the Chukyo University Economics Research Book Series No. 5, by Keiso Shobo. There are several reasons for this first new publication in 20 years: nonlinear models are still relevant as a tool in economic issues today, and many areas of the original text required revision. In this publication, I have changed the title somewhat from the Japanese version to "Eigenvalue Problem and Nonlinear Programming Problem."

The definition of the eigenvalue of a matrix is as follows: "When a linear transformation using a square matrix is represented by a simple constant multiple, the constant is called the eigenvalue of the square matrix". The readers just may not understand what this is all about. However, in such multi-dimensional economic models as economic growth, business cycles, and input-output analysis, the eigenvalue of the matrix plays an important role in checking the conditions for stability of paths or feasibility of equilibrium solutions, although the meaning of the eigenvalue is slightly different from the above-mentioned definition. The eigenvalue problem dealt with in this book refers to various problems when applying the eigenvalues and eigenvectors of a matrix to theoretical economics. Specifically speaking, this book includes the condition for achieving balanced growth in the von Neumann model and the condition on the input coefficients for obtaining a positive equilibrium output given a positive final demand in the static input-output analysis. And the Nonlinear Programming Problem is an optimization problem in which nonlinear expressions are included in the objective function or constraints. Both these have been researched and developed in various fields beyond mathematics, engineering, and economics. In this publication, therefore, I aim to describe both the close relationship between the two problems and introduce useful reference material in the final chapter of this book for those readers who wish to study this further.

I would like to express my deep gratitude to Mr. Shozo Miyamoto of Keiso Shobo for allowing me to contribute to this work. Additionally, I would like to thank Professor Emeritus Yoshiro Higano from Tsukuba University, Professor Kiyoshi Fujikawa from Aichi Gakuen University, and Professor Emeritus Akio Matsumoto from Chuo University for their many suggestions during the writing process, as well as Professor Emeritus Nozomu Matsubara who offered encouragement and advice when I wrote the initial version, and Mr. Yutaka Hirachi editor of Springer Japan, for his support while I was writing this book.

Nagoya, Aichi, Japan
January 2024

Keiko Nakayama

Contents

Chapter 1
Possibilities in Linear Inequality Systems

1.1 Introduction

Knowledge relating to the severability of linear inequality systems is important in discussing constrained optimization problems, which is one of the primary topics of this book, as shown in subsequent chapters. Thus, in this chapter, the basic theorems relating to the solvability of many forms of linear inequality systems will be systematically discussed and proven.[1]

1.2 Tucker's Theorem

In this chapter, Tucker's theorem, which is useful in deriving multiple theorems relating to the solvability of linear inequality equation systems, is shown.

[Lemma 1.1] Let $\boldsymbol{P}$ be a given $m \times n$ matrix. Herein, let the set of solutions of the following inequality equation systems

$$\text{I} \quad [\boldsymbol{Px} \geqq \boldsymbol{0}]$$

and

[1] The names of theorems in this chapter are mainly based on Mangasarian (1969).

© Springer Nature Singapore Pte Ltd. 2024
K. Nakayama, *Eigenvalue Problem and Nonlinear Programming Problem*,
New Frontiers in Regional Science: Asian Perspectives 70,
https://doi.org/10.1007/978-981-99-2943-6_1

$$\text{II} \quad [\boldsymbol{P}^t\boldsymbol{\eta} = \boldsymbol{0}, \ \ \boldsymbol{\eta} \geqq \boldsymbol{0} \ ]$$

be $\boldsymbol{S} = \left\{ \begin{pmatrix} \boldsymbol{x} \\ \boldsymbol{\eta} \end{pmatrix} \in \boldsymbol{R}^{n+m} | \text{I and II} \right\}$.[2] There exists $\boldsymbol{x}^1 \in \boldsymbol{R}^n$ and $\boldsymbol{\eta}^1 \in \boldsymbol{R}^m$ that belong to $\boldsymbol{S}$ and

$$[\boldsymbol{p}_1\boldsymbol{x}^1 + (\boldsymbol{\eta}^1)_1 > 0]$$

holds, where $\boldsymbol{p}_1 = [p_{11}, \ldots, p_{1n}]$ is the first row of $\boldsymbol{P}$ and $(\boldsymbol{\eta}^1)_1$ is the first element of $\boldsymbol{\eta}^1$.

(Proof) Prove using the induction method in respect of $m \times n$ matrix $\boldsymbol{P}$'s matrix m.

(A) If we let $m = 1$, $\boldsymbol{P} = \boldsymbol{p}_1$.

In response to $\boldsymbol{p}_1 = \boldsymbol{0}$ or $\boldsymbol{p}_1 \neq \boldsymbol{0}$, $\begin{pmatrix} \boldsymbol{p}_1{}^t \\ 1 \end{pmatrix}$ or $\begin{pmatrix} \boldsymbol{p}_1{}^t \\ 0 \end{pmatrix}$ are the corresponding sought-after solutions.

(B) Let $m \geqq 2$. To demonstrate the two stages induction method, split the $(m + 1) \times n$ matrix $\overline{\boldsymbol{P}}$ into

$$\bar{\boldsymbol{P}} = \begin{bmatrix} \boldsymbol{P} \\ \boldsymbol{p}_{m+1} \end{bmatrix} = \begin{bmatrix} \boldsymbol{p}_1 \\ \vdots \\ \boldsymbol{p}_m \\ \boldsymbol{p}_{m+1} \end{bmatrix}.$$

If the assumptions of the induction method are applied to $\boldsymbol{P}$, there exists $\boldsymbol{x}^1$ and $\boldsymbol{\eta}^1$ that satisfy

$$\boldsymbol{P}\boldsymbol{x}^1 \geqq \boldsymbol{0},\ \boldsymbol{P}^t\boldsymbol{\eta}^1 = \boldsymbol{0},\ \boldsymbol{\eta}^1 \geqq \boldsymbol{0} \text{ and } \boldsymbol{p}_1\boldsymbol{x}^1 + (\boldsymbol{\eta}^1)_1 > 0. \tag{1.1}$$

Next, consider whether $\boldsymbol{p}_{m+1}\boldsymbol{x}^1 \geqq \boldsymbol{0}$ in respect of $\boldsymbol{x}^1$.

(B-1) Letting $\boldsymbol{p}_{m+1}\boldsymbol{x}^1 \geqq \boldsymbol{0}$ in respect of $\overline{\boldsymbol{\eta}}^t = \left((\boldsymbol{\eta}^1)^t, 0\right)$ and $\boldsymbol{x}^1$ by (1.1),

$$\overline{\boldsymbol{P}}\boldsymbol{x}^1 \geqq \boldsymbol{0}, \boldsymbol{P}^t\overline{\boldsymbol{\eta}} = \boldsymbol{P}^t\boldsymbol{\eta}^1 = \boldsymbol{0}, \overline{\boldsymbol{\eta}} \geqq \boldsymbol{0} \text{ and } \boldsymbol{p}_1\boldsymbol{x}^1 + (\boldsymbol{\eta}^1)_1 > 0.$$

Thus, the sought after conclusion has been obtained in respect of $\overline{\boldsymbol{P}}$.

(B-2) When $\boldsymbol{p}_{m+1}\boldsymbol{x}^1 < 0$, define $\lambda_j = -\dfrac{\boldsymbol{p}_j\boldsymbol{x}^1}{\boldsymbol{p}_{m+1}\boldsymbol{x}^1} \geqq 0$ $(j = 1, \ldots, m)$, and set $m \times n$ matrix $\boldsymbol{Q}$ to be

---

[2] $\boldsymbol{S} = \left\{ \begin{pmatrix} \boldsymbol{x} \\ \boldsymbol{\eta} \end{pmatrix} \in \boldsymbol{R}^{n+m} | \boldsymbol{P}\boldsymbol{x} \geqq \boldsymbol{0}, \boldsymbol{P}^t\boldsymbol{\eta} = \boldsymbol{0}, \ \ \boldsymbol{\eta} \geqq \boldsymbol{0} \right\}$. $\boldsymbol{x} = \boldsymbol{0}$ and $\boldsymbol{\eta} = \boldsymbol{0}$ clearly satisfy I and II, and thus $\boldsymbol{S} \neq \phi$.

$$Q = \begin{bmatrix} q_1 \\ \vdots \\ q_m \end{bmatrix} = \begin{bmatrix} p_1 + \lambda_l p_{m+1} \\ \vdots \\ p_m + \lambda_m p_{m+1} \end{bmatrix}$$

concerning all $j \in \{1, \ldots, m\}$,

$$q_j x^1 = p_j x^1 + \lambda_j p_{m+1} x^1 = 0,$$

i.e.,

$$Qx^1 = \mathbf{0}.$$

Herein, if the assumptions in the induction method are applied to Q, there exist v^1 and u^1 and

$$Qv^1 \geqq \mathbf{0},\ Q^t u^1 = \mathbf{0},\ u^1 \geqq \mathbf{0} \text{ and } q_1 v^1 + (u^1)_1 > 0 \tag{1.2}$$

are satisfied. Finally, if we set $(\overline{u}^1)^t = \left((u^1)^t, \sum_{j=1}^m \lambda_j (u^1)_j\right)$, $w^1 = v^1 - \frac{p_{m+1} v^1}{p_{m+1} x^1} x^1$, by (1.2) and definition of w^1,

$$\overline{u}^1 \geqq \mathbf{0}$$

and

$$p_{m+1} w^1 = p_{m+1} v^1 - p_{m+1} v^1 = 0.$$

Furthermore, from (1.2) and the direct calculation, the following is obtained.

$$\begin{aligned} \overline{P}^t \overline{u^1} &= P^t u^1 + p_{m+1}^t \textstyle\sum_{j=1}^m \lambda_j (u^1)_j \\ &= \left(Q^t u^1 - \textstyle\sum_{j=1}^m \lambda_j p_{m+1}^t (u^1)_j\right) + p_{m+1}^t \textstyle\sum_{j=1}^m \lambda_j (u^1)_j \\ &= Q^t u^1 \\ &= 0, \end{aligned}$$

$$\begin{aligned} \overline{P} w^1 &= \begin{bmatrix} P \\ p_{m+1} \end{bmatrix} w^1 \\ &= \begin{bmatrix} P w^1 \\ 0 \end{bmatrix} \\ &= \begin{bmatrix} (q_1 - \lambda_1 p_{m+1}) w^1 \\ (q_m - \lambda_m p_{m+1}) w^1 \\ \vdots \\ 0 \end{bmatrix} \end{aligned}$$

$$
\begin{aligned}
&= \begin{bmatrix} Qw^1 \\ 0 \end{bmatrix} \\
&= \begin{bmatrix} Qv^1 - \dfrac{p_{m+1}v^1}{p_{m+1}x^1} Qx^1 \\ 0 \end{bmatrix} \\
&= \begin{bmatrix} Qv^1 \\ 0 \end{bmatrix} \\
&\geqq \mathbf{0}
\end{aligned}
$$

and

$$
\begin{aligned}
p_1 w^1 + (u^1)_1 &= (q_1 - \lambda_1 p_{m+1}) w^1 + (u^1)_1 \\
&= q_1 w^1 + (u^1)_1 \\
&= q v^1 - \frac{p_{m+1}v^1}{p_{m+1}x^1} q_1 x^1 + (u^1)_1 \\
&= q_1 v^1 + (u^1)_1 \\
&> 0.
\end{aligned}
$$

This concludes the proof. QED

This lemma can be generalized to [Theorem 1.1].

[Theorem 1.1] Tucker's Theorem[3] For a given $m \times n$ matrix $\boldsymbol{P}$,

$$\text{I} \quad [\boldsymbol{Px} \geqq \mathbf{0}]$$

and

$$\text{II} \quad [\boldsymbol{P}^t \boldsymbol{\eta} \geqq \mathbf{0}, \boldsymbol{\eta} \geqq \mathbf{0}]$$

have solutions $\boldsymbol{x} \in \boldsymbol{R}^n$, $\boldsymbol{\eta} \in \boldsymbol{R}^m$ that satisfy

$$[\boldsymbol{Px} + \boldsymbol{\eta} > \mathbf{0}].$$

[3] See Tucker (1956).

(Proof) For any row number k of $\boldsymbol{P}$, define $\boldsymbol{Q}$ to be a matrix with m-order unit matrix with the first row and kth row switched and apply the [Lemma 1.1] to the matrix ($\boldsymbol{QP}$), there exist $\boldsymbol{y}$ and $\boldsymbol{\xi}$ are

$$(\boldsymbol{QP})\boldsymbol{y} \geqq \boldsymbol{0},\ (\boldsymbol{QP})^t\boldsymbol{\xi} \geqq \boldsymbol{0},\ \boldsymbol{\xi} \geqq \boldsymbol{0} \text{ and } (\boldsymbol{QP})_1\boldsymbol{y} + \left(\boldsymbol{\xi}^1\right)_1 > \boldsymbol{0}. \qquad (1.3)$$

From the definition of $\boldsymbol{Q}$,

$$(\boldsymbol{QP})_i = \boldsymbol{p}_k \text{ and } (\boldsymbol{Q}^t\boldsymbol{\xi})_i = \xi_k, \quad (i = 1)$$
$$(\boldsymbol{QP})_i = \boldsymbol{p}_1 \text{ and } (\boldsymbol{Q}^t\boldsymbol{\xi})_i = \xi_1, \quad (i = k)$$

and

$$(\boldsymbol{QP})_i = \boldsymbol{p}_i \text{ and } (\boldsymbol{Q}^t\boldsymbol{\xi})_i = \xi_i, \quad (i \neq 1, k).$$

Thus, $\boldsymbol{Q}^t\boldsymbol{\xi} = \boldsymbol{\eta}^k$ and letting $\boldsymbol{y} = \boldsymbol{x}^k$, (1.3) can be rewritten as follows:

$$\boldsymbol{Px}^k \geqq \boldsymbol{0},\ \boldsymbol{P}^t\boldsymbol{\eta}^k = \boldsymbol{0},\ \boldsymbol{\eta}^k \geqq 0 \text{ and } \boldsymbol{p}_k\boldsymbol{x}^k + \left(\boldsymbol{\eta}^k\right)_k > 0.$$

Hence, if we let $\boldsymbol{x} = \sum_{k=1}^{m} \boldsymbol{x}^k$ and $\boldsymbol{\eta} = \sum_{k=1}^{m} \boldsymbol{\eta}^k$, $\begin{pmatrix} \boldsymbol{x} \\ \boldsymbol{\eta} \end{pmatrix} \in \boldsymbol{S}$ and $\boldsymbol{px} + \boldsymbol{\eta} > \boldsymbol{0}$ are obtained. QED

If [Theorem 1.1] is written down as per the following series, it is useful for future use.

[Corollary 1.1] Let $\boldsymbol{A}, \boldsymbol{B}, \boldsymbol{C}, \boldsymbol{D}$ be a given $m_1 \times n$, $m_2 \times n$, $m_3 \times n$, $m_4 \times n$ matrix. Here,

$$\text{I} \quad [\boldsymbol{Ax} \geqq \boldsymbol{0}, \boldsymbol{Bx} \geqq \boldsymbol{0}, \boldsymbol{Cx} \geqq \boldsymbol{0}, \boldsymbol{Dx} = \boldsymbol{0}]$$

and

$$\text{II} \quad \begin{bmatrix} \boldsymbol{A}^t\boldsymbol{y}^1 + \boldsymbol{B}^t\boldsymbol{y}^2 + \boldsymbol{C}^t\boldsymbol{y}^3 + \boldsymbol{D}^t\boldsymbol{y}^4 = \boldsymbol{0} \\ \boldsymbol{y}^1 \geqq \boldsymbol{0},\ \boldsymbol{y}^2 \geqq \boldsymbol{0},\ \boldsymbol{y}^3 \geqq \boldsymbol{0} \end{bmatrix}$$

has solution $\boldsymbol{x} \in \boldsymbol{R}^n$, $\boldsymbol{y}^1 \in \boldsymbol{R}^{m_1}$, $\boldsymbol{y}^2 \in \boldsymbol{R}^{m_2}$, $\boldsymbol{y}^3 \in \boldsymbol{R}^{m_3}$, and $\boldsymbol{y}^4 \in \boldsymbol{R}^{m_4}$, which satisfy

$$\begin{bmatrix} \boldsymbol{Ax} + \boldsymbol{y}^1 > \boldsymbol{0} \\ \boldsymbol{Bx} + \boldsymbol{y}^2 > \boldsymbol{0} \\ \boldsymbol{Cx} + \boldsymbol{y}^3 > \boldsymbol{0} \end{bmatrix}.$$

(Proof) Let us define $m \times n$ matrix $\boldsymbol{P}$, $\boldsymbol{\eta} \in \boldsymbol{R}^m$ as follows.

$$P = \begin{bmatrix} A \\ B \\ C \\ D \\ -D \end{bmatrix}, \ \eta = \begin{bmatrix} z^1 \\ z^2 \\ z^3 \\ z^4 \\ z^5 \end{bmatrix},$$

where $z^i \in R^{m_i}$ $(i = 1, 2, 3)$, $z^4\ z^5 \in R^{m_4}$, $m_1 + m_2 + m_3 + 2m_4 = m$.
From [Theorem 1.1],

$$\text{I}' \quad [Ax \geqq 0, Bx \geqq 0, Cx \geqq 0, Dx = 0, \ -Dx \geqq 0]$$

and

$$\text{II}' \quad \begin{bmatrix} A^t z^1 + B^t z^2 + C^t z^3 + D^t z^4 - D^t z^5 = 0 \\ z^1 \geqq 0, z^2 \geqq 0, \ z^3 \geqq 0, z^4 \geqq 0, z^5 \geqq 0 \end{bmatrix}$$

have solutions x and z^i $(i = 1, 2, 3, 4, 5)$ that satisfy

$$\begin{bmatrix} Ax + z^1 > 0 \\ Bx + z^2 > 0 \\ Cx + z^3 > 0 \\ Dx + z^4 > 0 \\ -Dx + z^5 > 0 \end{bmatrix}.$$

Thus, while noting that I′s $Dx \geqq 0$ and $-Dx \geqq 0$ have the same value as $Dx = 0$, define $y^i = z^i \geqq 0 (i = 1, 2, 3)$ and $y^4 = z^4 - z^5$, the conclusion is immediately obtained. QED

(Notes relating to [Corollary 1.1]): In future, if this corollary is to be used unless there is a particular reason not to, the solution of a corollary will be expressed as x_0, $y_0{}^i$ $(i = 1, 2, 3, 4)$.

1.3 Theorem of Alternatives in Respect of Linear Inequality Systems

Generally, the alternative theorem indicates that two prepositions I and II are exclusive and that at least one of the prepositions always holds. Accordingly, to demonstrate the theorem of alternatives,

(a) I ⇒∼ II[4] (equivalent to ∼I⇐II)
and
(b) ∼I ⇒ II (equivalent to I ⇐∼ II)
must be proven.

[4](a) indicates that prepositions I and II do not both hold together. "⇒" is a logical symbol that shows implication and "∼" is a sign that shows non-implication.

Originally, the description of [Theorem 1.2] shown below is correct. Tucker However, in the various theorems below ([Theorem 1.3] ~ [Theorem 1.11]), for simplification of notation, when only one of the prepositions I or II hold, it is simply expressed as "I or II." Moreover, matrices $\boldsymbol{A}$, $\boldsymbol{B}$, $\boldsymbol{C}$, and $\boldsymbol{D}$ are considered to be real matrices with n columns and m_1, m_2, m_3, m_4 rows, respectively, and $\boldsymbol{y}^i$ is defined to be a $m_i(i = 1, 2, 3, 4)$-dimension real vector. The number of order of a vector and the number of columns and rows of matrices other than this will be clarified only if they cannot be understood from the context.

Firstly, [Corollary 1.1] is used to explain [Theorem 1.2] and [Theorem 1.3]. In particular, [Theorem 1.2] is a versatile theorem of alternatives.

[Theorem 1.2] Slater's Theorem of the Alternative[5] Let $\boldsymbol{A}$, $\boldsymbol{B}$, $\boldsymbol{C}$ and $\boldsymbol{D}$ be given matrices, with $\boldsymbol{A}$ and $\boldsymbol{B}$ being nonvacuous. Then either

$$\text{I} \quad [\boldsymbol{Ax} > \boldsymbol{0}, \boldsymbol{Bx} \geq \boldsymbol{0}, \boldsymbol{Cx} \geqq \boldsymbol{0}, \boldsymbol{Dx} = \boldsymbol{0}] \text{ has a solution } \boldsymbol{x},$$

or

$$\text{II} \quad \begin{bmatrix} \boldsymbol{A}^t\boldsymbol{y}^1 + \boldsymbol{B}^t\boldsymbol{y}^2 + \boldsymbol{C}^t\boldsymbol{y}^3 + \boldsymbol{D}^t\boldsymbol{y}^4 = \boldsymbol{0} \\ \text{However,(i) } \boldsymbol{y}^1 \geq \boldsymbol{0}, \ \boldsymbol{y}^2 \geqq \boldsymbol{0}, \ \boldsymbol{y}^3 \geqq \boldsymbol{0} \\ \text{or} \\ \text{(ii) } \boldsymbol{y}^1 \geqq \boldsymbol{0}, \ \boldsymbol{y}^2 > \boldsymbol{0}, \ \boldsymbol{y}^3 \geqq \boldsymbol{0} \end{bmatrix} \text{ has a solution } \boldsymbol{y}^1, \boldsymbol{y}^2, \boldsymbol{y}^3, \boldsymbol{y}^4.$$

But never both.

(Proof)

(a) If the conclusion is contradicted and it is assumed that I and II holds,

$$(\boldsymbol{x}^t\boldsymbol{C}^t)\boldsymbol{y}^3 \geqq 0 \text{ and } (\boldsymbol{x}^t\boldsymbol{D}^t)\boldsymbol{y}^4 = 0.$$

Furthermore, in response to II (i) or (ii),

$$(\boldsymbol{x}^t\boldsymbol{A}^t)\boldsymbol{y}^1 > 0 \text{ and } (\boldsymbol{x}^t\boldsymbol{B}^t)\boldsymbol{y}^2 \geqq 0$$

Moreover,

$$(\boldsymbol{x}^t\boldsymbol{A}^t)\boldsymbol{y}^1 \geqq 0 \text{ and } (\boldsymbol{x}^t\boldsymbol{B}^t)\boldsymbol{y}^2 > 0$$

are established. Thus, in any case,

[5] See Slater (1951).

$$(x'A')y^1 + (x'B')y^2 + (x'C')y^3 + (x'D')y^4 > 0.$$

However, the left-hand side of the above inequality is

$$x'\left(A'y^1 + B'y^2 + C'y^3 + D'y^4\right) = \mathbf{0}$$

Accordingly, both these two equations do not hold together.
(b) If I is denied, in respect of any x, at least one of the four equations in I no longer holds. However, if [Corollary 1.1] is considered, $Cx_0 \geqq \mathbf{0}$ and $Dx_0 = \mathbf{0}$. Thus, I may become untrue only in the case of (A) $Ax_0 \ngtr \mathbf{0}$ or (B) $Bx_0 \ngeq \mathbf{0}$.

(A) $Ax_0 \ngtr \mathbf{0}$ and $Ax_0 \geqq \mathbf{0}$. Hence, there exists a number $i_0 \in \{1, \ldots, m_4\}$ that satisfies $(Ax_0)_{ia} = \sum_{j=1}^{n} a_{io}x_0 = 0$. Moreover, from $y_0^1 \geqq \mathbf{0}$ and $Ax_0 + y_0^1 > \mathbf{0}$, $y_0^1 \geq \mathbf{0}$. $y_0^i \geqq \mathbf{0}$ $(i = 2, 3)$ is guaranteed, and therefore II(i) has been demonstrated.
(B) $Bx_0 \ngeq \mathbf{0}$ and $Bx_0 \geqq \mathbf{0}$. Therefore, $Bx_0 = \mathbf{0}$. If this is substituted into $Bx_0 + y_0^2 > \mathbf{0}$, $y_0^2 > \mathbf{0}$ is immediately obtained and $y_0^i \geqq \mathbf{0}$ $(i = 1, 3)$ is self-apparent. Accordingly, in this case, II(ii) holds. QED

(Notes on [Theorem 1.2]): If the solution of I in [Theorem 1.2] is considered to be $x = -z$, I can be rewritten as follows:

$$\text{I}' \quad [Az < \mathbf{0}, Bz \leq \mathbf{0}, Cz \leqq \mathbf{0}, Dz = \mathbf{0}] \text{ has a solution } z.$$

It is the same for the different theorems below. Thus, when using the theorem, a more useful approach from the inequality system I or I′ must be selected.

Below, the situation when a specific matrix as per [Theorem 1.2] hence a vector which corresponds to it does not exist is considered. Thus, a matrix or a vector, according to Mangasarian (1969), when A or y^1 does not exist, is expressed as $A = [\phi]$ or $y^1 = [\phi]$ and it is said that A is a **vacuous matrix** or that y^1 is an **empty vector**. Henceforth, nonvacuous matrix A shall be written as $A \neq [\phi]$.

[Theorem 1.3] Tucker's Theorem of the Alternative[6] Concerning the given matrices $B(\neq[\phi])$, C and D

$$\text{I} \quad [Bx \geq \mathbf{0}, Cx \geqq \mathbf{0}, Dx = \mathbf{0}] \text{ has a solution } x,$$

or

$$\text{II} \quad \begin{bmatrix} B'y^2 + C'y^3 + D'y^4 = \mathbf{0} \\ y^2 > \mathbf{0}, \ \ y^3 \geqq \mathbf{0} \end{bmatrix} \text{has solutions } y^2, y^3, y^4.$$

---

[6] See Tucker (1956).

(Proof) In the inequality system of [Theorem 1.2], $\boldsymbol{A}$, accordingly, if there exists no $\boldsymbol{y}^1$, the inequality system of the theorem is obtained.

(a) It is evident from proof (a) of [Theorem 1.2].
(b) Shown in (B) in proof (b) of [Theorem 1.2]. QED

Furthermore, when there are no $\boldsymbol{C}$ and $\boldsymbol{D}$ for [Theorem 1.3], the following theorem can be obtained.

[Theorem 1.4] Stiemke's Theorem[7] For a given $\boldsymbol{B}$,

$$\text{I} \quad [\boldsymbol{Bx} \geq \boldsymbol{0}] \text{ has a solution,}$$

or

$$\text{II} \quad [\boldsymbol{B}^t\boldsymbol{y}^2 = \boldsymbol{0},\ \ \boldsymbol{y}^2 > \boldsymbol{0}] \text{ has a solution } \boldsymbol{y}^2.$$

(Proof) It is evident by letting [Theorem 1.3]'s $\boldsymbol{C} = [\phi]$ and $\boldsymbol{D} = [\phi]$. QED

[Theorem 1.5] Motzkin's Theorem of the Alternative[8] Let $\boldsymbol{A}(\neq[\phi])$, $\boldsymbol{C}$ and $\boldsymbol{D}$ be given matrices. Then

$$\text{I} \quad [\boldsymbol{Ax} > \boldsymbol{0},\ \ \boldsymbol{Cx} \geqq \boldsymbol{0}, \boldsymbol{Dx} = \boldsymbol{0}] \text{ has a solution } \boldsymbol{x},$$

or

$$\text{II} \quad \begin{bmatrix} \boldsymbol{A}^t\boldsymbol{y}^1 + \boldsymbol{C}^t\boldsymbol{y}^3 + \boldsymbol{D}^t\boldsymbol{y}^4 = \boldsymbol{0} \\ \boldsymbol{y}^1 \geq \boldsymbol{0}, \boldsymbol{y}^3 \geqq \boldsymbol{0} \end{bmatrix} \text{ has a solution } \boldsymbol{y}^1, \boldsymbol{y}^3, \boldsymbol{y}^4.$$

(Proof) Just require to ignore [Theorem 1.2]'s $\boldsymbol{B} = [\phi]$ and thus $\boldsymbol{y}^2$.

(a) It can be shown in the same as Proof (a) in [Theorem 1.2].
(b) It gets back to proof (b) of [Theorem 1.2]. QED

Next, let us rewrite this theorem such that it can be easily applied to an inhomogeneous system.

[Corollary 1.5] Let β, γ, ξ, and $\eta \in \boldsymbol{R}$ and define $\widetilde{\boldsymbol{x}} = \begin{bmatrix} \boldsymbol{x} \\ \xi \end{bmatrix} \in \boldsymbol{R}^{n+1}, \widetilde{\boldsymbol{y}_1} = \begin{bmatrix} \eta \\ \gamma \end{bmatrix} \in \boldsymbol{R}^2$. Here, corresponding to [Theorem 1.5]'s $\boldsymbol{C} = [\phi]$ or $\boldsymbol{D} = [\phi]$, (i) and (ii) hold.

---

[7] See Stiemke (1915).

[8] See Motzkin (1936, 1951). Regarding Motzkin's transposition theorem, there are many papers, but for example, Slater (1951), Ben (2001) and Nermirovski and Roos (2009) are recommended.

(i) Given $\widetilde{A} = \begin{bmatrix} b^t & -\beta \\ 0 & 1 \end{bmatrix}$, $\widetilde{D} = [-A, c]$,

$$\text{I} \quad \left[\widetilde{A}\widetilde{x} = \begin{bmatrix} b^t x - \beta\xi \\ \xi \end{bmatrix} > 0, \widetilde{D}\widetilde{x} = [-Ax + \xi c] = 0\right] \text{ has a solution } \widetilde{x} = \begin{bmatrix} x \\ \xi \end{bmatrix},$$

or

$$\text{II} \quad \begin{bmatrix} \widetilde{A}^t\widetilde{y}^1 + \widetilde{D}^t y^4 = \begin{bmatrix} \eta b - A^t y^4 \\ -\beta\eta + \gamma + C^t y^4 \end{bmatrix} = 0 \\ \widetilde{y}^1 = \begin{bmatrix} \eta \\ \gamma \end{bmatrix} \geq 0 \end{bmatrix} \text{ has a solution } \widetilde{y}^1, y^4.$$

(ii) Given $\widetilde{A} = \begin{bmatrix} b^t & -\beta \\ 0 & 1 \end{bmatrix}$, $\tilde{C} = [-A, c]$,

$$\text{I} \quad \left[\widetilde{A}\widetilde{x} = \begin{bmatrix} b^t x - \beta\xi \\ \xi \end{bmatrix} > 0, \; \widetilde{C}\widetilde{x} = [-Ax + \xi c] \geqq 0\right] \text{ has a solution } \widetilde{x} = \begin{bmatrix} x \\ \xi \end{bmatrix},$$

or

$$\text{II} \quad \begin{bmatrix} \widetilde{A}^t\widetilde{y}^1 + \widetilde{C}^t y^3 = \begin{bmatrix} \eta b - A^t y^3 \\ -\beta\eta + \gamma + C^t y^3 \end{bmatrix} = 0 \\ \widetilde{y}_1 = \begin{bmatrix} \eta \\ \gamma \end{bmatrix} \geq 0, \; y^3 \geqq 0 \end{bmatrix} \text{ has a solution } \widetilde{y}^1, y^3.$$

Now, let us show below the various theorems that can be derived from [Theorem 1.5]. First, the case of [Theorem 1.5]'s $C = [\phi]$ will be discussed.

[Theorem 1.6] Gale's Theorem for Linear Equalities[9] For a given A and c,

$$\text{I} \quad [Ax = c] \text{ has a solution } x,$$

or

$$\text{II} \quad [A^t y = 0, \;\; c^t y = 1] \text{ has a solution } y.$$

(Proof) Let [Theorem 1.5]'s $C = [\phi]$. Accordingly, by letting it be system [1.5; (i)]'s $b = 0 \in R^n$, $\beta = -1$, $\eta = 0$; in other words, it can be rewritten as follows:

[9] See Gale (1960).

$$\widetilde{A} = \begin{bmatrix} \mathbf{0} & 1 \\ \mathbf{0} & 1 \end{bmatrix},\ \widetilde{D} = [A, c],\ \widetilde{x} = \begin{bmatrix} x \\ \xi \end{bmatrix},\ \widetilde{y}^{\,1} = \begin{bmatrix} 0 \\ \gamma \end{bmatrix}$$

[Corollary 1.5; (i)],

$$\text{I}' \quad [\xi > 0,\ \ Ax = c\xi\,] \text{ has a solution } \begin{bmatrix} x \\ \xi \end{bmatrix},$$

or

$$\text{II}' \quad \begin{bmatrix} A^t y^4 = \mathbf{0},\ \gamma + c^t y^4 = 0 \\ \begin{bmatrix} 0 \\ \gamma \end{bmatrix} \geq \mathbf{0} \end{bmatrix} \text{ has solution } \begin{bmatrix} 0 \\ \gamma \end{bmatrix} \text{ and } y^4.$$

Clearly, I′ and II′ and II are mutually equivalent, and hence this concludes the proof. QED

[Theorem 1.7]–[Theorem 1.9] corresponds to the case when D does not exist in [Theorem 1.5]; thus, it is more directly obtained by applying [Corollary 1.5; (ii)].

[Theorem 1.7] Farkas's Theorem[10] For a given A and b,

$$\text{I} \quad [Ax \leqq \mathbf{0},\ \ b^t x > 0] \text{ has a solution } x,$$

or

$$\text{II} \quad [A^t y = b, y \geqq \mathbf{0}] \text{ has a solution } y.$$

(Proof)

If we define $\widetilde{A} = \begin{bmatrix} b^t & 0 \\ \mathbf{0} & 1 \end{bmatrix}, \widetilde{C} = [-A, \mathbf{0}], \widetilde{x} = \begin{bmatrix} x \\ \xi \end{bmatrix},\ \widetilde{y}^{\,1} = \begin{bmatrix} \eta \\ 0 \end{bmatrix}$

it is no other than the case when we set [Series 1.5; (ii)]'s $c = \mathbf{0}$, $\beta = \gamma = 0$. Accordingly, I or II is equivalent to:

$$\text{I}' \quad [b^t x > 0,\ \ \xi > 0,\ \ Ax \leqq \mathbf{0}\,] \text{ has a solution } x,$$

or

$$\text{II}' \quad \begin{bmatrix} \eta b - A^t y^3 = \mathbf{0} \\ \eta > 0, y^3 \geqq \mathbf{0} \end{bmatrix} \text{ has a solution } \eta, y^3.$$

QED

[10] See Farkas (1902).

[Theorem 1.8] Nonhomogeneous Farkas Theorem (Duffin's Theorem[11]) For a given $\boldsymbol{A}$, $\boldsymbol{b}$, and $\boldsymbol{c}$,

$$\text{I} \quad [\boldsymbol{b}^t\boldsymbol{x} > \beta,\ \boldsymbol{A}\boldsymbol{x} \leqq \boldsymbol{c}\,] \text{ has a solution } \boldsymbol{x},$$

or

$$\text{II} \quad \begin{bmatrix} \text{(i)} \quad \boldsymbol{A}^t\boldsymbol{y} = \boldsymbol{b},\ \boldsymbol{C}^t\boldsymbol{y} \leqq \beta,\ \boldsymbol{y} \geqq \boldsymbol{0} \\ \text{or} \\ \text{(ii)} \quad \boldsymbol{A}^t\boldsymbol{y} = \boldsymbol{0},\ \boldsymbol{C}^t\boldsymbol{y} < 0,\ \boldsymbol{y} \geqq \boldsymbol{0} \end{bmatrix} \text{ has a solution } \boldsymbol{y}.$$

(Proof) If [Corollary 1.5; (ii)] is applied without changing anything,

$$\text{I}' \quad \left[\begin{bmatrix} \boldsymbol{b}^t\boldsymbol{x} - \beta\xi \\ \xi \end{bmatrix} > \boldsymbol{0},\ [-\boldsymbol{A}\boldsymbol{x} + \xi\boldsymbol{c}] \geqq \boldsymbol{0}\right] \text{ has a solution} \begin{bmatrix} \boldsymbol{x} \\ \xi \end{bmatrix}.$$

or

$$\text{II}' \quad \begin{bmatrix} \begin{bmatrix} \eta\boldsymbol{b} - \boldsymbol{A}^t\boldsymbol{y}^3 \\ -\beta\eta + \gamma + \boldsymbol{C}^t\boldsymbol{y}^3 \end{bmatrix} = \boldsymbol{0}, \begin{bmatrix} \eta \\ \gamma \end{bmatrix} \geq \boldsymbol{0}, \\ \boldsymbol{y}^3 \geqq \boldsymbol{0} \end{bmatrix} \text{ has a solution } \begin{bmatrix} \eta \\ \gamma \end{bmatrix}, \boldsymbol{y}^3.$$

I and I′ are equivalent and therefore only the equivalence between II and II′ must be observed.

(II′ ⇒ II) $[\eta, \gamma\,]^t \geq \boldsymbol{0}$ in II′ indicates that one of (A)$\eta > 0$ (thus, $\gamma \geqq 0$) or (B)$\eta = 0$ (thence, $\gamma > 0$) always holds. Consequently, each case will be considered.

(A) $\boldsymbol{A}^t\boldsymbol{y}^3 = \eta\boldsymbol{b}$ and $\boldsymbol{C}^t\boldsymbol{y}^3 = \beta\eta - \gamma \leqq \beta\eta$. Thus, by selecting $\boldsymbol{y} = \boldsymbol{y}^3/\eta \geqq \boldsymbol{0}$, II(i) is derived.
(B) $\boldsymbol{A}^t\boldsymbol{y}^3 = \boldsymbol{0}$ and $\boldsymbol{C}^t\boldsymbol{y}^3 = -\gamma < 0$. Accordingly, by defining $\boldsymbol{y} = \boldsymbol{y}^3$, $\boldsymbol{y} \geqq \boldsymbol{0}$ is satisfied and II(ii) is obtained.

(II′ ⇐ II) Prove for each of (i) or (ii).

(i) If (i), in respect of $\eta > 0$, $\boldsymbol{A}^t(\eta\boldsymbol{y}) = \eta\boldsymbol{b}$, $\boldsymbol{C}^t(\eta\boldsymbol{y}) \leqq \beta\eta$ and $\eta\boldsymbol{y} \geqq \boldsymbol{0}$. Thus, if we define $\gamma = \beta\eta - \boldsymbol{C}^t(\eta\boldsymbol{y}) \geqq 0$ and $\boldsymbol{y}^3 = \eta\boldsymbol{y}$, (A) in the proof of (II′ ⇒ II) is obtained.
(ii) If we let $\boldsymbol{y} = \boldsymbol{y}^3$ and $-\boldsymbol{c}^t\boldsymbol{y} = \gamma > 0$, (B) holds.

Accordingly, II and II′ become equivalent. QED

[11] See Duffin (1956).

[Theorem 1.9] Gale's Theorem for Linear Inequalities[12] For a given $\boldsymbol{A}$ and $\boldsymbol{c}$,

$$\text{I} \quad [\boldsymbol{Ax} \leqq \boldsymbol{c}] \text{ has a solution } \boldsymbol{x},$$

or

$$\text{II} \quad [\boldsymbol{A}^t\boldsymbol{y} = \boldsymbol{0}, \boldsymbol{c}^t\boldsymbol{y} = -1, \boldsymbol{y} \geqq \boldsymbol{0}] \text{ has a solution } \boldsymbol{y}.$$

(Proof)

$$\text{Let } \widetilde{\boldsymbol{A}} = \begin{bmatrix} \boldsymbol{0} & 1 \\ \boldsymbol{0} & 1 \end{bmatrix}, \ \widetilde{\boldsymbol{C}} = [-\boldsymbol{A}, \boldsymbol{c}], \ \widetilde{\boldsymbol{x}} = \begin{bmatrix} \boldsymbol{x} \\ \xi \end{bmatrix}, \ \widetilde{\boldsymbol{y}}^1 = \begin{bmatrix} 0 \\ \gamma \end{bmatrix}$$

and if [Corollary 1.5; (ii)] is applied, the conclusion is evident. QED

If neither $\boldsymbol{C}$ nor $\boldsymbol{D}$ in [Theorem 1.5] exists, the following is obtained.

[Theorem 1.10] Gordan's Theorem[13] For a given $\boldsymbol{A}$,

$$\text{I} \quad [\boldsymbol{Ax} > \boldsymbol{0}] \text{ has a solution } \boldsymbol{x},$$

or

$$\text{II} \quad [\boldsymbol{A}^t\boldsymbol{y}^1 = \boldsymbol{0}, \boldsymbol{y}^1 \geq \boldsymbol{0}] \text{ has a solution } \boldsymbol{y}^1.$$

(Proof) If we let $\boldsymbol{C} = [\phi]$ and $\boldsymbol{D} = [\phi]$ in [Theorem 1.5], the conclusion is immediately obtained. QED

Finally, let us show one conclusion, which can be obtained from [Corollary 1.1] by considering $\boldsymbol{A}$, $\boldsymbol{B}$, $\boldsymbol{C}$ and $\boldsymbol{D}$ as not being empty.

[Theorem 1.11] For a given $\boldsymbol{A}(\neq[\phi])$, $\boldsymbol{B}(\neq[\phi])$, $\boldsymbol{C}$ and $\boldsymbol{D}$,

$$\text{I} \quad \begin{bmatrix} \text{(i)} \quad \boldsymbol{Ax} \geq \boldsymbol{0}, \boldsymbol{Bx} \geqq \boldsymbol{0}, \boldsymbol{Cx} \geqq \boldsymbol{0}, \boldsymbol{Dx} = \boldsymbol{0} \\ \text{or} \\ \text{(ii)} \quad \boldsymbol{Ax} \geqq \boldsymbol{0}, \boldsymbol{Bx} > \boldsymbol{0}, \boldsymbol{Cx} \geqq \boldsymbol{0}, \boldsymbol{Dx} = \boldsymbol{0} \end{bmatrix} \text{ has a solution } \boldsymbol{x},$$

or

$$\text{II} \quad \begin{bmatrix} \boldsymbol{A}^t\boldsymbol{y}^1 + \boldsymbol{B}^t\boldsymbol{y}^2 + \boldsymbol{C}^t\boldsymbol{y}^3 + \boldsymbol{D}^t\boldsymbol{y}^4 = \boldsymbol{0} \\ \boldsymbol{y}^1 > \boldsymbol{0}, \boldsymbol{y}^2 \geq \boldsymbol{0}, \boldsymbol{y}^3 \geqq \boldsymbol{0} \end{bmatrix} \text{ has a solution } \boldsymbol{y}^1, \boldsymbol{y}^2, \boldsymbol{y}^3, \boldsymbol{y}^4.$$

---

[12] See Gale (1960).

[13] See Gordan (1873).

(Proof)
(a) If I and II are both true,

$$\left(x^t C^t y^3\right) \geqq \mathbf{0} \quad \text{and} \quad (x^t D^t) y^4 = \mathbf{0}.$$

Moreover, with (i) and (ii) holding,

$$(x^t A^t) y^1 > \mathbf{0} \text{ and } (x^t B^t) y^2 \geqq \mathbf{0},$$

and

$$(x^t A^t) y^1 \geqq \mathbf{0} \quad \text{and} \quad (x^t B^t) y^2 > \mathbf{0}.$$

Accordingly, the following self-contradiction

$$\begin{aligned} 0 &< (x^t A^t) y^1 + (x^t B^t) y^2 + (x^t C^t) y^3 + (x^t D^t) y^4 \\ &= x^t \left(A^t y^1 + B^t y^2 + C^t y^3 + D^t y^4\right) \\ &= 0 \end{aligned}$$

is obtained.
(b) Rather than (~I ⇒ II), show its opposite, (I ⇐~ II). From the assumption, in respect of any y^1, y^2, y^3, y^4,

$$A^t y^1 + B^t y^2 + C^t y^3 + D^t y^4 \neq \mathbf{0},$$

or

$$y^1 \ngtr \mathbf{0} \text{ or } y^2 \ngeq \mathbf{0} \text{ or } y^3 \ngeqq \mathbf{0}.$$

Here, if we consider [Corollary 1.1],

$$A^t y_0^1 + B^t y_0^2 + C^t y_0^3 + D^t y_0^4 = \mathbf{0} \quad \text{and} \quad y_0^3 \geqq \mathbf{0}$$

hold; hence, the case (A) $y_0^1 \ngtr \mathbf{0}$ or (B) $y_0^2 \ngeq \mathbf{0}$ must be considered.

(A) From $y_0^1 \ngtr \mathbf{0}$ and $y_0^1 \geqq \mathbf{0}$, there exists $i_0 \in \{1, \ldots, m_1\}$ such that it is the i_0th element $= 0$ of $\left(y_0^1\right)_{i_0} = y_0^1$. Combining this with $Ax_0 + y_0^1 > \mathbf{0}$establishes $(Ax_0)_{i_0} > 0$ in respect of i_0 but from $Ax_0 + y_0^1 > \mathbf{0}$, $Ax_0 \geq \mathbf{0}$ is obtained; however, $Bx_0 \geqq \mathbf{0}$, $Cx_0 \geqq \mathbf{0}$ and $Dx_0 = \mathbf{0}$ are self-apparent. Thus, I(i) has a solution x_0.
(B) $y_0^2 \geqq \mathbf{0}$ is already guaranteed, and hence $y_0^2 \ngeq \mathbf{0}$ indicates $y_0^2 = \mathbf{0}$. Accordingly, $Bx_0 + y_0^2 > \mathbf{0}$ indicates $Bx_0 > \mathbf{0}$ and ultimately I(ii) is obtained. QED

Figure 1.1 shows the mutual relationships between the abovementioned theorems.

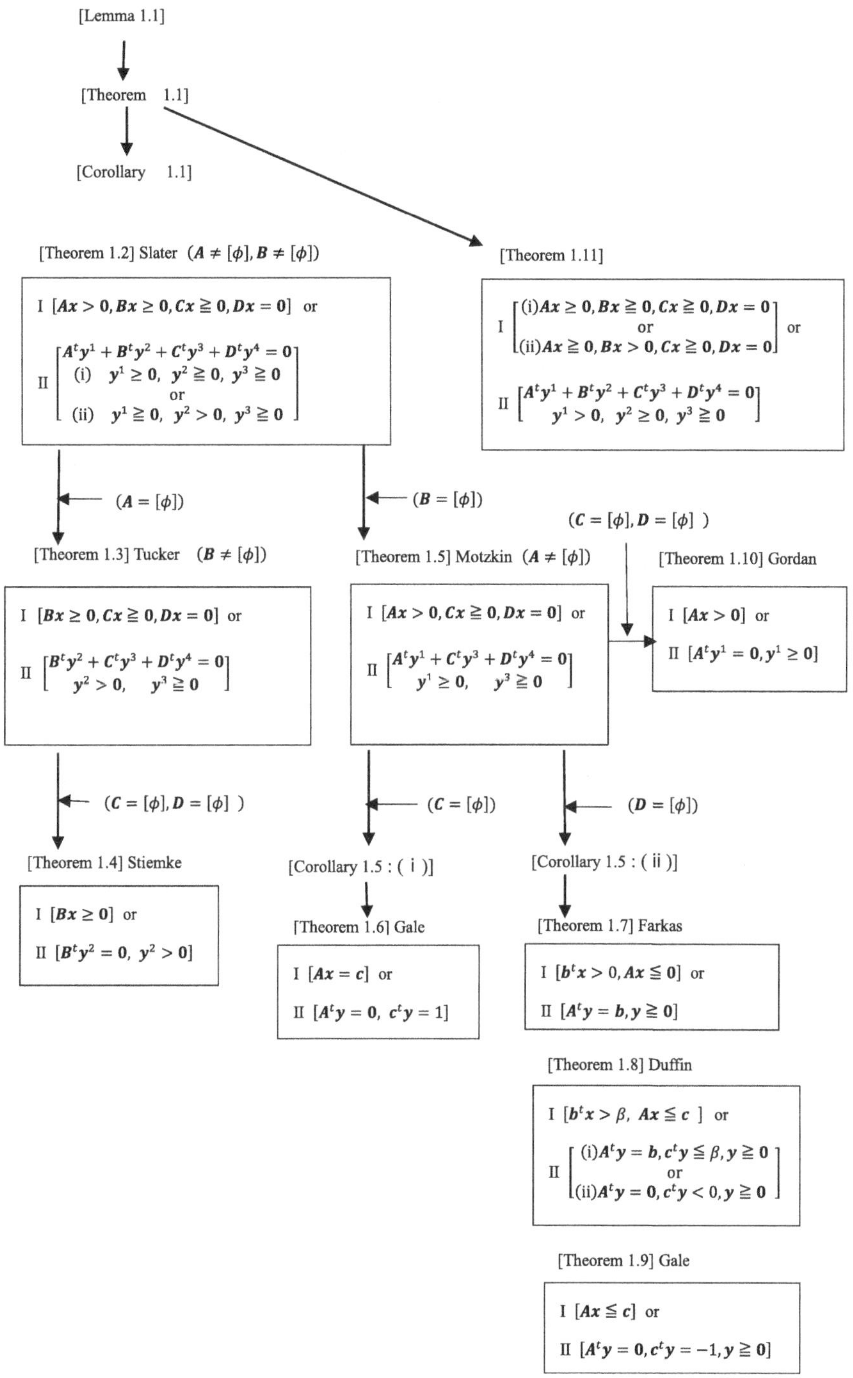

Fig. 1.1 Theorems of the alternative

References

Ben, A. (2001). Motzkin's transposition theorem, and the related theorems of Farkas, Gordan and Stiemke. *Encyclopaedia of Mathematics, Supplement III*, 1–4.

Duffin, R. J. (1956). Infinite programs. *Linear Inequalities and Related Systems, 38*, 157–170.

Farkas, J. (1902). Über die Theorie der einfachen Ungleichungen. *Journal für Reine und AngewandteMathematik, 124*, 1–27.

Gale, D. (1960). *The theory of linear economic models*. McGraw-Hill. ISBN-13: 978-0070227286. Reprinted by University of Chicago Press (New edition), 1989. ISBN-13: 978-0226278841.

Gordan, P. (1873). Über die Auflösungen linearer Gleichungen mit reelen Coefficienten. *Mathematische Annalen, 6*(1), 23–28.

Mangasarian, O. L. (1969). *Nonlinear programming*. McGraw-Hill. ISBN-13: 978-0898713412.

Motzkin, T. S. (1936). *Beiträge zur Theorie der Linearen Ungleichungen*. Inaugural Dissertation. Basel.

Motzkin, T. S. (1951). Two consequences of the transposition theorem on linear inequalities. *Econometrica, 19*, 184–185.

Nemirovski, A., & Roos, K. (2009). Motskin Transposition theorem. In C. A. Floudas & P. M. Pardalos (Eds.), *Encyclopedia of optimization* (2nd ed., pp. 1537–1540). Springer. ISBN-13: 978-0387747583.

Slater, M. L. (1951). A note on Motzkin's transposition theorem. *Econometrica, 19*(2), 185–187.

Stiemke, E. (1915). Über positive lösungen homogener linearer gleichugen. *Mathematische Annalen, 76*, 340–342.

Tucker, A. W. (1956). Dual systems of homogeneous linear relations. *Linear Inequalities and Related Systems, 38*, 3–18.

Chapter 2
Concave Functions, Convex Functions, and Their Periphery

2.1 Introduction

Generally, for an optimization problem to be solvable, the function under consideration must satisfy certain conditions. This is evident from the fact that for the real-valued function $f(\boldsymbol{x})$ of a variable to be at a local maximum value at an inner point x^* within its domain, f must not be a strictly monotonically increasing function in any vicinity of x^*. In particular, in relation to constrained maximization problems, functions considered tend to be concave functions or a function, which is a slight strengthening or expansion of them, i.e., strictly concave function, strictly quasiconcave function, semistrictly quasiconcave function, quasiconcave function, strictly pseudoconcave function, and pseudoconcave function.

In this chapter, the characteristics and mutual relationships between these functions will be explained as preparation for discussing nonlinear programming problems.

2.2 Assumptions and Definitions

In this chapter, the assumptions and glossary used are as follows. Furthermore, throughout this chapter, (2.A1) is assumed

(Assumptions)

(2.A1) $f(\boldsymbol{x}) = f(x_1, \ldots, x_n)^t$ is a real-valued function defined on the convex set $\boldsymbol{D}$ of $\boldsymbol{R}^n$.

(2.A2) $f(\boldsymbol{x})$ is differentiable on the open convex set $\boldsymbol{D}$ of $\boldsymbol{R}^n$.

© Springer Nature Singapore Pte Ltd. 2024

K. Nakayama, *Eigenvalue Problem and Nonlinear Programming Problem*,
New Frontiers in Regional Science: Asian Perspectives 70,
https://doi.org/10.1007/978-981-99-2943-6_2

(2.A3) $f(\boldsymbol{x})$ is twice differentiable on the open convex set $\boldsymbol{D}$ of $\boldsymbol{R}^n$.

⟨**Definition 2.1**⟩ When certain point $\bar{\boldsymbol{x}} \in \boldsymbol{D}$ satisfies

$$^{\exists}\varepsilon > 0|^{\forall}\boldsymbol{x} \in \boldsymbol{B}(\bar{\boldsymbol{x}};\varepsilon) \cap \boldsymbol{D}, f(\boldsymbol{x}) \leqq (\geqq) f(\bar{\boldsymbol{x}})$$

f is said to achieve a maximum (minimum) on $\boldsymbol{x} = \bar{\boldsymbol{x}}$ and $\bar{\boldsymbol{x}} \in \boldsymbol{D}$ is called a **local maximum (local minimum)**. Moreover, the local maximum and local minimum are collectively called **extremum**, although $\boldsymbol{B}(\bar{\boldsymbol{x}};\varepsilon)$ is the $\bar{\varepsilon}$ neighborhood of $\bar{\boldsymbol{x}}$.

When the $\bar{\varepsilon}$ neighborhood of $\bar{\boldsymbol{x}}$ in formula above can be replaced by the domain $\boldsymbol{D}$, in other words, when some $\bar{\boldsymbol{x}} \in \boldsymbol{D}$ satisfies:

$$^{\exists}\varepsilon > 0|^{\forall}\boldsymbol{x} \in \boldsymbol{D}, f(\boldsymbol{x}) \leqq (\geqq) f(\bar{\boldsymbol{x}})$$

f is said to satisfy a **maximum** (**minimum**) on $\boldsymbol{x} = \bar{\boldsymbol{x}}$ and $\bar{\boldsymbol{x}} \in \boldsymbol{D}$ and $f(\bar{\boldsymbol{x}})$ are called a **maximum** (**minimum**)and the **maximum value** (**minimum value**), respectively. In particular, if previous formula holds with a strong inequality $<$ ($>$) for any $\boldsymbol{x} \in \boldsymbol{D}$, which is different to $\bar{\boldsymbol{x}}$, $\bar{\boldsymbol{x}} \in \boldsymbol{D}$ is said to be a **unique maximum** (**minimum**).

(Notes on ⟨Definition 2.1⟩): Fig. 2.1's x^1, x^3 are both a local maximum and $f(x^1)$ and $f(x^3)$ are local maxima but x^1, $f(x^1)$ are simultaneously the maximum point and maximum value over $\boldsymbol{D}$. The same is true for a local minimum.

However, depending on the shape's definition of the function or domain $\boldsymbol{D}$, there may not be a maximum or a minimum point, even if they did, it may not be unique and many may be in existence.

⟨**Definition 2.2**⟩ **Strictly Concave Function** When there exists certain $\boldsymbol{x}^1 \in \boldsymbol{D}$ and the following is satisfied for any $\boldsymbol{x}^2 \in \boldsymbol{D}$ different from $\boldsymbol{x}^1$ and $\lambda \in (0, 1)$,

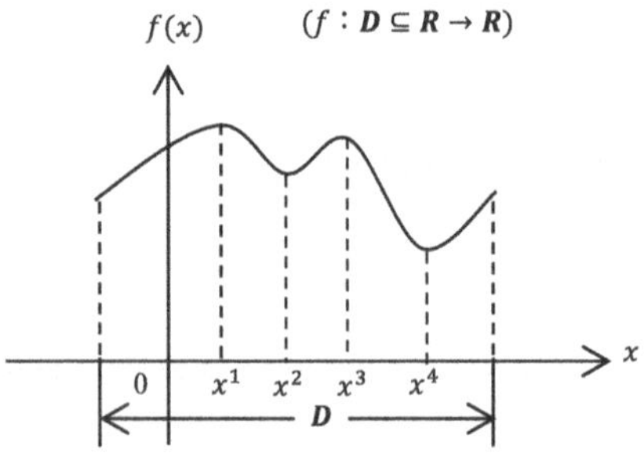

Fig. 2.1 Local maximum and muximum

$$f\left[(1-\lambda)\boldsymbol{x}^1+\lambda\boldsymbol{x}^2\right]>(1-\lambda)f\left(\boldsymbol{x}^1\right)+\lambda f\left(\boldsymbol{x}^2\right)$$

f is said to be a **strictly concave function** on $\boldsymbol{x}^1 \in \boldsymbol{D}$.[1] Moreover, if *f* is a strictly concave function for any $\boldsymbol{x}^1 \in \boldsymbol{D}$, *f* is said to be a strictly concave function on $\boldsymbol{D}$.

(Notes on ⟨Definition 2.2⟩): Here, for simplification, the definition of *f* at certain point will be abbreviated to only write down the definition of *f* on $\boldsymbol{D}$.

⟨Definition 2.3⟩ Concave Function For $\boldsymbol{x}^1, \boldsymbol{x}^2 \in \boldsymbol{D}$ and $\lambda \in [0, 1]$, if

$$f\left[(1-\lambda)\boldsymbol{x}^1+\lambda\boldsymbol{x}^2\right]\geqq(1-\lambda)f\left(\boldsymbol{x}^1\right)+\lambda f\left(\boldsymbol{x}^2\right)$$

holds, *f* is said to be a **concave function** on $\boldsymbol{D}$.

⟨Definition 2.4⟩ Strictly Quasiconcave Function For any different $\boldsymbol{x}^1, \boldsymbol{x}^2 \in \boldsymbol{D}$ and $\lambda \in (0, 1)$,

$$f\left[(1-\lambda)\boldsymbol{x}^1+\lambda\boldsymbol{x}^2\right]>min\left\{f\left(\boldsymbol{x}^1\right),f\left(\boldsymbol{x}^2\right)\right\}$$

When (the above inequality) holds, *f* is said to be a strictly quasiconcave function on $\boldsymbol{D}$.

⟨Definition 2.5⟩ Semistrictly Quasiconcave Function If for any different $\boldsymbol{x}^1$, $\boldsymbol{x}^2 \in \boldsymbol{D}$ and $\lambda \in (0, 1)$ that satisfy $f(\boldsymbol{x}^1) \neq f(\boldsymbol{x}^2)$,

$$f\left[(1-\lambda)\boldsymbol{x}^1+\lambda\boldsymbol{x}^2\right]>min\left\{f\left(\boldsymbol{x}^1\right),f\left(\boldsymbol{x}^2\right)\right\}$$

holds, *f* is said to be semistrictly quasiconcave on $\boldsymbol{D}$.

⟨Definition 2.6⟩ Quasiconcave Function For $\boldsymbol{x}^1, \boldsymbol{x}^2 \in \boldsymbol{D}$ and $\lambda \in [0, 1]$, if

$$f\left[(1-\lambda)\boldsymbol{x}^1+\lambda\boldsymbol{x}^2\right]\geqq min\left\{f\left(\boldsymbol{x}^1\right),f\left(\boldsymbol{x}^2\right)\right\}$$

holds, *f* is said to be a quasiconcave function on $\boldsymbol{D}$.

⟨Definition 2.7⟩ to ⟨Definition 2.9⟩ will assume (2.A2), in addition to (2.A1).

⟨Definition 2.7⟩ Strictly Pseudoconcave Function If for any different $\boldsymbol{x}^1$ and $\boldsymbol{x}^2 \in \boldsymbol{D}$,

$$\nabla f\left(\boldsymbol{x}^1\right)\left(\boldsymbol{x}^2-\boldsymbol{x}^1\right)\leqq 0 \Rightarrow f\left(\boldsymbol{x}^2\right)<f\left(\boldsymbol{x}^1\right)$$

---

[1] As long as the definition involves two points, $\boldsymbol{x}^1$ and $\boldsymbol{x}^2$, within *f* ' s domain $\boldsymbol{D}$ and function values involving their linear combination, $(1-\lambda)\boldsymbol{x}^1+\lambda\boldsymbol{x}^2$; in other words, $f(\boldsymbol{x}^1)$, $f(\boldsymbol{x}^2)$, $f[(1-\lambda)\boldsymbol{x}^1+\lambda f(\boldsymbol{x}^2)]$, in order for the definition to be meaningful, it requires to be $(1-\lambda)\boldsymbol{x}^1+\lambda\boldsymbol{x}^2 \in \boldsymbol{D}$. This is the reason that, in (2.A1), $\boldsymbol{D}$ was assumed to be a convex set.

holds, f is said to be strictly pseudoconcave on $\boldsymbol{D}$, where $\nabla f(\boldsymbol{x}^1) = \left(\dfrac{\partial f}{\partial x_1^1}, \ldots, \dfrac{\partial f}{\partial x_n^1}\right)$.

⟨Definition 2.8⟩ Pseudoconcave Function In for any $\boldsymbol{x}^1, \boldsymbol{x}^2 \in \boldsymbol{D}$, when

$$\nabla f(\boldsymbol{x}^1)(\boldsymbol{x}^2 - \boldsymbol{x}^1) \leqq 0 \Longrightarrow f(\boldsymbol{x}^2) \leqq f(\boldsymbol{x}^1)$$

holds, f is said to be a **pseudoconcave function** on $\boldsymbol{D}$.

⟨Definition 2.9⟩

$$\text{When } -f \text{ is } \begin{bmatrix} \text{Strictly concave} \\ \text{Concave} \\ \text{Strictly quasiconcave} \\ \text{Semistrictly quasiconcave} \\ \text{Quasiconcave} \\ \text{Strictly pseudoconcave} \\ \text{Pseudoconcave} \end{bmatrix},$$

$$f \text{ is said to be } \begin{bmatrix} \textbf{Strictly convex} \\ \textbf{Convex} \\ \textbf{Strictly quasiconvex} \\ \textbf{Semistrcitly quasiconvex} \\ \textbf{Quasiconvex} \\ \textbf{Strictly pseudoconvex} \\ \textbf{Pseudoconvex} \end{bmatrix}.$$

However, there is a corresponding relationship between each row on the left and right.[2]

(Notes on ⟨Definition 2.9⟩): The linear function $f(\boldsymbol{x}) = \boldsymbol{a}'\boldsymbol{x} + b$ (although, $\boldsymbol{a}$ and $\boldsymbol{x}$ are each a given n-order real column vector and b is a real number) is a concave function and a convex function at the same time. In reality, for any $\lambda \in [0, 1]$,

$$\begin{aligned} f\left[(1-\lambda)\boldsymbol{x}^1 + \lambda x^2\right] &= a'\left[(1-\lambda)\boldsymbol{x}^1 + \lambda x^2\right] + b[(1-\lambda) + \lambda] \\ &= (1-\lambda)f(\boldsymbol{x}^1) + \lambda f(\boldsymbol{x}^2). \end{aligned}$$

(Notes on ⟨Definition 2.7⟩ and ⟨Definition 2.8⟩): There are times when <Defnition 2.7> and <Defnition 2.8> are defined as their contraposition:

$$\nabla f(\boldsymbol{x}^1)(\boldsymbol{x}^2 - \boldsymbol{x}^1) > 0 \Longleftarrow f(\boldsymbol{x}^2) \geqq f(\boldsymbol{x}^1) \qquad \text{<Defnition 2.7}'\text{>}$$

$$\nabla f(\boldsymbol{x}^1)(\boldsymbol{x}^2 - \boldsymbol{x}^1) > 0 \Longleftarrow f(\boldsymbol{x}^2) > f(\boldsymbol{x}^1) \qquad \text{<Defnition 2.8}'\text{>}$$

The difference between the above definitions can be easily understood from the following diagram (Figs. 2.2–2.9).

[2] For example, Mangasarian (1965).

However, in Figs. 2.2, 2.3, 2.4, 2.5 and 2.6, $f : \boldsymbol{D} \subseteq \boldsymbol{R} \to \boldsymbol{R}$ and in Figs. 2.7, 2.8 and 2.9, $f : \boldsymbol{D} \subseteq \boldsymbol{R}^2 \to \boldsymbol{R}$. This is because the differences between a strictly quasiconcave function, a semistrictly quasiconcave function, and a quasiconcave function can be shown more clearly in a single view drawing.

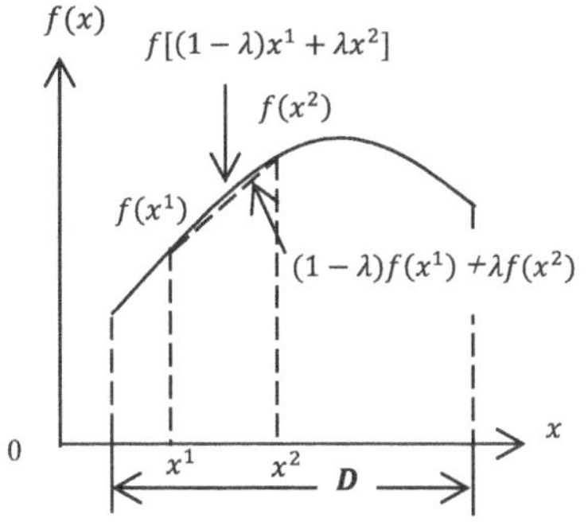

Fig. 2.2 f is a strictly concave function on $\boldsymbol{R}$

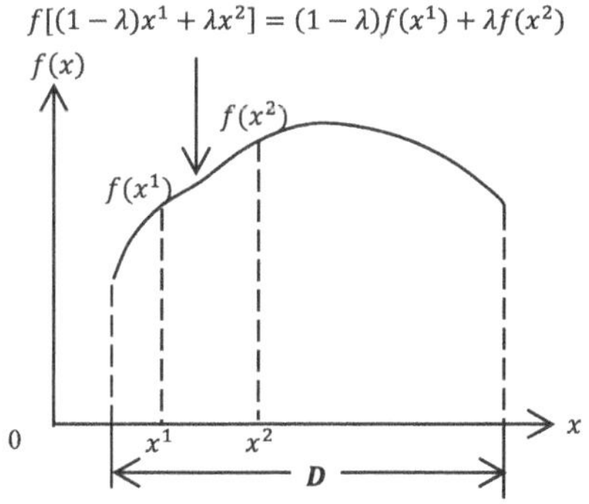

Fig. 2.3 f is not a strictly concave function on (x^1, x^2) but a concave function on $\boldsymbol{R}$

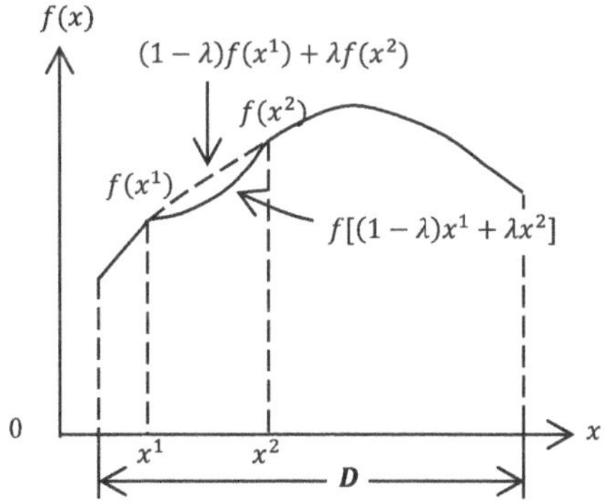

Fig. 2.4 f is not a concave function on (x^1, x^2) but a semistrictly quasiconcave function on $\boldsymbol{D}$

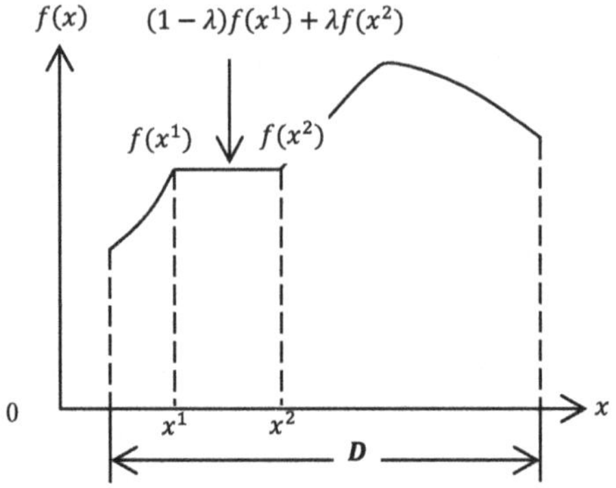

Fig. 2.5 f is not a strictly quasiconcave function on (x^1, x^2) but a quasiconcave function on $\boldsymbol{D}$

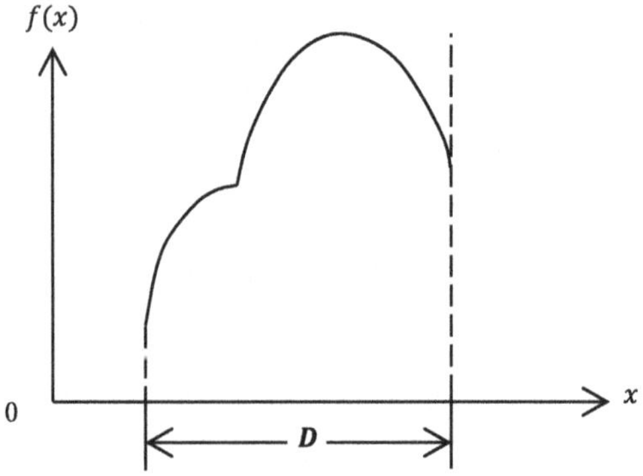

Fig. 2.6 f is a pseudconcave function on $\boldsymbol{D}$

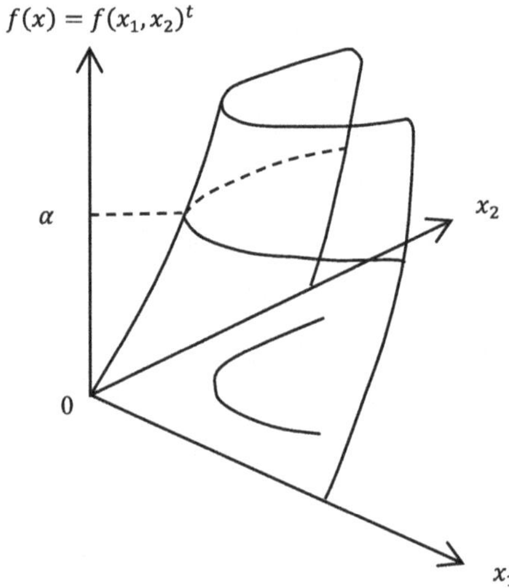

Fig. 2.7 f is a strictly quasiconcave, semistrictly quasiconcave and quasiconcavefunction

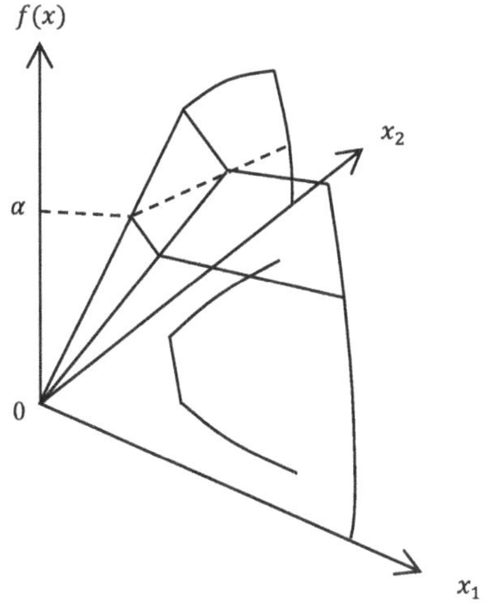

Fig. 2.8 f is a semistrictly quasiconcave and a quasiconcave function, but not a strictly quasiconcave function

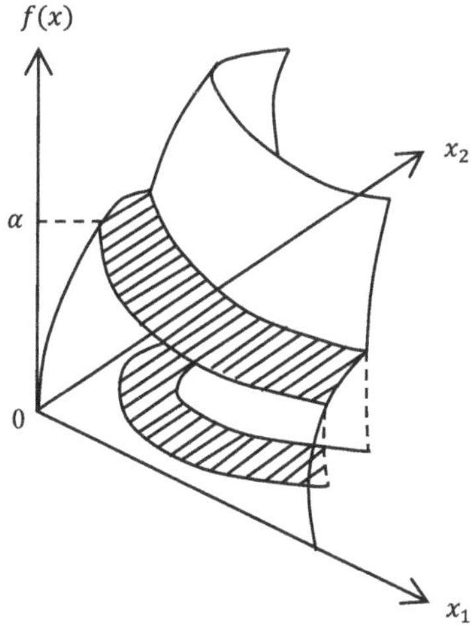

Fig. 2.9 f is a quasiconcave function, but neither a semistrictly quasiconcave function nor a strictly quasiconcave function

(Notes on ⟨Definition 2.2⟩ to ⟨Definition 2.8⟩): The fact that f is a strictly concave function on $\boldsymbol{D}$ (⟨Definition 2.2⟩) is sometimes expressed as f being a strictly concave function or f being a strictly concave function. The same shall apply with ⟨Definition 2.3⟩ to ⟨Definition 2.8⟩.

(Notes on ⟨Definition 2.2⟩, ⟨Definition 2.3⟩): The fact that the real-valued function f defined on a set $\boldsymbol{D}$ of convex functions in $\boldsymbol{R}^n$ is a strictly concave function on $\boldsymbol{D}$ indicates that for any different $\boldsymbol{x}^1, \boldsymbol{x}^2 \in \boldsymbol{D}$ (any $\boldsymbol{x}^1, \boldsymbol{x}^2 \in \boldsymbol{D}$), any point from ($\boldsymbol{x}^1$,

$x^2) = \{(1 - \lambda)x^1 + x^2 | 0 < \lambda < 1\}$ ($[x^1, x^2] = \{(1 - \lambda)x^1 + x^2 | 0 \leqq \lambda \leqq 1\}$) and a function of $(1 - \lambda)x^1 + \lambda x^2$, $=f[(1 - \lambda)x^1 + \lambda x^2]$, is greater (is not smaller) than the values on the open line connecting $f(x^1)$ and $f(x^2) = (1 - \lambda)f(x^1) + \lambda f(x^2)$. Thus, an advantage of a strictly concave (concave) function is that its local maximum value is the unique maximum (maximum value) (Figs. 2.2 and 2.3).

(Notes on ⟨Definition 2.4⟩ and ⟨Definition 2.5⟩): f is a point to point mapping hence $f(x^1) \neq f(x^2)$ indicates $x^1 \neq x^2$ but even if $x^1 \neq x^2$, it can be the case that $f(x^1) = f(x^2)$. Therefore, ⟨Definition 2.4⟩, as long as $x^1 \neq x^2$, irrespective of the difference between $f(x^1)$ and $f(x^2)$, demand that the following hold for all $\lambda \in (0, 1)$.

$$f\left[(1-\lambda)x^1 + x^2\right] > min\{f(x^1), f(x^2)\}$$

However, ⟨Definition 2.5⟩ requires that the aforementioned inequality holds only when $x^1 \neq x^2$ and $f(x^1) \neq f(x^2)$. A semistrictly quasiconcave function is broader than a strictly quasiconcave function.

(Notes on function names in ⟨Definition 2.4⟩ and ⟨Definition 2.5⟩): We will note here that a unified function name is as yet given for ⟨Definition 2.4⟩ and ⟨Definition 2.5⟩ similar to a quasiconcave function in ⟨Definition 2.6⟩.

⟨Definition 2.4⟩ Generally, in economics, a strictly quasiconcave function often refers to ⟨Definition 2.4⟩. Ortega and Rheinboldt (1970), Kemp and Kimura (1978), Avriel et al. (1988), and Takayama (1993) show real examples of this. However, there are other names such as Ponstein (1967)'s unnamed function, Thompson and Parke (1973) as X-concave function, Avriel (1976) and Bazaraa et al. (1993) as a strongly quasiconcave function.

⟨Definition 2.5⟩ Mangasarian (1969) conducted analyses relating to ⟨Definition 2.2⟩, ⟨Definition 2.3⟩, ⟨Definition 2.5⟩, ⟨Definition 2.6⟩ and ⟨Definition 2.8⟩; hence, it is probably the case that he referred to ⟨Definition 2.5⟩ as a strongly quasiconcave function in the sense that it is a stronger quasiconcave function than in ⟨Definition 2.6⟩. Avriel (1976), Bazaraa et al. (1993) followed in his footsteps. Moreover, Martos (1975), Thompson and Parke (1973), Takayama (1993) called⟨Definition 2.5⟩ as an explicitly quasiconcave function and Hanson (1964) referred to it as functionally concave. In this book, because ⟨Definition 2.5⟩ call in between ⟨Definition 2.4⟩ and ⟨Definition 2.6⟩, it is been referred to as semistrictly quasiconcave in accordance with Avriel et al. (1988).

(Notes on ⟨Definition 2.7⟩, ⟨Definition 2.8⟩): ⟨Definition 2.8⟩ was presented by Mangasarian (1965) as a sort of an intermediary concept in between ⟨Definition 2.3⟩ and ⟨Definition 2.5⟩. Similarly, ⟨Definition 2.7⟩, which is a slightly strengthened version of ⟨Definition 2.8⟩, can be said to position in between ⟨Definition 2.2⟩ and ⟨Definition 2.4⟩. The various theorems in the next section will confirm these points.

2.3 Basic Characteristics and Mutual Relationships

Firstly, let us explain a lemma as a preparation to [Theorem 2.1] and [Theorem 2.2].

[Lemma 2.1] (2.A1) is considered as a premise to think about the following seven propositions.

① f is a strictly concave function.
② $\boldsymbol{H}_f \equiv \{(\boldsymbol{x}, \alpha))|\boldsymbol{x} \in \boldsymbol{D}, \alpha \in \boldsymbol{R}, f(\boldsymbol{x}) \geqq \alpha\} \subseteq \boldsymbol{R}^{n+1}$ is a convex set.
However, $\boldsymbol{H}_f$ is an abbreviation of $\bigcup_{\alpha \in \boldsymbol{R}} \{(\boldsymbol{x}, \alpha)|\boldsymbol{x} \in \boldsymbol{D}, f(\boldsymbol{x}) \geqq \alpha\}$.
③ f is a concave function.
④ f is a quasiconcave function.
⑤ For any $\alpha \in \boldsymbol{R}$, $L_f(a) \equiv \{\boldsymbol{x} \in \boldsymbol{D}|f(\boldsymbol{x}) \geqq \alpha\} \subseteq \boldsymbol{R}^n$ is a convex set.
⑥ f is a strictly quasiconcave function.
⑦ f is a semistrictly quasiconcave function.

In this case, the following agreement holds.

(a) ①⇒②⇒③⇒④⇒⑤
(b) ⑥⇒⑤⇒④
(c) ①⇒⑥
(d) ③⇒⑦
(e) ⑥⇒⑦

(Proof)

(a)
① ⇒ ② Let $(\boldsymbol{x}^i, \alpha_i)(i = 1, 2)$ be any two points of $\boldsymbol{H}_f$. If $\boldsymbol{x}^1 = \boldsymbol{x}^2$, for any $\lambda \in [0, 1]$,

$$\begin{aligned} f[1-\lambda]x^1 + \lambda x^2 &= f(x^i) \quad (i = 1, 2) \\ &= (1-\lambda)f(\boldsymbol{x}^1) + \lambda f(\boldsymbol{x}^2) \\ &\geqq (1-\lambda)\alpha_1 + \lambda\alpha_2. \end{aligned}$$

From $\boldsymbol{x}^1 = \boldsymbol{x}^2 \in \boldsymbol{D}$ and $(1 - \lambda)\alpha_1 + \lambda\alpha_2 \in \boldsymbol{R}$, $(1 - \lambda)(\boldsymbol{x}^1, \alpha_1) + \lambda(\boldsymbol{x}^2, \alpha_2) \in \boldsymbol{H}_f$ is evident. Next, let $\boldsymbol{x}^1 \neq \boldsymbol{x}^2$. For $\lambda = 0$ (or 1), $(1 - \lambda)\boldsymbol{x}^1 + \lambda\boldsymbol{x}^2 = \boldsymbol{x}^1$ (or $\boldsymbol{x}^2$) and this gets to the proof of $\boldsymbol{x}^1 = \boldsymbol{x}^2$. Therefore, if you consider any $\lambda \in (0, 1)$, for this λ,

$$\begin{aligned} f\left[(1-\lambda)\boldsymbol{x}^1 + \lambda\boldsymbol{x}^2\right] &> (1-\lambda)f\boldsymbol{x}^1 + \lambda f(\boldsymbol{x}^2) \\ &\geqq (1-\lambda)\alpha_1 + \lambda\alpha_2 \end{aligned} \tag{2.1}$$

holds. Moreover, from $\boldsymbol{x}^i \in \boldsymbol{D}$ and $\alpha_i \in \boldsymbol{R}$ $(i = 1, 2)$,

$$(1-\lambda)\boldsymbol{x}^1 + \lambda\boldsymbol{x}^2 \in \boldsymbol{D} \text{ and } (1-\lambda)\alpha_1 + \lambda\alpha_2 \in \boldsymbol{R}. \tag{2.2}$$

Thus, from (2.1) and (2.2), $(1 - \lambda)(\boldsymbol{x}^i, \alpha_i) + \lambda(\boldsymbol{x}^2, \alpha_2) \in \boldsymbol{H}_f$. Therefore $\boldsymbol{H}_f$ is certainly a convex set.
② ⇒ ③ From the definition of $\boldsymbol{H}_f$, for any $\boldsymbol{x}^i \in \boldsymbol{D}$ $(i = 1, 2)$, $(\boldsymbol{x}^i, f(\boldsymbol{x}^i)) \in \boldsymbol{H}_f$. Moreover, $\boldsymbol{H}_f$ is a convex set; hence, for any $\lambda \in [0, 1]$, $(1 - \lambda)(\boldsymbol{x}^1, f(\boldsymbol{x}^1)) + \lambda(\boldsymbol{x}^2, f(\boldsymbol{x}^2)) \in \boldsymbol{H}_f$. Therefore,

$$f\left[(1-\lambda)\boldsymbol{x}^1+\lambda\boldsymbol{x}^2\right] \geqq (1-\lambda)f\left(\boldsymbol{x}^1\right)+\lambda f\left(\boldsymbol{x}^2\right) \tag{2.3}$$

holds and f becomes a concave function.
③ ⇒ ④ If f is a concave function, for any $\boldsymbol{x}^1$, $\boldsymbol{x}^2 \in \boldsymbol{D}$ and $\lambda \in [0, 1]$, (2.3) holds. Hence, if we let $min\{f(\boldsymbol{x}^1), f(\boldsymbol{x}^2)\} = \alpha$,

$$\begin{aligned} f\left[(1-\lambda)\boldsymbol{x}^1+\lambda\boldsymbol{x}^2\right] &\geqq (1-\lambda)\alpha+\lambda\alpha \\ &= \alpha, \end{aligned}$$

The quasiconvexity of f has been shown.
④ ⇒ ⑤ If we let α be any real numbers and $\boldsymbol{x}^1$ and $\boldsymbol{x}^2$ be any two points of $\boldsymbol{L}_f(\alpha)$, from the assumed quasiconcaveness of f and the definition of $\boldsymbol{L}_f(\alpha)$, for all $\lambda \in [0, 1]$,

$$\begin{aligned} f\left[(1-\lambda)\boldsymbol{x}^1+\lambda\boldsymbol{x}^2\right] &\geqq min\left\{f\left(\boldsymbol{x}^1\right), f\left(\boldsymbol{x}^2\right)\right\} \\ &\geqq \alpha. \end{aligned}$$

Therefore, for any $\lambda \in [0, 1]$. $(1 - \lambda)\boldsymbol{x}^1 + \lambda\boldsymbol{x}^2 \in \boldsymbol{L}_f(\alpha)$ can be derived and $\boldsymbol{L}_f(\alpha)$ becomes a convex set.
(b)
⑥ ⇒ ⑤ Let α be any real number and $\boldsymbol{x}^1$ and $\boldsymbol{x}^2$ any two points from $\boldsymbol{L}_f(\alpha)$. If $\boldsymbol{x}^1 = \boldsymbol{x}^2$, for of any $\lambda \in [0, 1]$,

$$\begin{aligned} f\left[(1-\lambda)\boldsymbol{x}^1+\lambda\boldsymbol{x}^2\right] &= f\boldsymbol{x}^i \quad (i=1,2) \\ &= min\left\{f\left(\boldsymbol{x}^1\right), f\left(\boldsymbol{x}^2\right)\right\} \\ &\geqq \alpha \end{aligned}$$

holds. When $\lambda = 0(1)$, for all $\boldsymbol{x}^1$, $\boldsymbol{x}^2 \in \boldsymbol{L}_f(\alpha)$,

$$\begin{aligned} f\left[(1-\lambda)\boldsymbol{x}^1+\lambda\boldsymbol{x}^2\right] &= f\left(\boldsymbol{x}^1\right)\left(f\left(\boldsymbol{x}^2\right)\right) \\ &\geqq min\left\{f\left(\boldsymbol{x}^1\right), f\left(\boldsymbol{x}^2\right)\right\} \\ &\geqq \alpha. \end{aligned}$$

Moreover, for any different $\boldsymbol{x}^1, \boldsymbol{x}^2 \in \boldsymbol{L}_f(\alpha)$ and $\lambda \in (0, 1)$,

$$\begin{aligned} f\left[(1-\lambda)\boldsymbol{x}^1+\lambda\boldsymbol{x}^2\right] &> min\left\{f\left(\boldsymbol{x}^1\right), f\left(\boldsymbol{x}^2\right)\right\} \\ &\geqq \alpha. \end{aligned}$$

Therefore, for any $\boldsymbol{x}^1, \boldsymbol{x}^2 \in \boldsymbol{L}_f(\alpha)$ and $\lambda \in [0, 1]$, $(1 - \lambda)\boldsymbol{x}^1 + \lambda\boldsymbol{x}^2 \in \boldsymbol{L}_f(\alpha)$ holds.
⑤ ⇒ ④ Let us define $\alpha = min\ \{f(\boldsymbol{x}^1), f(\boldsymbol{x}^2)\}$ for any $\boldsymbol{x}^1$ and $\boldsymbol{x}^2 \in \boldsymbol{D}.\boldsymbol{L}_f(\alpha)$ is a convex set. Hence, for all $\lambda \in [0, 1]$, $(1 - \lambda)\boldsymbol{x}^1 + \lambda\boldsymbol{x}^2 \in \boldsymbol{L}_f(\alpha)$. Thus, from the definition of $\boldsymbol{L}_f(\alpha)$,

$$f\left[(1-\lambda)\boldsymbol{x}^1+\lambda\boldsymbol{x}^2\right] \geqq \alpha = min\{f(\boldsymbol{x}^1), f(\boldsymbol{x}^2)\}$$

is derived.
(c) ① ⇒ ⑥ f is a strictly concave function. Hence, for any different $\boldsymbol{x}^1$, $\boldsymbol{x}^2 \in \boldsymbol{D}$ and $\lambda \in (0, 1)$,

$$\begin{aligned} f\left[(1-\lambda)\boldsymbol{x}^1+\lambda\boldsymbol{x}^2\right] &> (1-\lambda)f(\boldsymbol{x}^1)+\lambda f(\boldsymbol{x}^2) \\ &\geqq min\{f(\boldsymbol{x}^1), f(\boldsymbol{x}^2)\} \end{aligned}$$

holds and the conclusion is obtained.
(d) ③ ⇒ ⑦ If f is not a semistrictly concave function, there exist $\overline{\boldsymbol{x}}^1, \overline{\boldsymbol{x}}^2 \in \boldsymbol{D}$ and $\overline{\lambda} \in (0, 1)$ which are $f(\overline{\boldsymbol{x}}^1) \neq f(\overline{\boldsymbol{x}}^2)$ and

$$f\left[(1-\overline{\lambda})\overline{\boldsymbol{x}}^1+\lambda\overline{\boldsymbol{x}}^2\right] \leqq min\{f(\overline{\boldsymbol{x}}^1), f(\overline{\boldsymbol{x}}^2)\}. \tag{2.4}$$

Without losing generality, if $f(\overline{\boldsymbol{x}}^1) < f(\overline{\boldsymbol{x}}^2)$, from the assumed concaveness of f, for $\overline{\boldsymbol{x}}^i$ $(i = 1, 2)$ and $\overline{\lambda}$,

$$\begin{aligned} f\left[(1-\overline{\lambda})\overline{\boldsymbol{x}}^1+\lambda\overline{\boldsymbol{x}}^2\right] &\geqq (1-\overline{\lambda})f(\overline{\boldsymbol{x}}^1)+\overline{\lambda}f(\overline{\boldsymbol{x}}^2) \\ &> (1-\overline{\lambda})f(\overline{\boldsymbol{x}}^1)+\overline{\lambda}f(\overline{\boldsymbol{x}}^1) \\ &= f(\overline{\boldsymbol{x}}^1) = min\{f(\overline{\boldsymbol{x}}^1), f(\overline{\boldsymbol{x}}^2)\} \end{aligned}$$

but this contradicts (4).
(e) ⑥ ⇒ ⑦ Any $\boldsymbol{x}^1$, $\boldsymbol{x}^2 \in \boldsymbol{D}$ satisfying $f(\boldsymbol{x}^1) \neq f(\boldsymbol{x}^2)$ is such that clearly $\boldsymbol{x}^1 \neq \boldsymbol{x}^2$; hence, from the strict quasiconcaveness of f, the sought-after conclusion is immediately obtained. QED

This [Lemma 2.1] is useful for deriving [Theorem 2.1] and [Theorem 2.2] (Figs. 2.10 and 2.11).

[Theorem 2.1] Under (2.A1), the following holds.

(i) For f to be a concave function, the necessary and sufficient conditions is that

$$\boldsymbol{H}_f = \{(\boldsymbol{x}, \alpha) | \boldsymbol{x} \in \boldsymbol{D}, \alpha \in \boldsymbol{R}, f(\boldsymbol{x}) \geqq \alpha\} \subseteq \boldsymbol{R}^{n+1}$$

is a convex set.
(ii) For f to be a quasiconcave function, the necessary and sufficient condition is that, for any $\alpha \in \boldsymbol{R}$,

$$\boldsymbol{L}_f(\alpha) = \{\boldsymbol{x} \in \boldsymbol{D} | f(\boldsymbol{x}) > \alpha\}$$

is a convex set.

(Proof)

(i)
(Necessity) Let $(\boldsymbol{x}^1, \alpha_i)(i = 1, 2)$ be any two points of $\boldsymbol{H}_f$ and select λ arbitrarily from $[0, 1]$. f is a concave function, and thus (1) in the proof of [Lemma 2.1: (a); ①⇒②] holds with loose inequalities. Hence, the conclusion is evident.
(Sufficiency) It has been proven in [Lemma 2.1: (a); ②⇒③].
(ii) Necessity has been shown in [Lemma 2.1: (a); ④⇒⑤], and sufficiency in [Lemma 2.1: (b); ⑤⇒④]. QED

(Notes on [Theorem 2.1])
Notes on (i): The set $\boldsymbol{H}_f = \{(\boldsymbol{x}, \alpha) | \boldsymbol{x} \in \boldsymbol{D}, \alpha \in \boldsymbol{R}, f(\boldsymbol{x}) \geqq \alpha\} \subseteq \boldsymbol{R}^{n+1}$ in [Theorem 2.1; (i)] is known as a hypograph of f and the set $\boldsymbol{E}_f = \{(\boldsymbol{x}, \beta) | \boldsymbol{x} \in \boldsymbol{D}, \beta \in \boldsymbol{R}, f(\boldsymbol{x}) \leqq \beta\} \subseteq \boldsymbol{R}^{n+1}$ is known as an epigraph of f (Figs. 2.10 and 2.11).
Notes on (ii): $L_f(\alpha) = \{\boldsymbol{x} \in \boldsymbol{D} | f(\boldsymbol{x}) \geqq \alpha\}$ in [Theorem 2.1;(ii)] is called an ***α* upper-level set** and $L_f(\beta) = \{\boldsymbol{x} \in \boldsymbol{D} | f(\boldsymbol{x}) \leqq \beta\}$ is called a ***β* lower level set** (Figs. 2.12 and 2.13).

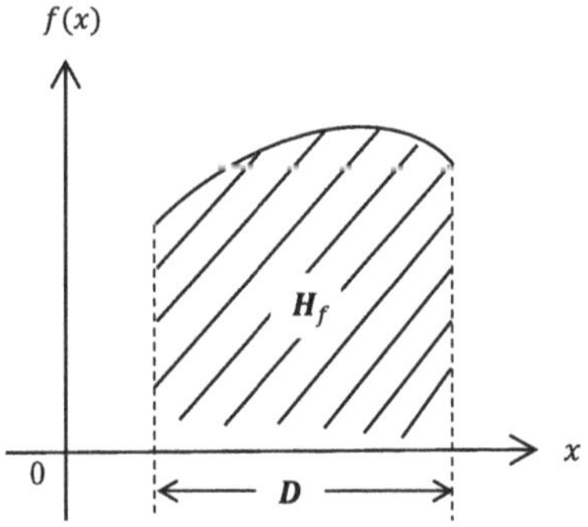

Fig. 2.10 Convex hypograph of a concave function $f(x)(f: \boldsymbol{D} \subseteq \boldsymbol{R} \to \boldsymbol{R})$

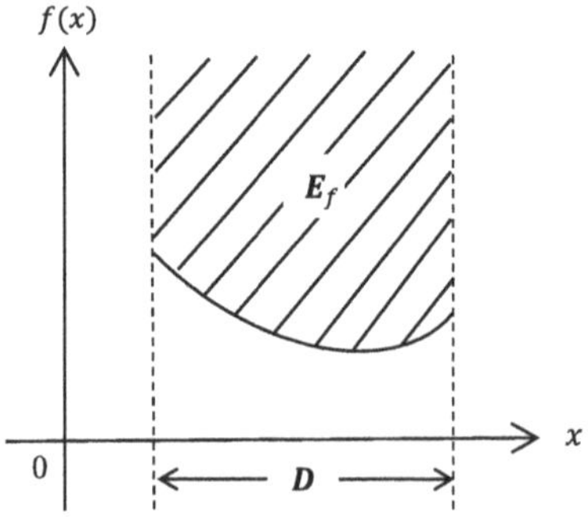

Fig. 2.11 Convex epigraph of a convex function $f(x)(f: \boldsymbol{D} \subseteq \boldsymbol{R} \to \boldsymbol{R})$

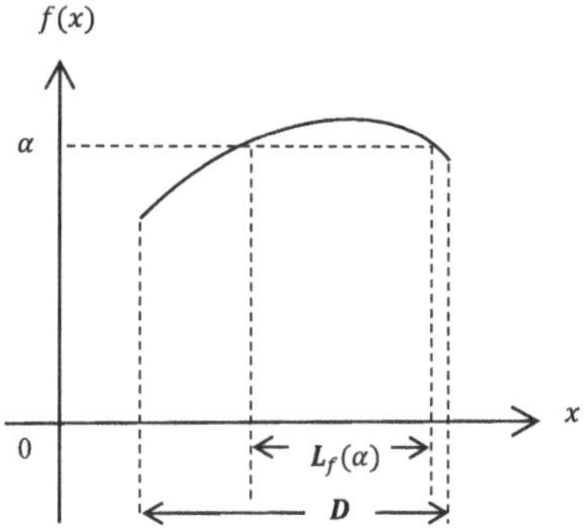

Fig. 2.12 $L_f(\alpha)$ of a convex function $f(x)(f: \boldsymbol{D} \subseteq \boldsymbol{R} \to \boldsymbol{R})$

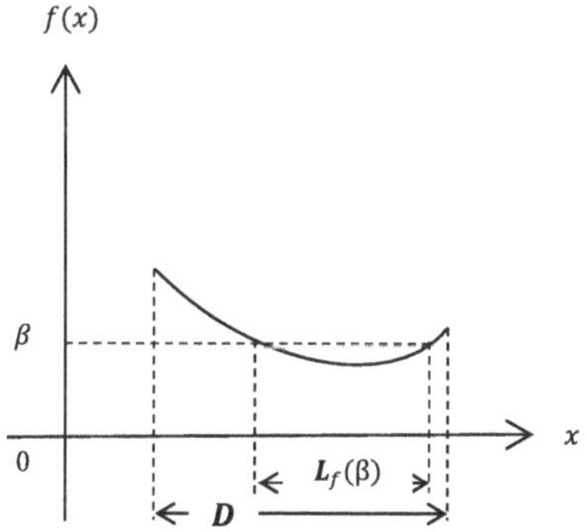

Fig. 2.13 $L_f(\beta)$ of a convex function $f(x)(f: \boldsymbol{D} \subseteq \boldsymbol{R} \to \boldsymbol{R})$

[Theorem 2.2] Under (2.A1), the following holds.

(i) [f:Strictly concave function] ⇒ [f:Concave function] ⇒ [f:Quasiconcave function]
(ii) [f:Strictly concave function] ⇒ [f:Strictly quasiconcave function] ⇒ [f: Quasiconcave function]
(iii) [f:□ function] ⇒ [f:Semistrictly quasiconcave function]
(iv) [f: strictly quasiconcave function] ⇒ [f:semi-strictly quasiconcave function]
(v) When f is upper semicontinuous,
[f: Semistrictly quasiconcave function] ⇒ [f: Quasiconcave function]
(vi) When $\bar{\boldsymbol{x}} \in \boldsymbol{D}$ is a local maximum of $f(\boldsymbol{x})$,
(vi-i) If f is a semistrictly quasiconcave function, $f(\boldsymbol{x})$ achieves its maximum value at $\boldsymbol{x} = \bar{\boldsymbol{x}}$.
and
(vi-ii) If f is a strictly quasiconcave function, $f(\bar{\boldsymbol{x}})$ is a unique maximum value.
(vii) When $\bar{\boldsymbol{x}} \in \boldsymbol{D}$ is a local maximum point of $f(\boldsymbol{x})$,

(vii-i) If f is a concave function, $f(\boldsymbol{x})$ achieves its maximum value on $\boldsymbol{x} = \overline{\boldsymbol{x}}$ and

(vii-ii) If f is a strictly concave function, $f(\overline{\boldsymbol{x}})$ becomes a unique maximum value.

(Proof)

(i) From [Lemma 2.1: (a); ①⇒②⇒③⇒④], it is self-apparent.
(ii) [Lemma 2.1: (c), (b)] should be used.
(iii), (iv) (iii) ((iv)) is shown in [Lemma 2.1: (d) ((e))].
(v) Let $\boldsymbol{x}^1$ and $\boldsymbol{x}^2$ be any two points in $\boldsymbol{D}$ and define for any $\lambda \in [0, 1]$, $x(\lambda) = (1 - \lambda)\boldsymbol{x}^1 + \lambda \boldsymbol{x}^2$, and then prove using the following groupings:

(1) $\boldsymbol{x}^1 = \boldsymbol{x}^2$ and $\lambda \in [0, 1]$
(2) When $\boldsymbol{x}^1 \neq \boldsymbol{x}^2$,
(2-1) $\lambda = 0$ or $\lambda = 1$
(2-2) $\lambda \in (0, 1)$ and $f(\boldsymbol{x}^1) \neq f(\boldsymbol{x}^2)$
(2-3) $\lambda \in (0, 1)$ and $f(\boldsymbol{x}^1) = f(\boldsymbol{x}^2)$
(1) If $\boldsymbol{x}^1 = \boldsymbol{x}^2$, for any $\lambda \in [0, 1]$,

$$f\left[(1-\lambda)\boldsymbol{x}^1 + \lambda \boldsymbol{x}^2\right] = f\left(\boldsymbol{x}^1\right) = f\left(\boldsymbol{x}^2\right).$$

(2-1) From the definition of $\boldsymbol{x}(\lambda)$, $\boldsymbol{x}(0) = \boldsymbol{x}^1$ and $\boldsymbol{x}(1) = \boldsymbol{x}^2$. Thus, $f[\boldsymbol{x}(0)] = f(\boldsymbol{x}^1)$ and $f[\boldsymbol{x}(1)] = f(\boldsymbol{x}^2)$. Hence, if $\lambda = 0$ or $\lambda = 1$,

$$f[\boldsymbol{x}(\lambda)] \geqq min\left\{f\left(\boldsymbol{x}^1\right), f\left(\boldsymbol{x}^2\right)\right\}.$$

(2-2) When $f(\boldsymbol{x}^1) \neq f(\boldsymbol{x}^2)$, from the assumed semistrictly quasiconcaveness of f, we immediately obtain the following for any $\lambda \in (0, 1)$:

$$f\left[(1-\lambda)\boldsymbol{x}^1 + \lambda \boldsymbol{x}^2\right] > min\left\{f\left(\boldsymbol{x}^1\right), f\left(\boldsymbol{x}^2\right)\right\}.$$

(2-3) If under $f(\boldsymbol{x}^1) = f(\boldsymbol{x}^2)$, for any different $\boldsymbol{x}^1$, $\boldsymbol{x}^2 \in \boldsymbol{D}$ and $\lambda \in (0, 1)$,

$$f\left[(1-\lambda)\boldsymbol{x}^1 + \lambda \boldsymbol{x}^2\right] \geqq f\left(\boldsymbol{x}^1\right) = f\left(\boldsymbol{x}^2\right) \tag{2.5}$$

is shown, the proof is complete. Here, if we contradict (2.5),

$$f\left[(1-\lambda)\boldsymbol{x}^1 + \lambda \boldsymbol{x}^2\right] < f\left(\boldsymbol{x}^1\right) = f\left(\boldsymbol{x}^2\right)$$

there exist $\lambda \in (0, 1)$ and $\boldsymbol{x}\left(\widehat{\lambda}\right) \equiv \left(1 - \widehat{\lambda}\right)\boldsymbol{x}^1 + \widehat{\lambda}\boldsymbol{x}^2 \in \boldsymbol{D}$ that satisfy the above. Furthermore, if we let:

$$Q=\{x\in(x^1,x^2)|f(x)<f(x^1)=f(x^2)\}$$

it is clear that $x(\widehat{\lambda})\in Q$, and thus $Q\neq\phi$. Moreover, f is upper semi-continuous, and hence Q is an open set in respect of (x^1,x^2).[3] Hence, there exists certain $\tilde{\lambda}\in(0,1)$ and $x(\tilde{\lambda})\equiv(1-\tilde{\lambda})x^1+\tilde{\lambda}x(\widehat{\lambda})\in Q$. Moreover, D is a convex set and hence $x(\tilde{\lambda})\in D$. Therefore, from $x(\widehat{\lambda})\in Q$ and from $f\left[x(\widehat{\lambda})\right]<f(x^1)$ and $x(\tilde{\lambda})\in Q$, and $f[x(\tilde{\lambda})]<f(x^1)=f(x^2)$. Furthermore, noting $x(\tilde{\lambda})\in\left(x^1,x(\widehat{\lambda})\right)\subset D$ and $x(\widehat{\lambda})\in(x(\tilde{\lambda}),x^2)\subset D$, f is a semistrictly quasiconcave function, and hence

$$\begin{aligned}f[x(\tilde{\lambda})]&>min\left\{f(x^1),f\left[x(\widehat{\lambda})\right]\right\}\\&=f\left[x(\widehat{\lambda})\right]\end{aligned}$$

and

$$\begin{aligned}f\left[x(\widehat{\lambda})\right]&>min\{f[x(\tilde{\lambda})],f(x^2)\}\\&=f[x(\tilde{\lambda})].\end{aligned}$$

On the other hand, combining both these equations, $f[x(\tilde{\lambda})]>f\left[x(\widehat{\lambda})\right]>f[x(\tilde{\lambda})]$, which is self-contradicting.

(vi)

(vi-i) $\bar{x}\in D$ is a local maximum point. Consequently, there exists a suitable $\delta>0$ and for any $x\in B\ (\bar{x};\delta)\cap D$, $f(x)\leqq f(\bar{x})$ holds. By denying the conclusion and considering $f(\bar{x})$ not be a maximum value, there exists $\widehat{x}\in D$, which is different from $\bar{x}$ and satisfies $f(\widehat{x})>f(\bar{x})$. If $\widehat{x}\in B\ (\bar{x};\delta)$ and $f(\widehat{x})\leqq f(\bar{x})$, and thus contradicts the assumption $(f(\widehat{x})>f(\bar{x}))$ of proof by contradiction, and hence let us $\widehat{x}\notin B\ (\bar{x};\delta)$. At this time, for any $\lambda\in(0,1)$,

$$\begin{aligned}f[(1-\lambda)\bar{x}+\lambda\widehat{x}]&>min\{f(\bar{x}),f(\widehat{x})\}\\&=f(\bar{x})\end{aligned}\tag{2.6}$$

is obtained. However, if $\lambda=\dfrac{\delta}{\|\widehat{x}-\bar{x}\|}>0$ is selected and consider the fact that D is a convex set, $(1-\lambda)\bar{x}+\lambda\widehat{x}\in B\ (\bar{x};\delta)\cap D$. Thus, for this $\lambda=\dfrac{\delta}{\|\widehat{x}-\bar{x}\|}$,

[3] If you consider the domain D of f in terms of only $(x^1,x^2)=\{(1-\lambda)x^1+\lambda x^2,0<\lambda<1\}$ if f is upper semi-continuous on D, f is upper semi-continuous on (x^1,x^2). Hence Q is an open set in respect of (x^1,x^2). See Rudin (1976, p.36, Theorem 2.30). For upper semi-continuity, please see ⟨Definition 14⟩(Appendix).

$$f[(1-\lambda)\bar{\boldsymbol{x}}+\lambda\hat{\boldsymbol{x}}] \leqq f(\bar{\boldsymbol{x}})$$

but this contradicts (2.6).

(vi-ii) If f is a strictly quasiconcave function, from (iv), f is a semistrictly quasiconcave function. Therefore, from (vi-i), $f(\bar{\boldsymbol{x}})$ is a maximum value. Hence, we will demonstrate that $\boldsymbol{x}=\bar{\boldsymbol{x}}$ is a unique maximum point. Here, if it is assumed that there exists a $\tilde{\boldsymbol{x}} \in \boldsymbol{D}$ different from $\bar{\boldsymbol{x}}$ satisfying $f(\tilde{\boldsymbol{x}})=f(\bar{\boldsymbol{x}})$, then for any $\lambda \in (0, 1)$,

$$\begin{aligned} f[(1-\lambda)\bar{\boldsymbol{x}}+\lambda\tilde{\boldsymbol{x}}] &> min\{f(\bar{\boldsymbol{x}}), f(\tilde{\boldsymbol{x}})\} \\ &= f(\bar{\boldsymbol{x}}) = f(\tilde{\boldsymbol{x}}). \end{aligned} \tag{2.7}$$

Moreover, $\boldsymbol{D}$ is a convex set, and thus $(1-\lambda)\bar{\boldsymbol{x}}+\lambda\tilde{\boldsymbol{x}} \in \boldsymbol{D}$. Accordingly,

$$f[(1-\lambda)\bar{\boldsymbol{x}}+\lambda\tilde{\boldsymbol{x}}] \leqq f(\bar{\boldsymbol{x}})$$

is derived but the above inequality contradicts (2.7). This demonstrated the uniqueness of $\bar{\boldsymbol{x}} \in \boldsymbol{D}$.

(vii) If f is a concave (strictly concave) function, from (iii) ((ii)), f is a semistrictly quasiconcave (strictly quasiconcave) function. Hence, conclusion is immediately obtained from (vi-i) (vi-ii). QED

(Notes on [Theorem 2.2])

Notes on (i)–(iv): The opposite implications of (i)–(iv) are not necessarily guaranteed (Figs. 2.3, 2.4, and 2.5).

Notes on (v): Even if f is a semistrictly quasiconcave function, if there is no hypothesis about the upper semicontinuity, there is a chance that it may not be a quasiconcave function, as demonstrated by Karamardian (1967)'s example.

Example 2.1

$$f(x)=\begin{cases} 0, & x=0 \\ 1, & x \neq 0 \end{cases}$$

$$(f: \boldsymbol{R} \rightarrow \boldsymbol{R})$$

f is a semistrictly quasiconcave function on $\boldsymbol{R}$ but when we let $x^1 = 1$, $x^2 = -1$, $\lambda = \frac{1}{2}$, we obtain:

$$\begin{aligned} f\left(\frac{1}{2}x^1+\frac{1}{2}x^2\right) &= f(0)=0<1 \\ &= min\{f(\boldsymbol{x}^1), f(\boldsymbol{x}^2)\}. \end{aligned}$$

Hence, it is not a quasiconcave function (See Fig. 2.14).

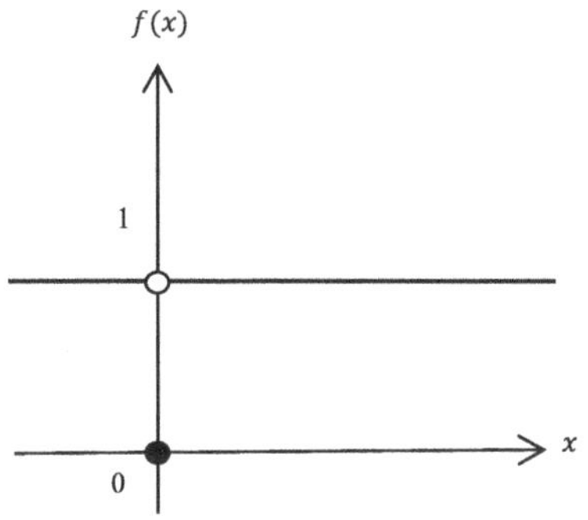

Fig. 2.14 Illustration of Example 2.1

Notes on (vi): Even if it is a quasiconcave function, there are cases when its local maximum value is not the maximum value. The example on the below shows this.

Example 2.2

$$f(x) = \begin{cases} x - 1, & x \geqq 1 \\ 0, & 0 < x < 1 \\ x, & x \leqq 0 \end{cases}$$

$$(f : \boldsymbol{R} \to \boldsymbol{R})$$

f is a quasiconcave function, for any $x \in (0, 1), f(x) = 0$ is a local maxima but is not a maximum value. However, f is a quasiconcave function. Thus, it should be understood at the same time that its local minimum $f(x) = 0$ does not become the minimum (See Fig. 2.15).

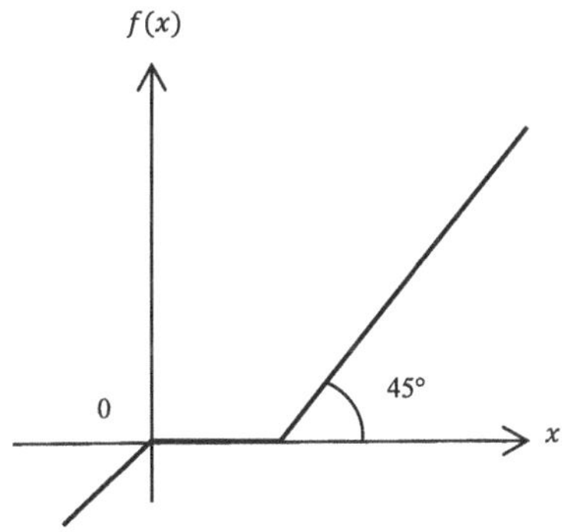

Fig. 2.15 Illustration of Example 2.2

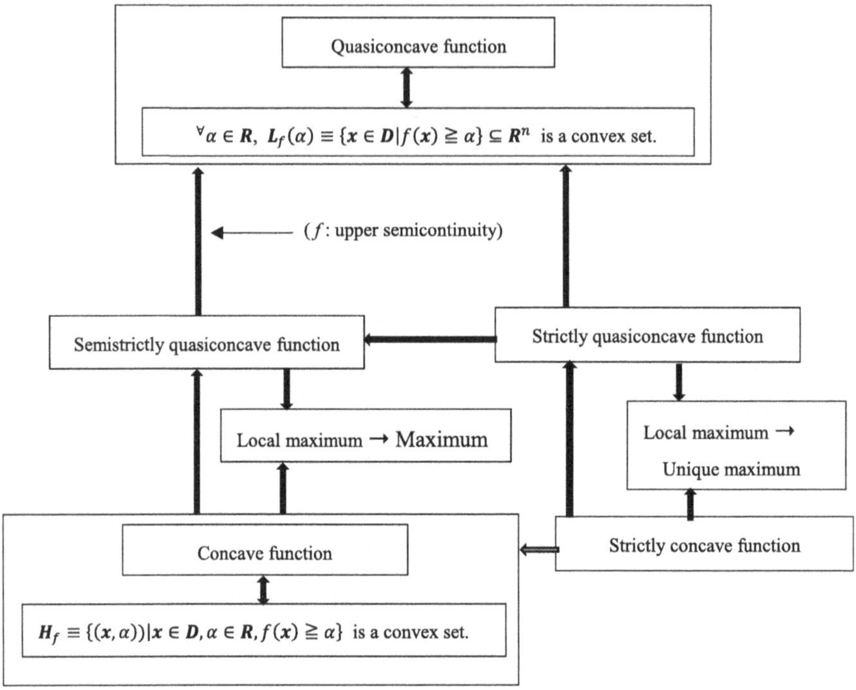

Fig. 2.16 The relationship shown by [Theorem 2.1] and [Theorem 2.2]

Figure 2.16 summarizes the results from [Theorem 2.1] and [Theorem 2.2].

[Theorem 2.3] The following hold under (2.A1) and (2.A2).

(i) For f to be a concave function, the necessary and sufficient condition is that

$$\text{for any } \boldsymbol{x}^1, \boldsymbol{x}^2 \in \boldsymbol{D},$$
$$\nabla f(\boldsymbol{x}^1)(\boldsymbol{x}^2 - \boldsymbol{x}^1) \geqq f(\boldsymbol{x}^2) - f(\boldsymbol{x}^1).$$

(ii) For f to be a strictly concave function, the necessary and sufficient condition is that

$$\text{for any different } \boldsymbol{x}^1, \boldsymbol{x}^2 \in \boldsymbol{D},$$
$$\nabla f(\boldsymbol{x}^1)(\boldsymbol{x}^2 - \boldsymbol{x}^1) > f(\boldsymbol{x}^2) - f(\boldsymbol{x}^1).$$

(iii) For f to be a quasiconcave function, the necessary and sufficient condition is that

$$\text{for any } \boldsymbol{x}^1, \boldsymbol{x}^2 \in \boldsymbol{D},$$
$$f(\boldsymbol{x}^2) \geqq f(\boldsymbol{x}^1) \Rightarrow \nabla f(\boldsymbol{x}^1)(\boldsymbol{x}^2 - \boldsymbol{x}^1) \geqq 0.$$

(Proof)

(i)
(Necessity) From the definition of a concave function, for any $\lambda \in [0, 1]$, $\boldsymbol{x}^1, \boldsymbol{x}^2 \in \boldsymbol{D}$,

$$\begin{aligned} f\left[(1-\lambda)\boldsymbol{x}^1 + \lambda\boldsymbol{x}^2\right] &= f\left[\boldsymbol{x}^1 + \lambda(\boldsymbol{x}^2 - \boldsymbol{x}^1)\right] \\ &\geqq (1-\lambda) f(\boldsymbol{x}^1) + \lambda f(\boldsymbol{x}^2) \\ &= f(\boldsymbol{x}^1) + \lambda\left[f(\boldsymbol{x}^2) - f(\boldsymbol{x}^1)\right] \end{aligned} \tag{2.8}$$

holds. Accordingly,

$$f\left[\boldsymbol{x}^1 + \lambda(\boldsymbol{x}^2 - \boldsymbol{x}^1)\right] - f(\boldsymbol{x}^1) \geqq \lambda\left[f(\boldsymbol{x}^2) - f(\boldsymbol{x}^1)\right]. \tag{2.8'}$$

If the differentiability of f and $\lambda > 0$ are considered,[4]

$$\begin{aligned} \frac{f[\boldsymbol{x}^1 + \lambda(\boldsymbol{x}^2 - \boldsymbol{x}^1)] - f(\boldsymbol{x}^1)}{\lambda} &= \nabla f(\boldsymbol{x}^1)(\boldsymbol{x}^2 - \boldsymbol{x}^1) + \alpha(\boldsymbol{x}^1,\ \lambda(\boldsymbol{x}^2 - \boldsymbol{x}^1))\left\|\boldsymbol{x}^2 - \boldsymbol{x}^1\right\| \\ &\geqq f(\boldsymbol{x}^2) - f(\boldsymbol{x}^1) \end{aligned}$$

is obtained. However, $\|\cdot\|$ is a vector norm. When $\lambda \to 0$, $\alpha(\boldsymbol{x}^1,\ \lambda(\boldsymbol{x}^2 - \boldsymbol{x}^1)) = 0$. Thus, [Theorem 2.3; (i)] is derived.
(Sufficiency) $\boldsymbol{D}$ is a convex set, and hence for any $\lambda \in [0, 1]$ and $\boldsymbol{x}^1, \boldsymbol{x}^2 \in \boldsymbol{D}$, $(1 - \lambda)\boldsymbol{x}^1 + \lambda\boldsymbol{x}^2 \in \boldsymbol{D}$. Therefore, by [Theorem 2.3; (i)],

$$f(\boldsymbol{x}^2) - f\left[(1-\lambda)\boldsymbol{x}^1 + \boldsymbol{x}^2\right] \leqq (1-\lambda)\nabla f\left[(1-\lambda)\boldsymbol{x}^1 + \boldsymbol{x}^2\right](\boldsymbol{x}^2 - \boldsymbol{x}^1) \tag{2.9}$$

$$f(\boldsymbol{x}^1) - f\left[(1-\lambda)\boldsymbol{x}^1 + \lambda\boldsymbol{x}^2\right] \leqq -\lambda\nabla f\left[(1-\lambda)\boldsymbol{x}^1 + \lambda\boldsymbol{x}^2\right](\boldsymbol{x}^2 - \boldsymbol{x}^1). \tag{2.10}$$

Here, if (2.9) is multiplied by λ and (2.10) by $(1 - \lambda)$, and each side are added together, for any $\lambda \in [0, 1]$,

$$(1-\lambda) f(\boldsymbol{x}^1) + \lambda f(\boldsymbol{x}^2) \leqq f\left[(1-\lambda)\boldsymbol{x}^1 + \lambda\boldsymbol{x}^2\right] \tag{2.11}$$

holds. Thus, f is a concave function.

[4] See ⟨Definition 16⟩ (Appendix).

(ii)
(Necessity) f is a strictly concave function. Hence, in (i)'s proof, $\lambda \in [0, 1], \boldsymbol{x}^1, \boldsymbol{x}^2 \in \boldsymbol{D}$ must be replaced with $\lambda \in (0, 1)$ and different $\boldsymbol{x}^1$, $\boldsymbol{x}^2 \in \boldsymbol{D}$, respectively, and (2.8) hence (2.8)′ must have the loose inequality replaced with a strict inequality.
(Sufficiency) For any different $\boldsymbol{x}^1$, $\boldsymbol{x}^2 \in \boldsymbol{D}$ and $\lambda \in (0, 1)$, the loose inequalities in (2.9)–(2.11) of (i)'s proof all hold with strong inequality. Thus, the conclusion is apparent.
(iii)
(Necessity) Choose any $\boldsymbol{x}^1, \boldsymbol{x}^2$. If $\boldsymbol{x}^1 = \boldsymbol{x}^2$, the conclusion is evident, and hence it only requires to be proven for the case except for $\boldsymbol{x}^1 \neq \boldsymbol{x}^2$. Here, for the proof's convenience, we let $min\{f(\boldsymbol{x}^1), f(\boldsymbol{x}^2)\} = f(\boldsymbol{x}^1)$. f is a quasiconcave function, and hence for any $\lambda \in [0, 1]$,

$$\begin{aligned} f\left[\boldsymbol{x}^1 + \lambda\left(\boldsymbol{x}^2 - \boldsymbol{x}^1\right)\right] &= f\left[(1-\lambda)\boldsymbol{x}^1 + \lambda\boldsymbol{x}^2\right] \\ &\geqq min\left\{f\left(\boldsymbol{x}^1\right), f\left(\boldsymbol{x}^2\right)\right\} \\ &= f\left(\boldsymbol{x}^1\right). \end{aligned}$$

If the differentiability of f is considered in the abovementioned equation, for any $\lambda \in [0, 1]$,

$$\begin{aligned} 0 &\leqq f\left[\boldsymbol{x}^1 + \lambda\left(\boldsymbol{x}^2 - \boldsymbol{x}^1\right)\right] - f\left(\boldsymbol{x}^1\right) \\ &= \lambda\left[\nabla f\left(\boldsymbol{x}^1\right)\left(\boldsymbol{x}^2 - \boldsymbol{x}^1\right) + \alpha\left(\boldsymbol{x}^1,\ \lambda\left(\boldsymbol{x}^2 - \boldsymbol{x}^1\right)\right)\left\|\boldsymbol{x}^2 - \boldsymbol{x}^1\right\|\right] \end{aligned} \tag{2.12}$$

is obtained but (2.12) hold for $\lambda \in (0, 1]$ too. Therefore, for any $\lambda \in (0, 1]$,

$$0 \leqq \nabla f\left(\boldsymbol{x}^1\right)\left(\boldsymbol{x}^2 - \boldsymbol{x}^1\right) + \alpha\left(\boldsymbol{x}^1,\ \lambda\left(\boldsymbol{x}^2 - \boldsymbol{x}^1\right)\right)\left\|\boldsymbol{x}^2 - \boldsymbol{x}^1\right\|. \tag{2.13}$$

Here, we contradict the conclusion and assume $0 > \nabla f(\boldsymbol{x}^1)(\boldsymbol{x}^2 - \boldsymbol{x}^1) \equiv -\varepsilon$. By the differentiability of f, when $\lambda \to 0$, $|\alpha(\boldsymbol{x}^1, \lambda(\boldsymbol{x}^2 - \boldsymbol{x}^1))| \to 0$. Thus, in respect of $\varepsilon_0 = \frac{\varepsilon}{\|\boldsymbol{x}^2 - \boldsymbol{x}^1\|}$, if there exists certain $\widehat{\lambda} > 0$ and $\lambda < \widehat{\lambda}$, $|\alpha(\boldsymbol{x}^1, \lambda(\boldsymbol{x}^2 - \boldsymbol{x}^1))| < \varepsilon$. Accordingly, for $\lambda \in \left(0, min\left\{\widehat{\lambda}, 1\right\}\right)$,

$$\begin{aligned} 0 &\leqq \nabla f\left(\boldsymbol{x}^1\right)\left(\boldsymbol{x}^2 - \boldsymbol{x}^1\right) + \alpha\left(\boldsymbol{x}^1,\ \lambda\left(\boldsymbol{x}^2 - \boldsymbol{x}^1\right)\right)\left\|\boldsymbol{x}^2 - \boldsymbol{x}^1\right\| \\ &\leqq \nabla f\left(\boldsymbol{x}^1\right)\left(\boldsymbol{x}^2 - \boldsymbol{x}^1\right) + \left|\alpha\left(\boldsymbol{x}^1,\ \lambda\left(\boldsymbol{x}^2 - \boldsymbol{x}^1\right)\right)\right|\left\|\boldsymbol{x}^2 - \boldsymbol{x}^1\right\| \\ &< -\varepsilon + \varepsilon \\ &= 0. \end{aligned}$$

As a result, this is self-contradictory.[5]

[5]Usually, the mean value theorem is used in the proof but here the differentiability of f was used.

(Sufficiency) If it is assumed that f is not a quasiconcave function, there exists $\widehat{\boldsymbol{x}}^1, \widehat{\boldsymbol{x}}^2 \in \boldsymbol{D}$ and $\lambda \in [0, 1]$, which satisfy

$$f\left[\left(1-\widehat{\lambda}\right)\widehat{\boldsymbol{x}}^1 + \widehat{\lambda}\widehat{\boldsymbol{x}}^2\right] < min\left\{f\left(\widehat{\boldsymbol{x}}^1\right), f\left(\widehat{\boldsymbol{x}}^2\right)\right\} \tag{2.14}$$

Furthermore, generality is not lost even as $f\left(\widehat{\boldsymbol{x}}^2\right) \geqq f\left(\widehat{\boldsymbol{x}}^1\right)$. Here, if for any $\lambda \in [0, 1]$, we set

$$g(\lambda) = f\left[(1-\lambda)\widehat{\boldsymbol{x}}^1 + \lambda\widehat{\boldsymbol{x}}^2\right] \quad \text{and} \quad \boldsymbol{x}(\lambda) = f(1-\lambda)\widehat{\boldsymbol{x}}^1 + \widehat{\lambda}\widehat{\boldsymbol{x}}^2, \tag{2.15}$$

$$\begin{aligned} g(0) &= f\left(\widehat{\boldsymbol{x}}^1\right) \\ g(1) &= f\left(\widehat{\boldsymbol{x}}^2\right) \end{aligned} \tag{2.16}$$

and

$$g'(\lambda) = \nabla f[\boldsymbol{x}(\lambda)]\widehat{\boldsymbol{x}}^2 - \widehat{\boldsymbol{x}}^1. \tag{2.17}$$

Thus,

$$\begin{aligned} g\left(\widehat{\lambda}\right) < g(0) &= f\left(\widehat{\boldsymbol{x}}^1\right) \\ &\leqq f\left(\widehat{\boldsymbol{x}}^2\right) = g(1) \end{aligned} \tag{2.18}$$

is obtained. If $\widehat{\boldsymbol{x}}^1 = \widehat{\boldsymbol{x}}^2$,

$$f[\boldsymbol{x}(\lambda)] = f\left(\widehat{\boldsymbol{x}}^1\right) = f\left(\widehat{\boldsymbol{x}}^2\right) = min\left\{f\left(\widehat{\boldsymbol{x}}^1\right), f\left(\widehat{\boldsymbol{x}}^2\right)\right\}. \tag{2.19}$$

Hence, it goes against (2.14). Moreover, depending on $\lambda = 0$ or $\lambda = 1$,

$$f[\boldsymbol{x}(0)] = f\left(\widehat{\boldsymbol{x}}^1\right) = min\left\{f\left(\widehat{\boldsymbol{x}}^1\right), f\left(\widehat{\boldsymbol{x}}^2\right)\right\}, \text{or} \tag{2.20}$$

$$f[\boldsymbol{x}(1)] = f\left(\widehat{\boldsymbol{x}}^2\right) \geqq min\left\{f\left(\widehat{\boldsymbol{x}}^1\right), f\left(\widehat{\boldsymbol{x}}^2\right)\right\} \tag{2.21}$$

Thus (2.19)–(2.21) all goes against (2.14). Here, let us think of $\widehat{\boldsymbol{x}}^1, \widehat{\boldsymbol{x}}^2 \in \boldsymbol{D}$ to be two different points and $\widehat{\lambda}$ as a real number in the open interval (0,1). From (2.18) and the continuity of g, there exists some $\lambda^* \in \left(\widehat{\lambda}, 1\right)$ and

$$g(0) > g(\lambda^*) \tag{2.22}$$

and

$$g'(\lambda^*) = \nabla f[\boldsymbol{x}(\lambda^*)]\left(\widehat{\boldsymbol{x}}^2 - \widehat{\boldsymbol{x}}^1\right) > 0 \tag{2.23}$$

are satisfied. If (2.15) and (2.16) are considered, (2.22) indicates $f\left(\widehat{\boldsymbol{x}}^1\right) > f[\boldsymbol{x}(\lambda^*)]$ and hence from assumption [Theorem 2.3;(iii)],

$$0 \leqq \nabla f[\boldsymbol{x}(\lambda^*)]\left(\widehat{\boldsymbol{x}}^1 - \boldsymbol{x}(\lambda^*)\right) = -\lambda^* \nabla f[\boldsymbol{x}(\lambda^*)]\left(\widehat{\boldsymbol{x}}^2 - \widehat{\boldsymbol{x}}^1\right)$$

holds. However, this contradicts (2.23). QED

(Notes on [Theorem 2.3; (i)]): [Theorem 2.3;(i)] can be rewritten as $f(\boldsymbol{x}^2) \leqq \nabla f(\boldsymbol{x}^1)(\boldsymbol{x}^2 - \boldsymbol{x}^1) + f(\boldsymbol{x}^1)$ but this inequality indicates that a linear approximation with the concave function f about $\boldsymbol{x}^1$, $(\nabla f(\boldsymbol{x}^1)(\boldsymbol{x}^2 - \boldsymbol{x}^1) + f(\boldsymbol{x}^1))$, are no smaller than $f(\boldsymbol{x}^2)$ for any $\boldsymbol{x}^2 \in \boldsymbol{D}$.

[Lemma 2.2] Assuming (2.A1) and (2.A2), consider the following six propositions.

① f is a strictly concave function.
② f is a strictly pseudoconcave function.
③ f is a strictly quasiconcave function.
④ f is a concave function.
⑤ f is a pseudoconcave function.
⑥ f is a semi-strictly quasiconcave function.

Here, the following implications hold.

(a) ① ⇒ ② ⇒ ③
(b) ④ ⇒ ⑤ ⇒ ⑥
(c) ② ⇒ ⑤

(Proof)

(a)
① ⇒ ② f is a strictly concave function and hence from [Theorem 2.3;(ii)],

$$\text{for any different } \boldsymbol{x}^1,\ \boldsymbol{x}^2 \in \boldsymbol{D},$$
$$\nabla f(\boldsymbol{x}^1)(\boldsymbol{x}^2 - \boldsymbol{x}^1) > f(\boldsymbol{x}^2) - f(\boldsymbol{x}^1).$$

Consequently, if $0 \geqq \nabla f(\boldsymbol{x}^1)(\boldsymbol{x}^2 - \boldsymbol{x}^1)$, $f(\boldsymbol{x}^1) > f(\boldsymbol{x}^2)$. Thus, f is a strictly pseudoconcave function.
② ⇒ ③ If f is considered not to be a strictly quasiconcave function, there exists two different points in $\boldsymbol{D}$, $\widehat{\boldsymbol{x}}^1$ and $\widehat{\boldsymbol{x}}^2$ and some $\widehat{\lambda} \in (0, 1)$, and:

$$f\left[\left(1-\widehat{\lambda}\right)\widehat{\boldsymbol{x}}^1+\widehat{\lambda}\widehat{\boldsymbol{x}}^2\right]\leqq min\left\{f\left(\widehat{\boldsymbol{x}}^1\right),f\left(\widehat{\boldsymbol{x}}^2\right)\right\}. \tag{2.24}$$

Here, for any $\lambda \in [0, 1]$, if we define

$$\boldsymbol{x}(\lambda)=(1-\lambda)\widehat{\boldsymbol{x}}^1+\lambda\widehat{\boldsymbol{x}}^2 \text{ and } g(\lambda)=f[\boldsymbol{x}(\lambda)], \tag{2.25}$$

$$g\left(\widehat{\lambda}\right)\leqq min\{g(0),g(1)\} \tag{2.26}$$

Moreover, $g(\lambda)$ is continuous on the compact set $[0, 1]$. Hence, from Weierstrass' theorem,[6] there exists some $\lambda^* \in (0, 1)$ and for all $\lambda \in [0, 1]$, $g(\lambda^*) \leqq g(\lambda)$ holds. Thus,

$$g(\lambda^*)\leqq g\left(\widehat{\lambda}\right)\leqq min\{g(0),g(1)\} \tag{2.27}$$

is obtained. If one of the inequalities in (2.27) holds with a strict inequality, $g(\lambda^*) < min\ \{g(0), g(1)\}$; therefore, $\lambda^* \in (0, 1)$. However, if all of the inequalities in (2.27) hold together, g becomes a minimum at $\widehat{\lambda} \in (0, 1)$. From the above, it has been demonstrated that $g(\lambda)$ considers its minimum value at $\lambda^* \in (0, 1)$. Accordingly,

$$\begin{aligned}\frac{dg(\lambda^*)}{d\lambda}&=\nabla f[x(\lambda^*)]\left(\widehat{\boldsymbol{x}}^2-\widehat{\boldsymbol{x}}^1\right)\\&=\frac{\nabla f[\boldsymbol{x}(\lambda^*)]}{1-\lambda^*}\left(\widehat{\boldsymbol{x}}^2-\boldsymbol{x}(\lambda^*)\right)\\&=0\end{aligned} \tag{2.28}$$

holds. From the assumed strict pseudoconcaveness of f, $f\left(\widehat{\boldsymbol{x}}^2\right)<f(x(\lambda^*))$ is derived from (2.28), however, this is inconsistent with

$$\begin{aligned}f[\boldsymbol{x}(\lambda^*)]\leqq f\left[\boldsymbol{x}\left(\widehat{\lambda}\right)\right]&\leqq min\left\{f\left(\widehat{\boldsymbol{x}}^1\right),f\left(\widehat{\boldsymbol{x}}^2\right)\right\}\\&\leqq f\left(\widehat{\boldsymbol{x}}^2\right)\end{aligned} \tag{2.29}$$

(b)
④ ⇒ ⑤ If we assume f to be a concave function, from [Theorem 2.3;(i)],

for any $\boldsymbol{x}^1,\ \boldsymbol{x}^2 \in \boldsymbol{D}$,

[6] See [Theorem 10] (Appendix).

$$f(\boldsymbol{x}^2) - f(\boldsymbol{x}^1) \leqq \nabla f(\boldsymbol{x}^1)(\boldsymbol{x}^2 - \boldsymbol{x}^1)$$

holds. Thus, if $\nabla f(\boldsymbol{x}^1)(\boldsymbol{x}^2 - \boldsymbol{x}^1) \leqq 0$, $f(\boldsymbol{x}^2) \leqq f(\boldsymbol{x}^1)$; in other words, f is a pseudoconcave function.

⑤ ⇒ ⑥ (a) It can be shown in the same way as for ②⇒③.

(c) ② ⇒ ⑤ If f is not a pseudoconcave function, there exists $\widehat{\boldsymbol{x}}^1, \widehat{\boldsymbol{x}}^2 \in \boldsymbol{D}$ that satisfies

$$\nabla f\left(\widehat{\boldsymbol{x}}^1\right)\left(\widehat{\boldsymbol{x}}^2 - \widehat{\boldsymbol{x}}^1\right) \leqq 0 \tag{2.30}$$

and

$$f\left(\widehat{\boldsymbol{x}}^2\right) > f\left(\widehat{\boldsymbol{x}}^1\right). \tag{2.31}$$

f is a point-to-point mapping and hence if $\widehat{\boldsymbol{x}}^1 = \widehat{\boldsymbol{x}}^2$, $f\left(\widehat{\boldsymbol{x}}^1\right) = f\left(\widehat{\boldsymbol{x}}^2\right)$; however, this contradicts (2.31). Consequently, $\widehat{\boldsymbol{x}}^1 \neq \widehat{\boldsymbol{x}}^2$. However, f is a strictly pseudoconcave function and hence under (2.30); this indicates the inequality $f\left(\widehat{\boldsymbol{x}}^2\right) < f\left(\widehat{\boldsymbol{x}}^1\right)$, but this again contradicts (2.31). QED

The next theorem is a direct conclusion from [Theorem 2.3] and [Lemma 2.2].

[Theorem 2.4] The following holds under (2.A1) and (2.A2).

(i) [f: Strictly concave function] ⇒ [f: Strictly quasiconcave] ⇒ [f: Strictly quasiconcave] ⇒ [f: Quasiconcave function]

(ii) [f: Strictly concave function] ⇒ [f: Concave function] ⇒ [f: Pseudoconcave function] ⇒ [f: Semistrictly quasiconcave function] ⇒ [f: Quasiconcave function]

(iii) [f: strictly pseudoconcave function] ⇒ [f: pseudoconcave function]

(iv) When there exists $\boldsymbol{x} = \overline{\boldsymbol{x}} \in \boldsymbol{D}$ that become $\nabla f(\overline{\boldsymbol{x}}) \neq 0$,

(iv-i) If f is a quasiconcave function, for any $\boldsymbol{x} \in \Omega_1(\overline{\boldsymbol{x}}) = \{\boldsymbol{x} \in \boldsymbol{D} | \nabla f(\overline{\boldsymbol{x}})(\boldsymbol{x} - \overline{\boldsymbol{x}}) = 0\}$, $f(\boldsymbol{x}) \leqq f(\overline{\boldsymbol{x}})$

and

(iv-ii) If f is a strictly quasiconcave function, for any $\boldsymbol{x} \in \Omega_2(\overline{\boldsymbol{x}}) = \{\boldsymbol{x} \in \boldsymbol{D} | \nabla f(\overline{\boldsymbol{x}})(\boldsymbol{x} - \overline{\boldsymbol{x}}) = 0, \boldsymbol{x} \neq \overline{\boldsymbol{x}}\}, f(\boldsymbol{x}) < f(\overline{\boldsymbol{x}})$.

(v) for any $\boldsymbol{x} \in \boldsymbol{D}$, when $\nabla f(\boldsymbol{x}) \neq 0$,

(v-i) [f: Quasiconcave function] ⇒ [f: Pseudoconcave]

and

(v-ii) [f: Strictly quasiconcave function] ⇒ [f: Strictly pseudoconcave function]

(vi) When there exists $\overline{\boldsymbol{x}} \in \boldsymbol{D}$ that is $\nabla f(\overline{\boldsymbol{x}}) = 0$,

(vi-i) If f is a pseudoconcave function, $f(\boldsymbol{x})$ attains its maximum value on $\boldsymbol{x} = \overline{\boldsymbol{x}}$. Furthermore, if f is considered to be a strictly pseudoconcave function, $f(\overline{\boldsymbol{x}})$ becomes a unique maximum value and

(vi-ii) If f is a concave function, $f(\boldsymbol{x})$ achieves its maximum value at $\boldsymbol{x} = \overline{\boldsymbol{x}}$. Furthermore, if f is considered to be a strictly concave function, $f(\overline{\boldsymbol{x}})$ becomes a unique maximum value.

(Proof)

(i) It can be derived from [Lemma 2.2; (a)] and [Theorem 2.2; (ii)].

(ii) Under (2.A2), f is continuous. Thus, the statement in [Theorem 2.2; (v)] holds without the assumption of upper semi-continuity of f. Accordingly, it is shown by [Lemma 2.2; (b)] and [Theorem 2.2; (i), (v)].

(iii) It is clear from [Lemma 2.2; (c)].

(iv)

(iv-i) Contradict the conclusion and assume (2.32) and (2.33).

$$\text{for some } \overline{\boldsymbol{x}} \in \boldsymbol{D},\ \nabla f(\overline{\boldsymbol{x}}) \neq 0 \tag{2.32}$$

and

$$\text{for some } \boldsymbol{x}^1 \in \boldsymbol{D},$$

$$\nabla f(\overline{\boldsymbol{x}})(\boldsymbol{x}^1 - \overline{\boldsymbol{x}}) = 0 \tag{2.33-1}$$

$$\text{and} \tag{2.33}$$

$$f(\boldsymbol{x}^1) > f(\overline{\boldsymbol{x}}). \tag{2.33-2}$$

However, f is differentiable on the open convex function $\boldsymbol{D}$. Hence, f becomes continuous and there exists certain $\delta_1 > 0$ and for any $\boldsymbol{x} \in \boldsymbol{B}\ (\boldsymbol{x}^1; \delta_1)$,

$$f(\boldsymbol{x}) > f(\overline{\boldsymbol{x}}) \tag{2.34}$$

holds. The fact that the quasiconcaveness of f indicates that (2.34) is

$$\nabla f(\overline{\boldsymbol{x}})(\boldsymbol{x} - \overline{\boldsymbol{x}}) \geqq 0 \tag{2.35}$$

is evident from [Theorem 2.3; (iii)]. Moreover, from (2.32), there exists certain $k \in \{1, \ldots, n\}$ that satisfy $f_k(\overline{\boldsymbol{x}}) = \dfrac{\partial f(\overline{\boldsymbol{x}})}{\partial \overline{\boldsymbol{x}}_k} \neq 0$. Here, corresponding to $f_k(\overline{\boldsymbol{x}}) > (<) 0$ for $\delta \in (0, \delta_1)$, let us define

$$\begin{aligned}\boldsymbol{x} &= \sum_{\substack{j\neq k\\ j=1}}^{n} x_j^1 \boldsymbol{e}^j + \left(x_k^1 \overset{-}{(+)}\ \delta\right)\boldsymbol{e}^k.\\ &= \boldsymbol{x}^1 \overset{-}{(+)}\ \delta \boldsymbol{e}^k,\end{aligned} \tag{2.36}$$

where $\boldsymbol{e}^j (j \in \{1, \ldots, n\})$ is the jth unit vector. Then,

$$\begin{aligned}\nabla f(\overline{\boldsymbol{x}})(\boldsymbol{x}-\overline{\boldsymbol{x}}) &= \nabla f(\overline{\boldsymbol{x}})\left(\boldsymbol{x}^1 \overset{-}{(+)}\ \delta\boldsymbol{e}^k - \overline{\boldsymbol{x}}\right)\\ &= \nabla f(\overline{\boldsymbol{x}})\left(\boldsymbol{x}^1-\overline{\boldsymbol{x}}\right) \overset{-}{(+)}\ \delta f_k(\overline{\boldsymbol{x}})\\ &< 0.\end{aligned} \tag{2.37}$$

$$f_k(\overline{\boldsymbol{x}}) > (<) 0 \Rightarrow \boldsymbol{x} = \boldsymbol{x}^1 \overset{-}{(+)}\ \delta\boldsymbol{e}^k$$

Thus,

$$\nabla f(\overline{\boldsymbol{x}})(\boldsymbol{x}-\overline{\boldsymbol{x}}) < 0. \tag{2.38}$$

(2.38) contradicts (2.35) and hence the conclusion sought after can be obtained.
(iv-ii) Contradict the conclusion and assume (2.39) and (2.40).

$$\text{for some } \overline{\boldsymbol{x}} \in \boldsymbol{D},\ \nabla f(\overline{\boldsymbol{x}}) \neq \boldsymbol{0}, \tag{2.39}$$

for some $\boldsymbol{x}^1 \in \boldsymbol{D}$ which is different to $\overline{\boldsymbol{x}}$,

$$\nabla f(\boldsymbol{x})\left(\boldsymbol{x}^1-\overline{\boldsymbol{x}}\right) = 0 \tag{2.40-1}$$

and (2.40)

$$f\left(\boldsymbol{x}^1\right) \geqq f(\overline{\boldsymbol{x}}). \tag{2.40-2}$$

For any $\lambda \in (0, 1)$, if we let $\tilde{\boldsymbol{x}} = (1-\lambda)\overline{\boldsymbol{x}} + \lambda\boldsymbol{x}^1$, $\boldsymbol{D}$ is a convex set and hence $\tilde{\boldsymbol{x}} = \boldsymbol{D}$. Therefore, from the strict quasiconcaveness of f, for any $\lambda \in (0, 1)$,

$$\begin{aligned}f(\tilde{\boldsymbol{x}}) &> min\left\{f(\overline{\boldsymbol{x}}), f\left(\boldsymbol{x}^1\right)\right\}\\ &= f(\overline{\boldsymbol{x}})\end{aligned} \tag{2.41}$$

and

$$\begin{aligned}\nabla f(\overline{\boldsymbol{x}})(\tilde{\boldsymbol{x}}-\overline{\boldsymbol{x}}) &= \lambda\nabla f(\overline{\boldsymbol{x}})\left(\boldsymbol{x}^1-\overline{\boldsymbol{x}}\right)\\ &= 0\end{aligned} \tag{2.42}$$

are obtained. Thus, $\tilde{\boldsymbol{x}}$ satisfies (2.40). Consequently, from the continuity of f, there exists a suitable $\tilde{\delta} > 0$ and for all $\boldsymbol{x} \in \boldsymbol{B}\left(\tilde{\boldsymbol{x}}; \tilde{\delta}\right)$,

$$f(\boldsymbol{x}) > f(\overline{\boldsymbol{x}}) \tag{2.43}$$

holds. Therefore, if the fact that f is a strictly quasiconcave function is used, from this theorem (i) and [Theorem 2.3; (iii)], and from,

$$\nabla f(\overline{\boldsymbol{x}})(\boldsymbol{x} - \overline{\boldsymbol{x}}) \geqq 0. \tag{2.44}$$

Although (2.39) guarantees the existence of $k \in \{1, \ldots, n\}$ that becomes $f_k(\overline{\boldsymbol{x}}) \neq 0$ and hence for $\delta \in \left(0, \tilde{\delta}\right)$ if

$$\boldsymbol{x} = \begin{cases} \tilde{\boldsymbol{x}} - \delta \boldsymbol{e}^k, f_k(\overline{\boldsymbol{x}}) > 0 \\ \tilde{\boldsymbol{x}} + \delta \boldsymbol{e}^k, f_k(\overline{\boldsymbol{x}}) < 0 \end{cases} \tag{2.45}$$

is defined depending on whether $f_k(\overline{\boldsymbol{x}})$ is positive or negative,

$$\begin{aligned} \nabla f(\overline{\boldsymbol{x}})(\boldsymbol{x} - \overline{\boldsymbol{x}}) &= \nabla f(\overline{\boldsymbol{x}})(\tilde{\boldsymbol{x}} - \overline{\boldsymbol{x}}) \overset{-}{(+)} \delta f_k(\overline{\boldsymbol{x}}) \\ &< 0. \end{aligned}$$

This contradicts (2.44).

(v)

(v-i) If f is considered to be a not pseudoconcave function, there exists a different $\widehat{\boldsymbol{x}}^1, \widehat{\boldsymbol{x}}^2 \in \boldsymbol{D}$ and

$$f\left(\widehat{\boldsymbol{x}}^2\right) > f\left(\widehat{\boldsymbol{x}}^1\right) \tag{2.46}$$

and

$$\nabla f\left(\widehat{\boldsymbol{x}}^1\right)\left(\widehat{\boldsymbol{x}}^2 - \widehat{\boldsymbol{x}}^1\right) \leqq 0 \tag{2.47}$$

hold. However, the quasiconcaveness of f and (2.46) indicate

$$\nabla f\left(\widehat{\boldsymbol{x}}^1\right)\left(\widehat{\boldsymbol{x}}^2 - \widehat{\boldsymbol{x}}^1\right) \geqq 0 \tag{2.48}$$

Thus, from (2.47) and (2.48), $\nabla f\left(\widehat{\boldsymbol{x}}^1\right)\left(\widehat{\boldsymbol{x}}^2 - \widehat{\boldsymbol{x}}^1\right) = 0$ is obtained but from the assumption, $\nabla f\left(\widehat{\boldsymbol{x}}^1\right) \neq \boldsymbol{0}$. Therefore,

$$f\left(\widehat{\boldsymbol{x}}^1\right) \geqq f\left(\widehat{\boldsymbol{x}}^2\right). \tag{2.49}$$

(2.46) and (2.49) are conflicting and hence f is a pseudoconcave function.

(v-ii) If f is a strictly quasiconcave function and note the fact that f is a quasiconcave function ((i)), (2.46) in the proof of (v-i) can be established as a weak inequality with (2.49) as a strong inequality, and then the conclusion is immediately obtained.

(vi)

(vi-i) If there exists $\overline{\boldsymbol{x}} \in \boldsymbol{D}$ that is $\nabla f(\overline{\boldsymbol{x}}) = 0$ and f is a pseudoconcave function, the fact that $f(\boldsymbol{x}) \leqq f(\overline{\boldsymbol{x}})$ holds for all $\boldsymbol{x} \in \boldsymbol{D}$ is evident from the definition of a pseudoconcave function. Furthermore, if f is a strictly pseudoconcave function, then from (iii), it is a pseudoconcave function. Thus, it should be demonstrated that $f(\overline{\boldsymbol{x}})$ is unique but from the definition of a strictly concave function, $f(\boldsymbol{x}) < f(\overline{\boldsymbol{x}})$ is shown for any $\boldsymbol{x} \in \boldsymbol{D}$ that is different from $\overline{\boldsymbol{x}}$.

(vi-ii) If f is a concave (strictly concave) function, from (ii) ((i)), f is a pseudoconcave (strictly pseudoconcave) function; therefore, the conclusion is apparent from (vi-i). QED

(Notes on concave function and strictly pseudoconcave function): There are no implications between the concave function and strictly pseudoconcave function. This fact can be easily verified based on the following two examples.

Example 2.3

$$f(\boldsymbol{x}) = 1 \ (f : \boldsymbol{R} \rightarrow 1)$$

For all $x^1, x^2 \in \boldsymbol{R}$ and $\lambda \in [0, 1]$, $f[(1 - \lambda)x^1 + \lambda x^2] = (1 - \lambda)f(x^1) + \lambda f(x^2) = 1$.

Hence, f is a concave function on $\boldsymbol{R}$. Although for any $x \in \boldsymbol{R}$, $f'(x) = 0$ and $f(x) = 1$, the definition of a strictly pseudoconcave function is not satisfied (Fig. 2.17).

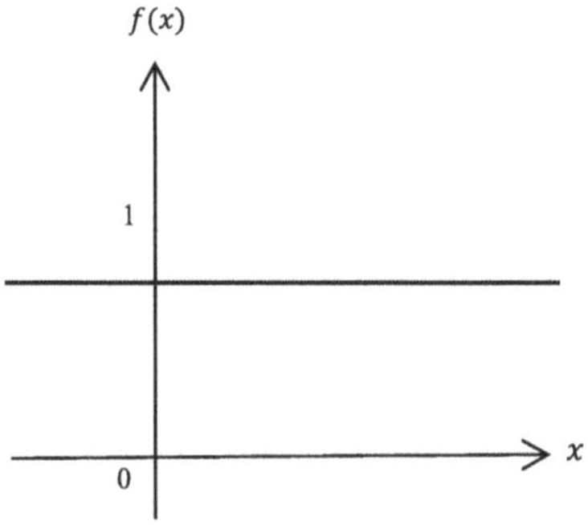

Fig. 2.17 Illustration of Example 2.3

Example 2.4 Let's define

$$f(\boldsymbol{x}) = \begin{cases} x^2 + 1, & x > 1 \\ 2x, & 0 < x \leqq 1 \end{cases} \quad (f : \boldsymbol{R}_{++}\boldsymbol{R}),$$

$$f'(x) = \begin{cases} 2x > 0, & x > 1 \\ 2, & 0 < x \leqq 1 \end{cases}$$

indicates that f satisfies the definition of strictly pseudoconcave function on $\boldsymbol{R}_{++}$ but f is not a concave function for all $x > 1$ (Fig. 2.18).

If (2.A2) is strengthened to be (2.A3), the properties of f with respect to a Hessian matrix and bordered Hessian are derived.

[Theorem 2.4] is summarized in the Fig. 2.19.

[Theorem 2.5] Consider (2.A1) and (2.A3) as a premise. Express the **Hessian matrix** with f evaluated at $\boldsymbol{x}$ as $H(x)$ and bordered Hessian matrix with f evaluated at x as $\boldsymbol{B}(\boldsymbol{x})$, and $k(k + 1)$-order small matrices obtained from the top left corner of $\boldsymbol{H}(\boldsymbol{x})$ and $(\boldsymbol{B}(\boldsymbol{x}))$ as $\boldsymbol{H}_k(\boldsymbol{x})$ $(\boldsymbol{B}_k(\boldsymbol{x}))$ and define the following:

$$\boldsymbol{H}(\boldsymbol{x}) = \begin{bmatrix} f_{11}(\boldsymbol{x}) & \cdots & f_{1n}(\boldsymbol{x}) \\ \vdots & & \vdots \\ f_{n1}(\boldsymbol{x}) & \cdots & f_{nn}(\boldsymbol{x}) \end{bmatrix},$$

$$\boldsymbol{B}(\boldsymbol{x}) = \begin{bmatrix} \boldsymbol{0} & f_1(\boldsymbol{x}) & & f_n(\boldsymbol{x}) \\ f_1(\boldsymbol{x}) & & & \\ \vdots & & \boldsymbol{H}(\boldsymbol{x}) & \\ f_n(\boldsymbol{x}) & & & \end{bmatrix}$$

$$\boldsymbol{H}_k(\boldsymbol{x}) = \begin{bmatrix} f_{11}(\boldsymbol{x}) & \cdots & f_{1k}(\boldsymbol{x}) \\ \vdots & & \vdots \\ f_{k1}(\boldsymbol{x}) & \cdots & f_{kk}(\boldsymbol{x}) \end{bmatrix}$$

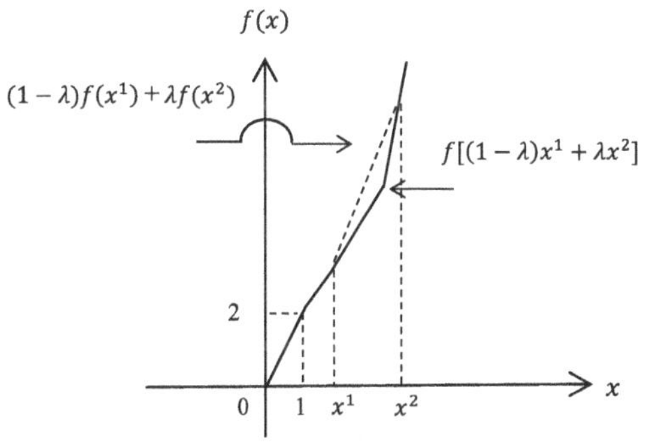

Fig. 2.18 Illustration of Example 2.4

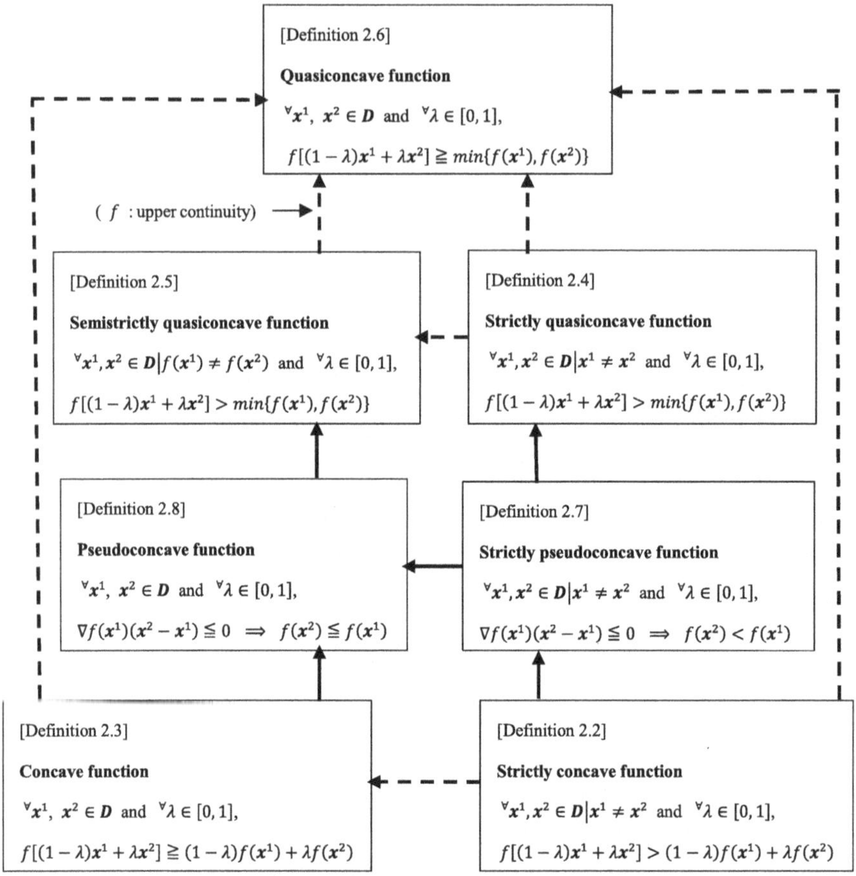

Fig. 2.19 The relation indicated by [Theorem 2.4]. The dotted line stands for that the relation holds under (2.A1). The solid line stands for that the relation holds under (2.A1) and (2.A2)

$$\boldsymbol{B}_k(\boldsymbol{x}) = \begin{bmatrix} \boldsymbol{0} & f_1(\boldsymbol{x}) & & f_k(\boldsymbol{x}) \\ f_1(\boldsymbol{x}) & & & \\ \vdots & & \boldsymbol{H}_k(\boldsymbol{x}) & \\ f_k(\boldsymbol{x}) & & & \end{bmatrix}, \quad (k = 1, \ldots, n)$$

Although $f_i(\boldsymbol{x}) = \dfrac{\partial f(\boldsymbol{x})}{\partial x_i}$, $(i = 1, \ldots, n)$,

$$f_{ij}(\boldsymbol{x}) = \frac{\partial}{\partial x_j}\left(\frac{\partial f(\boldsymbol{x})}{\partial x_i}\right) = \frac{\partial^2 f(\boldsymbol{x})}{\partial x_i \partial x_j}, \quad (i, j = 1, \ldots, n),$$

$det\ \boldsymbol{H}(\boldsymbol{x})$ is the matrix equation of $\boldsymbol{H}(\boldsymbol{x})$.

Here, the following holds.

(i) For f to be a concave function, the necessary and sufficient condition is that, for any $\boldsymbol{x} \in \boldsymbol{D}$, $\boldsymbol{H}(\boldsymbol{x})$ is a nonpositive definite,[7] i.e.,[8]

$$\text{For any } \boldsymbol{x} \in \boldsymbol{D} \text{ and } \boldsymbol{0} \neq \boldsymbol{y} \in \boldsymbol{R}^n, \boldsymbol{y}^t\boldsymbol{H}(\boldsymbol{x})\boldsymbol{y} \leqq 0.$$

(ii) For f to be a strictly concave function, the sufficient condition is that for any $\boldsymbol{x} \in \boldsymbol{D}$, $\boldsymbol{H}(\boldsymbol{x})$ has a negative sign, i.e.,

$$\text{for any } \boldsymbol{x} \in \boldsymbol{D} \text{ and } \boldsymbol{0} \neq \boldsymbol{y} \in \boldsymbol{R}^n, \boldsymbol{y}^t\boldsymbol{H}(\boldsymbol{x})\boldsymbol{y} < 0.$$

(iii) For f to be a concave function, the necessary and sufficient condition is that

$$\text{for any } \boldsymbol{x} \in \boldsymbol{D} \text{ and } k \in \boldsymbol{N} \equiv \{1, \ldots, n\},$$
$$(-1)^k \det(\boldsymbol{H}_k(\boldsymbol{x})) \geqq 0.$$

(iv) For f to be a strictly concave function, the sufficient condition is that:

$$\text{for any } \boldsymbol{x} \in \boldsymbol{D} \text{ and } k \in \boldsymbol{N}, (-1)^k \det(\boldsymbol{H}_k(\boldsymbol{x})) > 0.$$

(v) For f being a quasiconcave function, the necessary condition is that

$$\text{for any } \boldsymbol{x} \in \boldsymbol{D} \text{ and } k \in \boldsymbol{N}, (-1)^k \det(\boldsymbol{B}_k(\boldsymbol{x})) \geqq 0.$$

(vi) For f being a strictly quasiconcave function, the sufficient condition is that

$$\text{for any } \boldsymbol{x} \in \boldsymbol{D} \text{ and } k \in \boldsymbol{N}, (-1)^k \det(\boldsymbol{B}_k(\boldsymbol{x})) > 0.$$

(Proof)

(i)

(Necessity) It is thus assumed that $\boldsymbol{D}$ is an open set. Hence, for any $\boldsymbol{x} \in \boldsymbol{D}$ and $\boldsymbol{y}\ (\neq \boldsymbol{0}) \in \boldsymbol{R}^n$, there is a suitable $\bar{t} > 0$ and for all $t \in (0, \bar{t})$, $\boldsymbol{x} + t\boldsymbol{y} \in \boldsymbol{D}$. Moreover, f is a concave function, and hence for this $t > 0$,

[7] See ⟨Definition 5.11⟩ shown later.

[8] Under the assumption of this theorem, Young's theorem, in other words, for any $i, j \in N$, $f_{ij}(x) = f_{ji}(x)$ holds hence $H(x)$ becomes a real symmetic matrix. Moreover, if symmetry of $H(x)$ is not guaranteed, $H(x)$ is called a quasi non positive definite.

$$f(\boldsymbol{x}+t\boldsymbol{y})-f(\boldsymbol{x})-t\nabla f(\boldsymbol{x})\boldsymbol{y}\leqq 0. \tag{2.50}$$

From this and twice differentiability of f,[9]

$$\begin{aligned}0&\geqq f(\boldsymbol{x}+t\boldsymbol{y})-f(\boldsymbol{x})-t\nabla f(\boldsymbol{x})\boldsymbol{y}\\&=\frac{t^2}{2}\,\boldsymbol{y}^t\boldsymbol{H}(\boldsymbol{x})\boldsymbol{y}+t^2\beta(\boldsymbol{x}+t\boldsymbol{y})\|\boldsymbol{y}\|^2\end{aligned}$$

Accordingly,

$$0\geqq\frac{1}{2}\boldsymbol{y}^t\boldsymbol{H}(\boldsymbol{x})\boldsymbol{y}+\beta(\boldsymbol{x}+t\boldsymbol{y})\|\boldsymbol{y}\|^2 \tag{2.51}$$

are established. Here, if we let $t\to 0$, from the twice differentiability of f, $\beta(\boldsymbol{x},t\boldsymbol{y})\to 0$ and hence [Theorem 2.5](i) is obtained from (2.51).
(Sufficiency) From Taylor's theorem ([Theorem 15] (Appendix)), for any $\boldsymbol{x}^1$, $\boldsymbol{x}^2\in\boldsymbol{R}$, there exists $\theta\in(0,1)$ that satisfies

$$\begin{aligned}&f(\boldsymbol{x}^2)-f(\boldsymbol{x}^1)-\nabla f(\boldsymbol{x}^1)(\boldsymbol{x}^2-\boldsymbol{x}^1)\\&=\frac{1}{2}(\boldsymbol{x}^2-\boldsymbol{x}^1)^t\boldsymbol{H}\left[\boldsymbol{x}^1+\theta(\boldsymbol{x}^2-\boldsymbol{x}^1)\right](\boldsymbol{x}^2-\boldsymbol{x}^1).\end{aligned} \tag{2.52}$$

$\boldsymbol{D}$ is an open set, and so if θ is selected to be sufficiently small, $\boldsymbol{x}^1+\theta(\boldsymbol{x}^2-\boldsymbol{x}^1)\in\boldsymbol{D}$. Moreover, from [Theorem 2.5](i),

$$(\boldsymbol{x}^2-\boldsymbol{x}^1)^t\boldsymbol{H}\left[\boldsymbol{x}^1+\theta(\boldsymbol{x}^2-\boldsymbol{x}^1)\right](\boldsymbol{x}^2-\boldsymbol{x}^1)\leqq 0. \tag{2.53}$$

Thus,

$$f(\boldsymbol{x}^2)-f(\boldsymbol{x}^1)\leqq\nabla f(\boldsymbol{x}^1)(\boldsymbol{x}^2-\boldsymbol{x}^1) \tag{2.54}$$

is derived and using [Theorem 2.3; (i)], f becomes a concave function.
(ii) Proof of sufficiency of (i) holds for any different $\boldsymbol{x}^1$, $\boldsymbol{x}^2\in\boldsymbol{D}$. Combining this fact and (2.15), (2.53) and (2.54) hold with strong inequalities, and the conclusion sought after is obtained.
(iii), (iv) From [Theorem 5.8] ([Theorem 5.6]), (i) ((ii)) and (iii) ((iv)) are equivalent. Accordingly, (i) ((ii)) holding indicates (iii) ((iv)) holding is evident.

[9] For information about twice differentiability, please see ⟨Definition 16⟩(Appendix).

(v) $\boldsymbol{D}$ is an open set. Thus, for any $\boldsymbol{x} \in \boldsymbol{D}$, $\boldsymbol{y} \in \boldsymbol{R}^n$, there exists some real number $\widehat{\lambda} > 0$ and for all $\lambda \in \left(0, \widehat{\lambda}\right)$, $\boldsymbol{x} + \lambda \boldsymbol{y} \in \boldsymbol{D}$. Consequently, from Taylor's theorem,

$$f(\boldsymbol{x} + \lambda \boldsymbol{y}) = f(\boldsymbol{x}) + \lambda \nabla f(\boldsymbol{x})\boldsymbol{y} + \frac{\lambda^2}{2}\boldsymbol{y}^t \boldsymbol{H}(\boldsymbol{x} + \theta \lambda \boldsymbol{y})\boldsymbol{y}. \tag{2.55}$$

If $\nabla f(\boldsymbol{x}) = 0$, the fact that for any $k \in \boldsymbol{N}$, $\det(\boldsymbol{B}_k(\boldsymbol{x})) = 0$, is evident from the definition of $\boldsymbol{B}_k(\boldsymbol{x})$. Thus, if we assume $\nabla f(\boldsymbol{x}) \neq 0$, from [Theorem 2.4; (iv-i)], $f(\boldsymbol{x} + \lambda \boldsymbol{y}) \leqq f(\boldsymbol{x})$ holds for all $\boldsymbol{y} \in \boldsymbol{R}^n$ that satisfy $\nabla f(\boldsymbol{x})[(\boldsymbol{x} + \lambda \boldsymbol{y}) - \boldsymbol{x}] = \nabla f(\boldsymbol{x})\lambda \boldsymbol{y} = 0$. Therefore, $\boldsymbol{y}^t \boldsymbol{H}(\boldsymbol{x} + \theta \lambda \boldsymbol{y})\boldsymbol{y} \leqq 0$ is obtained from (2.55). Accordingly, if we set $\lambda \to 0$, for all $\boldsymbol{y} \in \boldsymbol{R}^n$ that are $\nabla f(\boldsymbol{x})\boldsymbol{y} = 0$, $\boldsymbol{y}^t \boldsymbol{H}(\boldsymbol{x})\boldsymbol{y} \leqq 0$. Thus, from [Corollary 5.10], [Theorem 2.5; (v)] holds.

(vi) If f is not a strictly quasiconcave function, as per the proof of [Lemma 2.2: (a); ②⇒③], the function $g(\lambda)$ achieves its minimum value on $\lambda^* \in (0, 1)$. Hence,

$$0 = g'(\lambda^*) = \nabla f[\boldsymbol{x}(\lambda^*)]\left(\widehat{\boldsymbol{x}}^2 - \widehat{\boldsymbol{x}}^1\right)$$

and

$$\left(\widehat{\boldsymbol{x}}^2 - \widehat{\boldsymbol{x}}^1\right)^t \boldsymbol{H}(\boldsymbol{x}(\lambda^*))\left(\widehat{\boldsymbol{x}}^2 - \widehat{\boldsymbol{x}}^1\right) = g''(x^*) \geqq 0. \tag{2.56}$$

On the other hand, $\boldsymbol{x}(\lambda^*) \in \boldsymbol{D}$. Therefore, condition [Theorem 2.5; (vi)], in respect of $\widehat{\boldsymbol{x}}^2 - \widehat{\boldsymbol{x}}^1 \neq 0$, which satisfies $\nabla f[\boldsymbol{x}(\lambda^*)]\left(\widehat{\boldsymbol{x}}^2 - \widehat{\boldsymbol{x}}^1\right) = 0$, means

$$\left(\widehat{\boldsymbol{x}}^2 - \widehat{\boldsymbol{x}}^1\right)^t \boldsymbol{H}(\boldsymbol{x}(\lambda^*))\left(\widehat{\boldsymbol{x}}^2 - \widehat{\boldsymbol{x}}^1\right) < 0$$

holds but this contradicts (2.56). QED

(Notes on [Theorem 2.5; (ii)]): From [Theorem 2.2; (i)], if it is a strictly concave function, then it is a concave function. Thus, from this theorem (i), for any $\boldsymbol{x} \in \boldsymbol{D}$, $\boldsymbol{H}(\boldsymbol{x})$ becomes a nonpositive definite. However, even if it is a strictly concave function, it is not necessarily the case that $\boldsymbol{H}(\boldsymbol{x})$ is a negative definite. This is demonstrated using the following example.[10]

Example 2.5

$$f(x) = -x^4 (f : \boldsymbol{R} \to \boldsymbol{R})$$

f is a strictly concave function on $\boldsymbol{R}$ but $\boldsymbol{H}(0) = f''(0) = 0$. Hence, even if $\boldsymbol{H}(0)$ is a nonpositive definite, it is not a negative definite.

[10] See Mangasarian (1969).

References

Avriel, M. (1976). *Nonlinear programming: analysis and methods*. Prentice Hall. ISBN-13: 978-0136236030. Reprinted by Dover Publications (Dover Books on Computer Science), 2003. ISBN-13: 978-0486432274.

Avriel, M., Diewert, W. E., Schaible, S., & Zang, I. (1988). *Generalized concavity*. Plenum Press. ISBN-13: 978-03064265568.

Bazaraa, M. S., Sherali, H. D., & Shetty, C. M. (1993). *Nonlinear programming theory and algorithms* (2nd ed.). John Wiley and Sons. ISBN-13: 978-0471557937. Reprinted by Wiley-Interscience (3rd ed.), 2006. ISBN-13: 978-0471486008.

Hanson, M. A. (1964). Bounds for functionally convex optimal control problems. *Journal of Mathematical Analysis and Applications, 8*, 84–89.

Karamadian, S. (1967). Duality in mathematical programming. *Journal of Mathematical Analysis and Applications, 20*, 344–358.

Kemp, M. C., & Kimura, Y. (1978). *Introduction to Mathematical Economics*. Springer-Verlag. ISBN-13: 978-1461262800.

Mangasarian, O. L. (1965). Pseudo-convex functions. *SIAM Journal on Control, 3*, 281–290.

Mangasarian, O. L. (1969). *Nonlinear programming*. McGraw-Hill. ISBN-13: 978-0472081356.

Martos, B. (1975). *Nonlinear programming theory and methods*. North-Holland. ISBN-13: 978-0-720428179.

Ortega, J. M., & Rheinboldt, W. C. (1970). *Interative solution of nonlinear equations in several variables*. Academic Press. ISBN-13: 978-0125285506.

Ponstein, J. (1967). Seven kinds of convexity. *SIAM Review, 9*, 115–119.

Rudin, W. (1976). *Principles of mathematical analysis* (3rd ed.). McGraw-Hill. ISBN-13: 978-0070856134.

Takayama, A. (1993). *Analytical methods in economics*. University of Michigan Press. ISBN-13: 978-0472081356.

Thompson, W. A., & Parke, D. W. (1973). Some properties of generalized concave functions. *Operations Research, 21*, 305–313.

Chapter 3
Nonlinear Programming Problems

3.1 Introduction

Mathematically, the majority of static optimization problems in economic theories fall under the category of nonlinear programming problems. Moreover, even when using Pontryagin's maximum principle to a dynamic optimization problem, nonlinear programming theories are used in the majority of the solutions. This is the primary reason for this chapter's discussion of nonlinear programming problems.

3.2 Nonlinear Programming Problems

Generally, a standard **nonlinear programming problem** refers to for a given real-valued function $f(\boldsymbol{x})$ and $g_1(\boldsymbol{x}), \cdots, g_m(\boldsymbol{x})$ defined on a subset D of R^n, seeking $\boldsymbol{x}^* = \left(x_1*, \cdots, x_n^*\right)^t$, which maximizes $f(\boldsymbol{x})$ under the constraint $g(\boldsymbol{x}) = (g_1(\boldsymbol{x}), \cdots, g_m(\boldsymbol{x}))^t \geqq 0$. Although throughout this book, we shall abbreviate this problem as

$$[\text{NLPX}]\left[\max_{\boldsymbol{x}\in \boldsymbol{D}} f(\boldsymbol{x}) \qquad \text{subject to} \quad g(\boldsymbol{x}) = (g_1(\boldsymbol{x}), \cdots, g_m(\boldsymbol{x}))^t \geqq \boldsymbol{0}\right].$$

Furthermore, standard nonlinear programming problems are sometimes formulated as constrained minimization problems as

$$[\text{NLPN}]\left[\min_{\boldsymbol{x}\in \boldsymbol{D}} F(\boldsymbol{x}) \qquad \text{subject to} \quad G(\boldsymbol{x}) = (G_1(\boldsymbol{x}), \cdots, G_m(\boldsymbol{x}))^t \leqq \boldsymbol{0}\right].$$

Although if each of the functions set as a problem in a standard format is multiplied by (-1) and max and min are switched, a new standard format is formed, and thus in this book only [NLPX] will be considered to be in scope.

© Springer Nature Singapore Pte Ltd. 2024

K. Nakayama, *Eigenvalue Problem and Nonlinear Programming Problem*,
New Frontiers in Regional Science: Asian Perspectives 70,
https://doi.org/10.1007/978-981-99-2943-6_3

Firstly, let us explain some of the terminology relating to [NLPX].

⟨**Definition 3.1**⟩

⟨D3.1-1⟩ $x_i (i = 1, \cdots, n)$ **Control variable**. A variable whose value can be controlled by the programmer.

$\boldsymbol{x} = (x_1, \cdots, x_n)^t$ **Control variable vector**. x can sometimes be simply referred to as the control variable.

⟨D3.1-2⟩ $f(\boldsymbol{x})$ $(\boldsymbol{D} \subseteq \boldsymbol{R}^n \to \boldsymbol{R})$ **Objective function**.

$g_i(\boldsymbol{x})$ $(\boldsymbol{D} \subseteq \boldsymbol{R}^n \to \boldsymbol{R})$ **Constraint function**.

$g(\boldsymbol{x}) = (g_1(\boldsymbol{x}), \cdots, g_m(\boldsymbol{x}))^t$ $(\boldsymbol{D} \subseteq \boldsymbol{R}^n \to \boldsymbol{R}^m)$

$\boldsymbol{D}(\subseteq \boldsymbol{R}^n)$ **Domain** of $f(\boldsymbol{x})$, $g_1(\boldsymbol{x})$, $\cdots$, $g_m(\boldsymbol{x})$.

$$\nabla f(\boldsymbol{x}) = \left(\frac{\partial f(\boldsymbol{x})}{\partial x_1}, \cdots, \frac{\partial f(\boldsymbol{x})}{\partial x_n}\right)$$

Gradient vector of $f(\boldsymbol{x})$.

⟨D3.1-3⟩ **Feasible solution** or **permissive solution**

A solution $\boldsymbol{x} \in \boldsymbol{D}$, which satisfies the constraint $g(\boldsymbol{x}) \geqq \boldsymbol{0}$.

$\boldsymbol{C} = \{\mathbf{x} \in \boldsymbol{D} | g_i(\mathbf{x}) \geqq 0, i = 1, \cdots, m\}$[1]

Constraint set. A set of feasible solutions.

<D3.1-4> **Global optimum**:

If $\boldsymbol{x}^* \in \boldsymbol{C}$ satisfies

$^\forall \boldsymbol{x} \in \boldsymbol{C}, f(\boldsymbol{x}) \leqq f(\boldsymbol{x}^*)$,

$\boldsymbol{x}^* \in \boldsymbol{C}$ is called a global optimum of [NLPX] or, if evident from the context, simply an **optimal solution** or an **optimum**.

Local optimum:

When in correspondence to $\boldsymbol{x}^* \in \boldsymbol{C}$, $\delta > 0$ is set and

$^\forall \boldsymbol{x} \in (\boldsymbol{B}(\boldsymbol{x}^*; \delta) \cap \boldsymbol{C}), f(\boldsymbol{x}) \leqq f(\boldsymbol{x}^*)$

this $\boldsymbol{x}^* \in \boldsymbol{C}$ is referred to as a local optimum of [NLPX].

<D3.1-5> $L(\boldsymbol{x}, \lambda) = f(\boldsymbol{x}) + \lambda g(\boldsymbol{x})$

$= f(\boldsymbol{x}) + \sum_{i=1}^{m} \lambda igi(x) (\boldsymbol{D} \times \boldsymbol{D} \to \boldsymbol{R})$

A **Lagrange function** corresponding to the [NLPX].

<D3.1-6> **Effective constraint** (or active constraint) and **effective set** (active set)

For of $\boldsymbol{x}^0 \in \boldsymbol{C}$, the constraint equation $g_i(\boldsymbol{x}^0) = 0$ is referred to as effective constraint and the numbered set of finite effective constraints

$\boldsymbol{E}(\boldsymbol{x}^0) = \{i \in \{1, \ldots, m\} | g_i x^0 = 0\}$

is referred to as an effective set on $\boldsymbol{x}^0$. When $i \notin \boldsymbol{E}(\boldsymbol{x}^0)$, $g_i(\boldsymbol{x}^0) > 0$ is called a **non-effective constraint** and the set of numbers of non-effective constraint is expressed as $\boldsymbol{F}(\boldsymbol{x}^0) = \{i \in \{1, \ldots, m\} | g_i x^0 > 0\}$.

[1] If a constraint set C is used, [NLPX] can be expressed as $\max_{x \in C} f(\boldsymbol{x})$.

(Notes on ⟨Definition 3.1⟩)

Note on ⟨D3.1-2⟩: For notation simplification, matrix $m \times n$ with $\nabla g_i(\boldsymbol{x})$ as the $i(i \in \{1, \cdots, m\})$ row can be expressed as $\nabla g(\boldsymbol{x})$.

Note on ⟨D3.1-5⟩**:** Henceforth, nonlinear programming problem with nonnegative constraints of $\boldsymbol{x}$ shall be abbreviated to

$$[\text{NLPX}]^{+} \quad \left[\max_{\boldsymbol{x}\in \boldsymbol{D}} f(\boldsymbol{x}) \qquad \text{subject to} \quad g(\boldsymbol{x}) = \begin{pmatrix} G(\boldsymbol{x}) \\ \boldsymbol{x} \end{pmatrix} \geqq \mathbf{0}\right]$$

or

$$\left[\max_{\boldsymbol{x}\in \boldsymbol{C}^{+}} f(\boldsymbol{x}) \qquad \boldsymbol{C}^{+} = \left\{ \boldsymbol{x} \in \boldsymbol{D} \,\middle|\, \begin{matrix} g(\boldsymbol{x}) \geqq \mathbf{0} \\ \boldsymbol{x} \geqq \mathbf{0} \end{matrix} \right\}\right]$$

However, $G(\boldsymbol{x}) = (g_1(\boldsymbol{x}), \cdots, g_{m-n}(\boldsymbol{x}))^t \geqq \mathbf{0}$ is a constraint equation in $g(\boldsymbol{x})$ other than $\boldsymbol{x} = (g_{m-n+1}(\boldsymbol{x}), \cdots, g_m(\boldsymbol{x}))^t \geqq \mathbf{0}$. When considering $[\text{NLPX}]^{+}$, regardless of whether $\boldsymbol{D}=\boldsymbol{R}^n_+$,

$$\Lambda(\boldsymbol{x}, \mu) = f(\boldsymbol{x}) + \mu G(\boldsymbol{x})$$

may be used as a Lagrange function.

Furthermore, let us define the **Lagrange condition** (hereinafter defined as [LC]) and **Kuhn–Tucker condition** (hereinafter defined as [KT]) corresponding to [NLPX]. Furthermore, in this section, unless otherwise specified, $f(\boldsymbol{x})$, $g_1(\boldsymbol{x})$, $\cdots$, $g_m(\boldsymbol{x})$ is considered to be differentiable.

⟨Definition 3.2⟩

⟨D3.2-1⟩ Lagrange condition

When for $\boldsymbol{x}^0 \in \boldsymbol{C}$, **Lagrange multiplier vector** $\boldsymbol{\lambda}^0 = (\lambda_1^0, \cdots, \lambda_m^0) \geqq \mathbf{0}$ is defined and

$$(\text{LC}-1) \quad \nabla L_x(\boldsymbol{x}^0, \boldsymbol{\lambda}^0) = \nabla f(\boldsymbol{x}^0) + \sum_{i=1}^{m} \lambda_i^0 \nabla g_i(\boldsymbol{x}^0) = 0$$

and

$$(\text{LC}-2) \quad \boldsymbol{\lambda}^0 \nabla L_\lambda(\boldsymbol{x}^0, \boldsymbol{\lambda}^0) = \boldsymbol{\lambda}^0 g(\boldsymbol{x}^0) = \sum_{i=1}^{m} \lambda_i^0 g_i(\boldsymbol{x}^0) = 0$$

are satisfied, [LC] forms on $\boldsymbol{x}^0 \in C$ or that $\boldsymbol{x}^0 \in C$ satisfies [LC].

Moreover, when on $\boldsymbol{x}^0 \in C$, [LC] holds whatever differentiable function $f(\boldsymbol{x})$ is selected, $g(\boldsymbol{x})$ is said to be **Lagrange regular** on $\boldsymbol{x}^0$.

⟨D 3.2-2⟩ Complementarity condition

If the nonnegativity of $g(\boldsymbol{x}^0)$ and $\boldsymbol{\lambda}^0$ are considered, (LC-2) is equivalent to

$$(\mathrm{LC}-2)' \quad \lambda_i^0 g_i(\boldsymbol{x}^0) = 0, (i = 1, \cdots, m).$$

(LC-2)′ indicates that if one of λ_i^0 or $g_i(\boldsymbol{x}^0)$ is positive, the other is always zero and hence it is called a **complementary slackness condition**. Furthermore, if we define $\boldsymbol{E}_+(\boldsymbol{x}^0) = \{i \in \{1, \cdots, m\} | \lambda_i^0 > 0\}$,[2] from complementary slackness condition, for any $i \in \boldsymbol{E}_+(\boldsymbol{x}^0)$, $g_i(\boldsymbol{x}^0) = 0$. Thus $\boldsymbol{E}_+(\boldsymbol{x}^0) \subseteq \boldsymbol{E}(\boldsymbol{x}^0)$. In particular, when $\boldsymbol{E}_+(\boldsymbol{x}^0) = \boldsymbol{E}(\boldsymbol{x}^0)$, it is said that a **narrow sense of complementary slackness condition**. If $\boldsymbol{E}_+(\boldsymbol{x}^0) = \boldsymbol{E}(\boldsymbol{x}^0)$, $i \notin \boldsymbol{E}_+(\boldsymbol{x}^0)$ and $i \notin \boldsymbol{E}(\boldsymbol{x}^0)$ are equivalent. Therefore, under the narrow sense of complementary slackness condition,

$$\lambda_i^0 = 0 \Rightarrow g_i(\boldsymbol{x}^0) > 0$$

or

$$g_i(\boldsymbol{x}^0) = 0 \Rightarrow \lambda_i^0 > 0.$$

Note that these implications are the opposite of the implications derived from the complementary slackness condition. Furthermore, if it is unnecessary to demonstrate the dependency of $\boldsymbol{E}(\boldsymbol{x}^0)$ on $\boldsymbol{x}^0$ or if there is a risk of making it over-complicated, $\boldsymbol{E}(\boldsymbol{x}^0)$ is simply expressed as $\boldsymbol{E}$.; this is the same as $\boldsymbol{E}_+(\boldsymbol{x}^0)$.

If $\boldsymbol{x} \geqq \boldsymbol{0}$ is included in $g(\boldsymbol{x}) \geqq \boldsymbol{0}$ ([NLPX]$^+$), [LC] and [KT] are equivalent. First, let us use the Lagrange function to determine [KT].

⟨Definition 3.3⟩ Kuhn–Tucker Condition For $\boldsymbol{x}^0 \in \boldsymbol{C}$, **Kuhn–Tucker's multiplier vector** $\boldsymbol{\mu}^0 = (\mu_1^0, \cdots, \mu_{m-n}^0) \geqq \boldsymbol{0}$ is defined, when

$$(\mathrm{KT}-1) \quad \nabla \Lambda_x(\boldsymbol{x}^0, \boldsymbol{\mu}^0) = \nabla f(\boldsymbol{x}^0) + \textstyle\sum_{i=1}^{m-n} \mu_i^0 \nabla G_i(\boldsymbol{x}^0) \leqq \boldsymbol{0},$$

$$(\mathrm{KT}-2) \quad \nabla \Lambda_x(\boldsymbol{x}^0, \boldsymbol{\mu}^0)\boldsymbol{x}^0 = [\nabla f(\boldsymbol{x}^0) + \textstyle\sum_{i=1}^{m-n} \mu_i^0 \nabla G_i(\boldsymbol{x}^0)]\boldsymbol{x}^0 = 0$$

and

$$(\mathrm{KT}-3) \quad \boldsymbol{\mu}^0 \nabla \Lambda_\mu(\boldsymbol{x}^0, \boldsymbol{\mu}^0) = \boldsymbol{\mu}^0 G(\boldsymbol{x}^0) = 0$$

hold, $\boldsymbol{x}^0 \in \boldsymbol{C}$ is said to satisfy [KT].

---

[2] The reason that the set of numbers of the positive elements of $\boldsymbol{\lambda}^0$ as $\boldsymbol{E}_+(\boldsymbol{x}^0)$ was to emphasize that $\boldsymbol{\lambda}^0$ itself is defined based on $\boldsymbol{x}^0$.

(Notes on ⟨Definition 3.3⟩): As long as [NLPX]$^+$ is being considered, [LC] and [KT] are equivalent. This is proven as shown below.

Let us assume that [LC] holds for $\boldsymbol{x}^0 \in \boldsymbol{C}$. Corresponding to $G(\boldsymbol{x}) = (g(x), \cdots, G_{m-n}(x))^t$ and $\boldsymbol{x}$ and divide in $\boldsymbol{\lambda}^0 = \left(\boldsymbol{\lambda}_I^0, \boldsymbol{\lambda}_{II}^0\right)$ (however, $\boldsymbol{\lambda}_I^0 \in R_+^{m-n}$, $\boldsymbol{\lambda}_{II}^0 \in R_+^n$), by (LC-1),

$$\begin{aligned}\nabla L_x\left(\boldsymbol{x}^0, \boldsymbol{\lambda}^0\right) &= \nabla f\left(\boldsymbol{x}^0\right) + \left(\boldsymbol{\lambda}_I^0, \boldsymbol{\lambda}_{II}^0\right)\begin{pmatrix}\nabla G\left(x^0\right) \\ \boldsymbol{I}_n\end{pmatrix} \\ &= 0,\end{aligned}$$

where, the ith row of $\nabla G(\boldsymbol{x}^0) = \nabla g_i(\boldsymbol{x})$ $(i = 1, \ldots, m - n)$.

Thus,

$$\begin{aligned}\nabla f\left(\boldsymbol{x}^0\right) + \boldsymbol{\lambda}_I^0 \nabla G\left(\boldsymbol{x}^0\right) &= -\boldsymbol{\lambda}_{II}^0 \\ &\leqq \boldsymbol{0}.\end{aligned} \tag{3.1}$$

By (LC-2),

$$\begin{aligned}0 &= \boldsymbol{\lambda}^0 g\left(\boldsymbol{x}^0\right) \\ &- \boldsymbol{\lambda}_I^0 G\left(\boldsymbol{x}^0\right) + \boldsymbol{\lambda}_{II}^0 \boldsymbol{x}^0.\end{aligned} \tag{3.2}$$

Using $\boldsymbol{\lambda}^0 \geqq \boldsymbol{0}$, $G(\boldsymbol{x}^0) \geqq \boldsymbol{0}$, (3.2) and (3.1)

$$\begin{aligned}\boldsymbol{0} &\leqq \boldsymbol{\lambda}_I^0 G\left(\boldsymbol{x}^0\right) \\ &= -\boldsymbol{\lambda}_{II}^0 \boldsymbol{x}^0 \\ &= \left[\nabla f\left(\boldsymbol{x}^0\right) + \lambda_{\mathrm{I}}^0 \nabla G\left(x^0\right)\right] x^0 \\ &\leqq \boldsymbol{0}.\end{aligned} \tag{3.3}$$

Hence, the following three equations

$$(\mathrm{KT}-1) \quad \nabla \Lambda_x\left(\boldsymbol{x}^0, \boldsymbol{\lambda}_I^0\right) = \nabla f\left(\boldsymbol{x}^0\right) + \boldsymbol{\lambda}_I^0 \nabla G\left(\boldsymbol{x}^0\right) \leqq \boldsymbol{0}$$

$$(\mathrm{KT}-2) \quad \nabla \Lambda_x\left(\boldsymbol{x}^0, \lambda_{\mathrm{I}}^0\right) x^0 = \left[\nabla f\left(x^0\right) + \boldsymbol{\lambda}_{\mathrm{I}}^0 \nabla G\left(x^0\right)\right] x^0 = \boldsymbol{0}$$

$$(\mathrm{KT}-3) \quad \lambda_{\mathrm{I}}^0 \nabla \Lambda_{\lambda \mathrm{I}}\left(\boldsymbol{x}^0, \lambda_I^0\right) = \boldsymbol{\lambda}_I^0 G\left(\boldsymbol{x}^0\right) = \boldsymbol{0}$$

hold and $\boldsymbol{x}^0 \in \boldsymbol{C}$ satisfies [KT].

Nevertheless, suppose that [KT] holds for $\boldsymbol{x}^0 \in \boldsymbol{C}$. Here, if we set

$$\boldsymbol{\lambda}^0 = \left\{\boldsymbol{\mu}^0, -\nabla \Lambda_x\left(\boldsymbol{x}^0, \boldsymbol{\mu}^0\right)\right\}$$

$$= (\boldsymbol{\mu}^0, -[\nabla f(\boldsymbol{x}^0) + \boldsymbol{\mu}^0 \nabla G(\boldsymbol{x}^0)])$$
$$\geqq \mathbf{0}$$

[LC] is satisfied over $\boldsymbol{x}^0 \in \boldsymbol{C}$. In fact, for this $\boldsymbol{\lambda}^0 \geqq \mathbf{0}$,

$$\begin{aligned}
\text{(LC}-1) \quad \nabla L_x(\boldsymbol{x}^0, \boldsymbol{\lambda}^0) &= \nabla f(\boldsymbol{x}^0) + \boldsymbol{\lambda}^0 \nabla g(\boldsymbol{x}^0) \\
&= \nabla f(\boldsymbol{x}^0) + \boldsymbol{\mu}^0 \nabla G(\boldsymbol{x}^0) - [\nabla f(\boldsymbol{x}^0) + \boldsymbol{\mu}^0 \nabla G(\boldsymbol{x}^0)] \\
&= \mathbf{0}
\end{aligned}$$

$$\begin{aligned}
\text{(LC}-2) \quad \boldsymbol{\lambda}^0 \nabla L_\lambda(\boldsymbol{x}^0, \boldsymbol{\lambda}^0) &= \boldsymbol{\mu}^0 G(\boldsymbol{x}^0) - [\nabla f(\boldsymbol{x}^0) + \mu^0 \nabla G(\boldsymbol{x}^0)]\boldsymbol{x}^0 \\
&= 0
\end{aligned}$$

is derived. QED

In this chapter, our primary interest is to consider the optimal solution for [NLPX] but firstly, as preparation, let us prove the following lemma.

[Lemma 3.1] Let $f(x)$, $g_1(x)$, $\cdots$, $g_m(x)$ be a real-valued function defined on a convex set D of R^n. Furthermore, let each $g_i(x)(i = 1, \cdots, m)$ be a quasi-concave function on D.

If for certain $\boldsymbol{x}^0 \in \boldsymbol{C}$, one of the following two conditions holds,

$$\text{[LCC]} \quad \left[\ {}^{\exists}\boldsymbol{\lambda}^0 \geqq \mathbf{0} \ \middle|\ \begin{array}{ll}
\text{(LCC}-1) & \nabla L_x(\boldsymbol{x}^0, \boldsymbol{\lambda}^0) \nabla f(\boldsymbol{x} - \boldsymbol{x}^0) \\
& = \{\nabla f(\boldsymbol{x}^0) + \boldsymbol{\lambda}^0 \nabla g(\boldsymbol{x}^0)\}(\boldsymbol{x} - \boldsymbol{x}^0) \\
& = \{\nabla f(\boldsymbol{x}^0) + \sum_{i=1}^m \lambda_i^0 g_i(\boldsymbol{x}^0)\}(\boldsymbol{x} - \boldsymbol{x}^0) \\
& \leqq 0 \qquad {}^{\forall}\boldsymbol{x} \in D \\
& \text{and} \\
\text{(LCC}-2) & \boldsymbol{\lambda}^0 \nabla L_x(\boldsymbol{x}^0, \boldsymbol{\lambda}^0) = \boldsymbol{\lambda}^0 g(\boldsymbol{x}^0) \\
& = \sum_{i=1}^m \lambda_i^0 g_i(\boldsymbol{x}^0) \\
& = 0
\end{array} \right]$$

or

$$\text{[LC]} \quad \left[\ {}^{\exists}\boldsymbol{\lambda}^0 \geqq \mathbf{0} \ \middle|\ \begin{array}{ll}
\text{(LC}-1) & \nabla L_x(\boldsymbol{x}^0, \boldsymbol{\lambda}^0) = \nabla f(\boldsymbol{x}^0) + \boldsymbol{\lambda}^0 \nabla g(\boldsymbol{x}^0) \\
& = \mathbf{0} \\
& \text{and} \\
\text{(LC}-2) & \boldsymbol{\lambda}^0 \nabla L_x(\boldsymbol{x}^0, \boldsymbol{\lambda}^0) = \boldsymbol{\lambda}^0 g(\boldsymbol{x}^0) \\
& = 0
\end{array} \right]$$

$$^{\forall}\boldsymbol{x} \in \boldsymbol{C}, \nabla f(\boldsymbol{x}^0)(\boldsymbol{x}-\boldsymbol{x}^0) \leqq 0.$$

Furthermore, if the domain $\boldsymbol{D}$ is an open convex set, [LCC] indicates [LC].

(Proof) If $\boldsymbol{x}$ is considered to be any vector of $\boldsymbol{C}$, the conclusion can be rewritten as follows:

$$\nabla f(\boldsymbol{x}^0)(\boldsymbol{x}-\boldsymbol{x}^0) = \left[\nabla f(\boldsymbol{x}^0) + \boldsymbol{\lambda}^0 \nabla g(\boldsymbol{x}^0)\right](\boldsymbol{x}-\boldsymbol{x}^0) - \boldsymbol{\lambda}^0 \nabla g(\boldsymbol{x}^0)(\boldsymbol{x}-\boldsymbol{x}^0) \quad (3.4)$$

Firstly, let us use (LC-2) which is common to the quasi-convexity of each g_i, [LCC] and [LC] to show [the second term on the right-hand side (3.4)]$\geqq 0$. From $\boldsymbol{\lambda}^0 \geqq \boldsymbol{0}$, if $\boldsymbol{E}_+(\boldsymbol{x}^0) = \left\{i \in \{1, \cdots, m\} \middle| \lambda_i^0 > 0\right\}$ is set, the second term on the right hand side becomes

$$\begin{aligned}\boldsymbol{\lambda}^0 \nabla g(\boldsymbol{x}^0)(\boldsymbol{x}-\boldsymbol{x}^0) &= \sum\nolimits_{i \in \boldsymbol{E}_+(\boldsymbol{x}^0)} \lambda_i^0 \left[\nabla g_i(\boldsymbol{x}^0)(\boldsymbol{x}-\boldsymbol{x}^0)\right] + \sum\nolimits_{i \notin \boldsymbol{E}_+(\boldsymbol{x}^0)} \lambda_i^0 \left[\nabla g_i(\boldsymbol{x}^0)(\boldsymbol{x}-\boldsymbol{x}^0)\right] \\ &= \sum\nolimits_{i \in \boldsymbol{E}_+(\boldsymbol{x}^0)} \lambda_i^0 \left[\nabla g_i(\boldsymbol{x}^0)(\boldsymbol{x}-\boldsymbol{x}^0)\right]\end{aligned}$$

Nevertheless, if the fact that for all $\boldsymbol{x} \in \boldsymbol{C}$ and $i \in \boldsymbol{E}_+(\boldsymbol{x}^0)$,

$$\begin{aligned} g_i(\boldsymbol{x}^0) &\geqq \boldsymbol{0} \\ &= g_i(\boldsymbol{x}^0)\end{aligned}$$

holds and on the quasi-concaveness of each g_i are noted, from [Theorem 2.3; (iii)], $\nabla g_i(\boldsymbol{x}^0)(\boldsymbol{x} - \boldsymbol{x}^0) \geqq \boldsymbol{0}$ is derived and [the second term on the right hand side] $\geqq 0$.

Moreover,

$$\begin{aligned}[\text{1st term on RHS}] = \left\{\nabla f(\boldsymbol{x}^0) + \boldsymbol{\lambda}^0 \nabla g(\boldsymbol{x}^0)\right\}(\boldsymbol{x}-\boldsymbol{x}^0) &= 0, \quad \text{when (LC} - 1) \\ &\leqq 0. \quad \text{when (LCC} - 1)\end{aligned}$$

Hence $^{\forall}\boldsymbol{x} \in \boldsymbol{C}, \nabla f(\boldsymbol{x}^0)(\boldsymbol{x}-\boldsymbol{x}^0) \leqq 0$.

Next, it demonstrates that when $\boldsymbol{D}$ is an open set, [LCC] indicates [LC]. Let us assume that [LCC] holds for some $\boldsymbol{x}^0 \in \boldsymbol{C}$. $\boldsymbol{D}$ is an open set and $\boldsymbol{x}^0 \in \boldsymbol{C} \subseteq \boldsymbol{D}$. Thus, for all $i \in \boldsymbol{N} \equiv \{1, \cdots, n\}$, $\delta_i^0 > 0$ exists and $\left(\boldsymbol{x}^0 \pm \delta_i^0 \boldsymbol{e}^i\right) \in \boldsymbol{D}$. Accordingly, if we let [LCC]'s $\boldsymbol{x} = \boldsymbol{x}^0 \pm \delta_i^0 \boldsymbol{e}^i$, for any $i \in \boldsymbol{N}$,

$$\begin{aligned} 0 &\geqq \nabla L_x\left(\boldsymbol{x}^0, \boldsymbol{\lambda}^0\right)(\boldsymbol{x}-\boldsymbol{x}^0) \\ &= \left[\nabla f(\boldsymbol{x}^0) + \boldsymbol{\lambda}^0 \nabla g(\boldsymbol{x}^0)\right]\left(\pm \delta_i^0 \boldsymbol{e}^i\right) \\ &= \pm \delta_i^0 \left[\nabla f(\boldsymbol{x}^0) + \boldsymbol{\lambda}^0 \nabla g(\boldsymbol{x}^0)\right]_i.\end{aligned}$$

Hence

$$0=\left[\nabla f\left(\boldsymbol{x}^0\right)+\boldsymbol{\lambda}^0\nabla g\left(\boldsymbol{x}^0\right)\right]_i. \quad (i=1,\cdots,n)$$

$$=\frac{\partial f}{\partial x_i}|_{\boldsymbol{x}^0}+\sum_{j=1}^{m}\lambda_j^0\frac{\partial g_j}{\partial x_i}|_{\boldsymbol{x}^0}$$

holds. However, $[\nabla f(\boldsymbol{x}^0) + \boldsymbol{\lambda}^0 \nabla g(\boldsymbol{x}^0)]_i$ is the *ith* element of $[\nabla f(\boldsymbol{x}^0) + \boldsymbol{\lambda}^0 \nabla g(\boldsymbol{x}^0)]$, which shows (LC-1). Then, the fact that (LC-2) holds is evident. Hence, $\boldsymbol{x}^0 \in C$ satisfies [LC]. QED

(Notes on [Lemma 3.1]): Even if the domain $\boldsymbol{D}$ is a closed set if it is an open set (example: $\boldsymbol{D}$ = universal set) or [LC] is established on $\boldsymbol{x}^0 \in \boldsymbol{C}$, there is no requirement to consider [LCC].

[Lemma 3.2] If a twice differentiable quasi-concave function $\varphi(u,v) : R_+^2 \rightarrow R$ satisfies the following three conditions,

$$\varphi(0,0)=0,$$

$$\varphi_u(0,0)=0$$

and

$$\varphi_v(0,0)>0,$$

then $\varphi(1,0) \leqq 0$.

(Proof) If we contradict the conclusion and assume $\varphi(1,0) > 0$, for any $u \in (0,1)$,

$$\begin{aligned}\varphi[u(1,0)+(1-u)(0,0)] &= \varphi(u,0)\\ &\geqq min\{\varphi(1,0),\varphi(0,0)\}\\ &=0.\end{aligned}$$

Accordingly, for any $u \in (0,1]$, $\varphi(u,0) \geqq 0$.

Either,

(i) there exists certain $\overline{u} \in (0,1]$ and $\varphi(\overline{u},0)=0$,

or

(ii) for all $u \in (0,1]$, $\varphi(u,0) > 0$

holds.

If (i) holds, $\overline{u} \in \boldsymbol{S}=\{u \in (0,1] | \varphi(u,0)=0\}$ indicates $\boldsymbol{S} \neq \varphi$. Moreover, $\boldsymbol{S}$ is bound from above hence if there exists $sup\boldsymbol{S} = u^*$ and $u \in \boldsymbol{S}$, $0 < u \leqq u^*$. Furthermore, the continuity of φ and $\varphi(1,0) > 0$ guarantees the existence of $\overline{\delta}>0$, which is $\varphi(u,0) > 0$ for all $u \in \left(1-\overline{\delta},1\right]$. Thus, because $1-\overline{\delta}$ is one of the upper bounds of S and u^* is the minimum upper bound of $\boldsymbol{S}$, $u^* \leqq 1-\overline{\delta} < 1$. Therefore, $0 < u^* < 1$, and then if we let $\varphi(u^*, 0) > 0$ to show $\varphi(u^*, 0) = 0$, there exists $\{u_n\} \subseteq \mathbf{S}$ that converge to $sup\mathbf{S} = u^*$. However, from the continuity of φ, for $\varepsilon = \frac{1}{2}\varphi(u^*, 0)$, there exists certain number n_ε, and hence if $n \geqq n_\varepsilon$,

$$
\begin{aligned}
|\varphi(u_n, 0) - \varphi(u^*, 0)| &= \varphi(u^*, 0) \\
&< \frac{1}{2}\varphi(u^*, 0)
\end{aligned}
$$

but this is self-contradictory. Thus,

$$\varphi(u^*, 0) = 0 \tag{3.5}$$

Hence, $u^* = \max_{u \in S} u$ is shown. Moreover, because of the third condition and potential twice differentiability of φ, when $v \in [0, \overline{v})$, there exists a $\overline{v} > 0$ that is $\varphi_v(0, v) > 0$. Therefore, for $v \in (0, \overline{v})$, $\varphi_v(0, v) > 0$. Accordingly, if $\tilde{v}$ is selected to be sufficiently small in $(0, \overline{v})$, as long as $v \in (0, \overline{v})$, the equation

$$\phi[u(v), 0] = \varphi(0, v) \tag{3.6}$$

has a solution $u(v)$ that is $u(v) \geqq u^*$ and

$$\lim_{v \to 0} u(v) = u^* \tag{3.7}$$

is established. If $\theta = 1 - u^*/u(v)$ is set, $\theta \in [0, 1]$ and $(1 - \theta)\ [u(v), 0] + \theta(0, v) = [(1 - \theta)u(v), \theta v]$. Thus

$$
\begin{aligned}
\varphi[(1 - \theta)u(v), \theta v] &= \varphi(u^*, \theta v) \\
&\geqq \min\ \{\varphi\{u(v), 0\}, \varphi(0, v)\} \\
&= \varphi(0, v).
\end{aligned}
\tag{3.8}
$$

From the mean value theorem, for certain $v^* \in (0, \theta v)$,

$$\varphi_v(u^*, v^*) = \frac{\varphi(u^*, \theta v) - \varphi(u^*, 0)}{\theta v} \tag{3.9}$$

holds. However, if the domain of the solution $u(v)$ and (3.5) are considered, (3.8) and (3.9) means

$$\varphi_v(u^*, v^*) \geqq \frac{1}{\theta}\frac{\varphi(0, v)}{v} \text{ for any } v \in (0,\ \tilde{v}). \tag{3.10}$$

Here, let us investigate the limit value of both sides of (3.10) when $v \to 0$. If from (3.7) and definition of θ, $v \to 0$, $\theta \to 0$. Moreover, from [Lemma 3.2; the third condition], $\lim_{v \to 0} \frac{\varphi(0,v)}{v} = \varphi_v(0,0) > 0$. Therefore, the right-hand side of (3.10) diverges to infinity. On the other hand, when $v \to 0$, $\theta v \to 0$, and hence $v^* \in (0, \theta v)$ converges to 0. Furthermore, φ_v is assumed to be differentiable and so φ_v is continuous. Thus, the left-hand side of (3.10), as per $v \to 0$, converges to the bound definitive value $\varphi_v(u^*, 0)$, but these contradict (3.10).

Next, if we focus on the case when (ii) is $u^* = 0$ in (i), there exists a function $u(v)$ defined over an appropriate section $(0, \tilde{v})$ and

$$\varphi(u(v), 0) = \varphi(0, v) \tag{3.11}$$

and

$$\lim_{v \to 0} u(v) = 0. \tag{3.12}$$

For the two points $(u(v), 0)$ and $(0, v)$, $\varphi(u, v)$ is a quasiconcave function, and hence by the application of [Theorem 2.3; (iii)],

$$\text{for any } v \in (0, \tilde{v}), u(v)\varphi_u(0, v) - v\varphi_v(0, v) \geqq 0. \tag{3.13}$$

If (3.13) is rearranged,

$$\text{for any } v \in (0, \tilde{v}), \quad u(v)\frac{\varphi_u(0, v)}{v} \geqq \varphi_v(0, v). \tag{3.14}$$

When $v \to 0$, the limit value of the right-hand side of (3.14), from [Lemma 3.2; the third condition], $\varphi_v(0, 0) > 0$. However, the left-hand side of (3.14), from (3.6) and the twice differentiability of φ, $\lim_{v \to 0} u(v)\varphi_u(0, v)/v = u(0)\varphi_{uv}(0, 0) = 0$, which contradicts (3.14).

Accordingly, if we let $\varphi(1, 0) > 0$, there are inconsistencies in all possible cases, and thus $\varphi(1, 0) \leqq 0$ must be true. QED

In this book, the first theorem clarifies the conditions under which $\boldsymbol{x}^0 \in \boldsymbol{C}$, which satisfies [LC] or [LCC] becomes the optimal solution for [NLPX].

[Theorem 3.1] Let $f(\boldsymbol{x})$, $g_1(\boldsymbol{x})$, $\cdots$, $g_m(\boldsymbol{x})$ be a real-valued function defined on the convex set $\boldsymbol{D}$ and each $g_i(\boldsymbol{x})(i = 1, \cdots, m)$ is to be a quasiconcave function on $\boldsymbol{D}$.

Here, if $\boldsymbol{x}^0 \in \boldsymbol{C}$ satisfies either [LLC] or [LC] and one of the following conditions hold, x_0 is the optimal solution of [NLPX].

(i) $f(\boldsymbol{x})$ is a quasiconcave function and
(i-i) for some $\boldsymbol{x}^1 \in D$, $\nabla f(\boldsymbol{x}^0)(\boldsymbol{x}^1 - \boldsymbol{x}^0) < 0$,

(i-ii) for some $\boldsymbol{x}^2 \in D$, $\nabla f(\boldsymbol{x}^0)(\boldsymbol{x}^2 - \boldsymbol{x}^0) \neq 0$, and $f(\boldsymbol{x})$ is twice differentiable near $\boldsymbol{x}^0$.
(ii) $f(\boldsymbol{x})$ is a pseudo concave function.

(Proof)

(i-i) If for an arbitrary $\boldsymbol{x} \in \boldsymbol{C}$, if we let $\boldsymbol{x}(\theta)$ and $\boldsymbol{x}^0(\theta)$ as

$$\boldsymbol{x}(\theta) = \theta \boldsymbol{x}^1 + (1-\theta)\boldsymbol{x}$$
$$\boldsymbol{x}^0(\theta) = \theta \boldsymbol{x}^1 + (1-\theta)\boldsymbol{x}^0$$

respectively, for $\theta > 0$

$$\begin{aligned}\nabla f(\boldsymbol{x}^0)\left[\boldsymbol{x}^0(\theta) - \boldsymbol{x}^0\right] &= \theta \nabla f(\boldsymbol{x}^0)(\boldsymbol{x}^1 - \boldsymbol{x}^0) \\ &< 0. \end{aligned} \tag{3.15}$$

However, from [Lemma 3.1], for all $\theta \in (0, 1]$,

$$\begin{aligned}\nabla f(\boldsymbol{x}^0)\left[\boldsymbol{x}(\theta) - \boldsymbol{x}^0(\theta)\right] &= (1-\theta)\nabla f(\boldsymbol{x}^0)(\boldsymbol{x} - \boldsymbol{x}^0) \\ &\leqq 0. \end{aligned} \tag{3.16}$$

Hence, by adding each side of (3.15) and (3.16), for $\theta \in (0, 1]$,

$$\nabla f(x^0)\left[x(\theta) - x^0\right] < 0 \tag{3.17}$$

is obtained. Furthermore, f is a quasiconcave function, and so from [Theorem 2.3 ; (iii)], (3.17) only holds when

$$f(x(\theta)) < f(x^0).$$

$\theta \to 0$ Then $\boldsymbol{x}(\theta) \to \boldsymbol{x}$. Here, if $\boldsymbol{x}^0$ is considered to be not an optimal solution, and there exists $\tilde{\boldsymbol{x}} \in \boldsymbol{C}$ that satisfies $f(\tilde{\boldsymbol{x}}) > f(\boldsymbol{x}^0)$ and it goes against the continuity of $f(\boldsymbol{x})$. Accordingly, for all $\boldsymbol{x} \in \boldsymbol{C}$, $f(\boldsymbol{x}) \leqq f(\boldsymbol{x}^0)$ holds and $\boldsymbol{x}^0 \in \boldsymbol{C}$ is optimal.
(i-ii) If condition (i-i) holds, there is no requirement for additional proof. Furthermore, it is fine to take it that condition (i-i) does not hold; in other words, for any $\boldsymbol{x} \in \boldsymbol{D}$,

$$\nabla f(\boldsymbol{x}^0)(\boldsymbol{x} - \boldsymbol{x}^0) \geq 0. \tag{3.18}$$

Here, the establishment of (i-ii), because of (3.18), guarantee the existence of $\boldsymbol{x}^2 \in \boldsymbol{D}$ that satisfy

$$\nabla f(x^0)(x^2 - x^0) > 0. \tag{3.19}$$

Here, if (3.18) is used again whilst noting [Lemma 3.1], for all $\boldsymbol{x}^1 \in \boldsymbol{C}$,

$$\nabla f(\boldsymbol{x}^0)(\boldsymbol{x}^1 - \boldsymbol{x}^0) = 0 \tag{3.20}$$

is derived. Here, if the abovementioned $\boldsymbol{x}^2 \in \boldsymbol{D}$ and for any $\boldsymbol{x}^1 \in \boldsymbol{C}$ and $\boldsymbol{0} \leqq (\alpha, \beta) \in R^2$, $\phi(\alpha, \beta)$ is defined to be

$$\phi(\alpha, \beta) = f\left[(1 - \alpha - \beta)\boldsymbol{x}^0 + \alpha\boldsymbol{x}^1 + \beta\boldsymbol{x}^2\right] - f(\boldsymbol{x}^0) \tag{3.21}$$

$\phi(\alpha, \beta)$ becomes a quasi-concave function. Actually, the following forms for any $\lambda \in [0, 1]$.

$$\begin{aligned}
&\phi[(1-\lambda)(\alpha_1, \beta_1) + \lambda(\alpha_1, \beta_1)] \\
&= \phi[(1-\lambda)\alpha_1 + \lambda\alpha_2, (1-\lambda)\beta_1 + \lambda\beta_2] \\
&= f\Big[(1 - ((1-\lambda)\alpha_1 + \lambda\alpha_2) - ((1-\lambda)\beta_1 + \lambda\beta_2))x^0 + ((1-\lambda)\alpha_1 + \lambda\alpha_2)x^1 \\
&\quad + ((1-\lambda)\beta_1 + \lambda\beta_2)x^2\Big] - f(x^0) \\
&= f\Big[(1-\lambda)(\boldsymbol{x}^0 + \alpha_1(\boldsymbol{x}^1 - \boldsymbol{x}^0) + \beta_1(\boldsymbol{x}^2 - \boldsymbol{x}^0)) \\
&\quad + \lambda(x^0 + \alpha_2(x^2 - x^0) + \beta_2(x^2 - x^0))\Big] - f(\boldsymbol{x}^0) \\
&\geqq min\{f[\boldsymbol{x}^0 + \alpha_1(\boldsymbol{x}^1 - \boldsymbol{x}^0) + \beta_1(\boldsymbol{x}^2 - \boldsymbol{x}^0)], f[\boldsymbol{x}^0 + \alpha_2(\boldsymbol{x}^1 - \boldsymbol{x}^0) + \beta_2(\boldsymbol{x}^2 - \boldsymbol{x}^0)]\} - f(\boldsymbol{x}^0) \\
&= min\{\phi(\alpha_1, \beta_1), \phi(\alpha_2, \beta_2)\}.
\end{aligned}$$

Furthermore, by (3.19) and (3.20),

$$\begin{aligned}
\phi(0,0) &= f(\boldsymbol{x}^0) - f(\boldsymbol{x}^0) = 0 \\
\phi_\alpha(0,0) &= \nabla f(\boldsymbol{x}^0)(\boldsymbol{x}^1 - \boldsymbol{x}^0) = 0 \\
\phi_\beta(0,0) &= \nabla f(\boldsymbol{x}^0)(\boldsymbol{x}^2 - \boldsymbol{x}^0) > 0
\end{aligned}$$

is evident, and so $\phi(\alpha, \beta)$ satisfies the conditions of [Lemma 3.2]. Hence, for any $\boldsymbol{x}^1 \in \boldsymbol{C}$,

$$\phi(1,0) = f(\boldsymbol{x}^1) - f(\boldsymbol{x}^0) \leqq 0$$

is derived.

(ii) From [Lemma 3.1: (3.2)] and the fact that f is a quasi convex function, the following conclusion is evident. QED

(Notes on [Theorem 3.1]):It was already shown in [Lemma 3.1] that when the domain $\boldsymbol{D}$ is an open set, [LC] is derived from [LCC]. Thus, when $\boldsymbol{D}$ is an open set, only [LC] requires to be assumed.

From the above theorem, the following two systems can be derived. [System 3.1.1] can be said to be a different expression of [Theorem 3.1; (i-i)].

[Corollary 3.1.1] The assumptions are similar for [Lemma 3.1]. Here, if $\boldsymbol{x}^1 \in \boldsymbol{C}$ and $\boldsymbol{x}^2 \in \boldsymbol{D}$ exist and $\nabla f(\boldsymbol{x}^0)\boldsymbol{x}^1 > \nabla f(\boldsymbol{x}^0)\boldsymbol{x}^2$ holds, $\boldsymbol{x}^0 \in \boldsymbol{C}$ becomes an optimal solution of [NLPX].[3]

(Proof) The above inequality is equivalent to

$$\nabla f(\boldsymbol{x}^0)(\boldsymbol{x}^1 - \boldsymbol{x}^0) > \nabla f(\boldsymbol{x}^0)(\boldsymbol{x}^2 - \boldsymbol{x}^0) \tag{3.22}$$

holding for certain $\boldsymbol{x}^1 \in \boldsymbol{C}$ and $\boldsymbol{x}^2 \in \boldsymbol{D}$. Moreover, from [Lemma 3.1], for any $\boldsymbol{x} \in \boldsymbol{C}$,

$$0 \geqq \nabla f(\boldsymbol{x}^0)(\boldsymbol{x} - \boldsymbol{x}^0). \tag{3.23}$$

Consequently, selecting $\boldsymbol{x} = \boldsymbol{x}^1$ guarantees the presence of $\boldsymbol{x}^2 \in \boldsymbol{D}$ that satisfies

$$\begin{aligned} 0 &\geqq \nabla f(\boldsymbol{x}^0)(\boldsymbol{x}^1 - \boldsymbol{x}^0) \\ &> \nabla f(\boldsymbol{x}^0)(\boldsymbol{x}^2 - \boldsymbol{x}^0) \end{aligned}$$

i.e., the [Theorem 3.1; (i-i)] holds. Thus, the optimality of $\boldsymbol{x}^0 \in \boldsymbol{C}$ is evident. QED

When the domain $\boldsymbol{D}$ in [Theorem 3.1] is $\boldsymbol{R}^n_+$ (one of the closed convex set of $\boldsymbol{R}^n$), the following system is obtained.

[Corollary 3.1.2] Let $\boldsymbol{D} = \boldsymbol{R}^n_+$ and $f(\boldsymbol{x})$, $g_1(\boldsymbol{x})$, $\cdots$, $g_m(\boldsymbol{x})$ be the same as in [Theorem 3.1]. Here, if $\boldsymbol{x}^0 \in \boldsymbol{C}$ satisfies [KT] and one of the following conditions are satisfied, $\boldsymbol{x}^0 \in \boldsymbol{C}$ becomes the optimal solution of [NLPX]$^+$

(i) $f(\boldsymbol{x})$ is a quasiconcave function and

(i-i) for some number $i_0 \in \{1, \cdots, m\}$, $\nabla f_{i_0}(\boldsymbol{x}^0) = \left(\left. \frac{\partial f}{\partial x_{i_0}} \right|_{\boldsymbol{x}^0} \right) < 0$,

(i-ii) $\nabla f(\boldsymbol{x}^0)\boldsymbol{x}^0 > 0$,

(i-iii) $\nabla f(\boldsymbol{x}^0) \neq \boldsymbol{0}$ and $f(\boldsymbol{x})$ are twice differentiable near $\boldsymbol{x}^0$.

(ii) $f(\boldsymbol{x})$ is a pseudo concave function.

---

[3] Arrow et al. (1961, p.185, Lemma (a)) discuss the optimality of $\boldsymbol{x}^0 \in \boldsymbol{C}$ under (3.6) without proof.

(Proof) When $\boldsymbol{D}=\boldsymbol{R}_+^n$, as shown in the note about ⟨Definition 3.3⟩, because [KT] and [LC] are equivalent, it is evident that the condition for [Theorem 3.1] is satisfied.

First, show that both system conditions (i-i) and (i-ii) return to the condition (i-i) of [Theorem 3.1].

(i-i) If $\boldsymbol{e}^i$ is considered to be a i ($i \in \{1, \cdots, n\}$) th unit column vector and $\boldsymbol{x}^1$ $\boldsymbol{x}^1 = \boldsymbol{x}^0 + \boldsymbol{e}^i$, $\boldsymbol{0} \leq \boldsymbol{x}^1 \in \boldsymbol{D} = \boldsymbol{R}_+^n$. Accordingly, (i-i) shows that there exists $\boldsymbol{x}^1 \in \boldsymbol{R}_+^n$ that satisfy

$$\frac{\partial f}{\partial x_i}|_{\boldsymbol{x}^0} = \nabla f(\boldsymbol{x}^0)\boldsymbol{e}^{i_0} = \nabla f(\boldsymbol{x}^0)(\boldsymbol{x}^1 - \boldsymbol{x}^0) < 0$$

(i-ii) By selecting $\boldsymbol{x}^1 = \boldsymbol{0}$, (i-ii) can be rewritten as follows

$$-\nabla f(\boldsymbol{x}^0)\boldsymbol{x}^0 = \nabla f(\boldsymbol{x}^0)(\boldsymbol{0} - \boldsymbol{x}^0) < 0$$

holding for this $\boldsymbol{x}^1 = \boldsymbol{0} \in \boldsymbol{R}_+^n$.

If either of the conditions (i-i) or (i-ii) hold, the argument in [Theorem 3.1 ; (i-i)] is applicable.

(i-iii) Let us assume that condition (i-i) and (i-ii) do not hold but (i-iii) holds. (i-i) and (i-ii) not holding implies

$$\nabla f(\boldsymbol{x}^0) \geqq \boldsymbol{0} \tag{3.24}$$

$$\nabla f(\boldsymbol{x}^0)\boldsymbol{x}^0 \leqq 0 \tag{3.25}$$

respectively. Accordingly the optimality of $\boldsymbol{x}^0 \in \boldsymbol{C}$ must be demonstrated under

$$\nabla f(\boldsymbol{x}^0) \geq \boldsymbol{0} \tag{3.26}$$

and

$$\nabla f(\boldsymbol{x}^0)\boldsymbol{x}^0 = 0 \tag{3.27}$$

If the set of numbers $\boldsymbol{J}$ is defined as $\boldsymbol{J} = \{i \in \{1, \cdots, n\} | \nabla f_i(\boldsymbol{x}^0) > 0\}$, from (3.26), $\boldsymbol{J} \neq \phi$. Thus, for any $\boldsymbol{x} \in \boldsymbol{C}$,

$$\begin{aligned} 0 &\geqq \nabla f(\boldsymbol{x}^0)(\boldsymbol{x} - \boldsymbol{x}^0) \\ &= \nabla f(\boldsymbol{x}^0)\boldsymbol{x} \\ &\geqq 0. \end{aligned}$$

In other words,

$$^{\forall}\boldsymbol{x} \in \boldsymbol{C}, \nabla f(\boldsymbol{x}^0)\boldsymbol{x} = 0. \tag{3.28}$$

However, if we consider the definition of $\boldsymbol{J}$, (28), for any $\boldsymbol{x} \in \boldsymbol{C}$ and $i \in \boldsymbol{J}$, indicates $x_i = 0$.

Thus, when $\boldsymbol{J} = \boldsymbol{N}$, $\boldsymbol{C}$ is constructed only of $\mathbf{0}$, thus making the optimality of $\boldsymbol{x}^0 \in \boldsymbol{C}$ obvious. Hence, if $\boldsymbol{J} \subset \boldsymbol{N}$ is assumed and we let $\#\boldsymbol{J} = (n - s_0)$, it is possible to set $\boldsymbol{J} = \{s_0 + 1, \cdots, n\}$ (however, $1 < s_0 < n$) without losing generality. Moreover, allocate $\boldsymbol{x}$ to the vector subspace $\boldsymbol{y}$ composed of the first s_0 elements of it and the vector subspace z composed of the remaining elements and express it as

$$\boldsymbol{x} = \begin{pmatrix} \boldsymbol{y} \\ z \end{pmatrix}$$

Furthermore, for this partition of $\boldsymbol{x}$, partition $\nabla f(\boldsymbol{x})$ in the following manner.

$$\nabla f(\boldsymbol{x}) = \left[\nabla f_y(\boldsymbol{x}) \nabla f_z(\boldsymbol{x})\right],$$

$$\text{where, } \nabla f_y(\boldsymbol{x}) = \left[\nabla f_1(\boldsymbol{x}), \cdots, \nabla f_{s_0}(\boldsymbol{x})\right],$$

$$\nabla f_z(\boldsymbol{x}) = \left[\nabla f_{s_0+1}(\boldsymbol{x}), \cdots, \nabla f_n(\boldsymbol{x})\right].$$

Here, in respect of $\boldsymbol{x}^0 = \begin{pmatrix} \boldsymbol{y}^0 \\ z^0 \end{pmatrix} \in \boldsymbol{C}$, $\nabla f_y(\boldsymbol{x}^0) = \mathbf{0}$, $\nabla f_z(\boldsymbol{x}^0) > \mathbf{0}$ and $z^0 = \mathbf{0}$. Moreover, from the definition of J, for all $\boldsymbol{x} = \begin{pmatrix} \boldsymbol{y} \\ z \end{pmatrix} \in \boldsymbol{C}$, $z = \mathbf{0}$. Thus,

$$^{\forall}\boldsymbol{y} \geqq \mathbf{0},\ f\begin{bmatrix} \boldsymbol{y}^0 \\ \mathbf{0} \end{bmatrix} \geqq f\begin{bmatrix} \boldsymbol{y} \\ \mathbf{0} \end{bmatrix}$$

must be proven. Here, if a function $\varphi(\alpha, \beta)$ is defined for all $\alpha \in [0, 1]$, $\beta \geqq 0$ and any $\boldsymbol{y} \geqq \mathbf{0}$ and any $\bar{z} \geqq \mathbf{0}$ such that $\bar{z}_{io} > 0$, is defined as follows:

$$\varphi(\alpha, \beta) = f\begin{bmatrix} (1-\alpha)\boldsymbol{y}^0 + \alpha\boldsymbol{y} \\ \beta\bar{z} \end{bmatrix} - f\begin{bmatrix} \boldsymbol{y}^0 \\ \mathbf{0} \end{bmatrix}$$

the following holds for all $\lambda \in [0, 1]$.

$$\begin{aligned}
&\varphi[(1-\lambda)(\alpha_1, \beta_1) + \lambda(\alpha_2, \beta_2)] \\
&= \varphi[(1-\lambda)\alpha_1 + \lambda\alpha_2, (1-\lambda)\beta_1 + \lambda\beta_2] \\
&= f\begin{bmatrix} (1-\lambda)(1-\alpha_1\boldsymbol{y}^0 + \alpha_1\boldsymbol{y}) + \lambda(1-\alpha_2\boldsymbol{y}^0 + \alpha_2\boldsymbol{y}) \\ ((1-\lambda)\beta_1 + \lambda\beta_2)\bar{z} \end{bmatrix} - f\begin{bmatrix} \boldsymbol{y}^0 \\ \mathbf{0} \end{bmatrix} \\
&\geqq min\left\{ f\begin{bmatrix} 1-\alpha_1\boldsymbol{y}^0 + \alpha_1\boldsymbol{y} \\ \beta_1\bar{z} \end{bmatrix}, f\begin{bmatrix} 1-\alpha_2\boldsymbol{y}^0 + \alpha_2\boldsymbol{y} \\ \beta_2\bar{z} \end{bmatrix} \right\} - f\begin{bmatrix} \boldsymbol{y}^0 \\ \mathbf{0} \end{bmatrix}
\end{aligned}$$

$$= min\{\varphi(\alpha_1,\beta_1),\varphi(\alpha_2,\beta_2)\}.$$

Consequently, $\varphi(\alpha,\beta)$ is a quasiconvex function and

$$\varphi(0,0)=f\begin{bmatrix}\boldsymbol{y}^0\\ \boldsymbol{0}\end{bmatrix}-f\begin{bmatrix}\boldsymbol{y}^0\\ \boldsymbol{0}\end{bmatrix}=0,$$

$$\varphi_\alpha(0,0)=\nabla f_y(\boldsymbol{x}^0)(\boldsymbol{y}-\boldsymbol{y}^0)=0$$

and

$$\varphi_\beta(0,0)=\nabla f_{\bar{z}}(\boldsymbol{x}^0)\bar{z}>0$$

can be easily confirmed. Thus, [Lemma 3.2] can be applied to $\varphi(\alpha,\beta)$ and for any $\boldsymbol{y} \geqq \boldsymbol{0}$,

$$\varphi(1,0)=f\begin{bmatrix}\boldsymbol{y}\\ \boldsymbol{0}\end{bmatrix}-f\begin{bmatrix}\boldsymbol{y}^0\\ \boldsymbol{0}\end{bmatrix}\leqq 0.$$

The above equation has shown the optimality of $\boldsymbol{x}^0=\begin{bmatrix}\boldsymbol{y}^0\\ \boldsymbol{0}\end{bmatrix}\in \boldsymbol{C}$.

(ii) It is exactly the same as the proof for [Theorem 3.1; (ii)]. QED

To date, we have been explaining the sufficient conditions for, when some $\boldsymbol{x}^0 \in \boldsymbol{C}$ satisfies [LC], $\boldsymbol{x}^0$ is the optimal solution for a nonlinear programming problem. Conversely, it is generally not that case that there exists $\boldsymbol{\lambda}^* \geqq \boldsymbol{0}$ that satisfies [LC] given an optimal solution $\boldsymbol{x}^* \in \boldsymbol{C}$ of a nonlinear programming problem. This fact can easily be understood using the following example from Arrow and Enthoven (1961, p.782).

$$\begin{bmatrix} \max\limits_{\boldsymbol{x}=(x_1,x_2)'\in \boldsymbol{D}} & f(\boldsymbol{x})=x_1x_2 \\ \text{subject to} & g(\boldsymbol{x})=\begin{pmatrix}(1-x_1-x_2)^3\\ x_1\\ x_2\end{pmatrix}\geqq\boldsymbol{0}\end{bmatrix}$$

The optimal solution of this problem is $\boldsymbol{x}^*=\left(\frac{1}{2},\frac{1}{2}\right)^t$ and $f(\boldsymbol{x}^*)=\frac{1}{4}$ is the maximum value of $f(\boldsymbol{x})$.

However, the Lagrange function corresponding to this maximization problem can be

$$L(\boldsymbol{x},\boldsymbol{\lambda}) = f(\boldsymbol{x}) + (\lambda_1, \lambda_2, \lambda_3)\begin{pmatrix} (1-x_1-x_2)^3 \\ x_1 \\ x_2 \end{pmatrix}$$

expressed as

$$\nabla L_x(\boldsymbol{x},\boldsymbol{\lambda}) = \left(x_2 - 3\lambda_1(1-x_1-x_2)^2 + \lambda_2, x_1 - 3\lambda_2(1-x_1-x_2)^2 + \lambda_3\right).$$

Substituting $\boldsymbol{x}^* = \left(\frac{1}{2}, \frac{1}{2}\right)^t$ in the above equation, as long as $(\lambda_2, \lambda_3) \geqq \mathbf{0}$, establishes $\nabla L_x(\boldsymbol{x}^*, \boldsymbol{\lambda}) > \mathbf{0}$, i.e., there are no $\boldsymbol{\lambda} \geqq \mathbf{0}$ that satisfies $\nabla L_x(\boldsymbol{x}^*, \boldsymbol{\lambda}) = \mathbf{0}$.

However, it is known that if the constraints are made to include appropriate additional conditions, a nonnegative Lagrange multiplier vector that satisfies [LC] in respect of the optimal solution is set or that [LC] becomes the necessary condition for optimality.

This additional condition is referred to as **constraint qualification** and numerous types of constraint qualifications have been put forward. Thus, in the next section, once definitions of several representative constraint qualifications are provided, the mutual relationships between them will be considered.

Problems relating to constraint qualifications are extremely important from the perspective of application to economic studies. This is because to examine the effect of a nonlinear programming problem on the exogenous variables and optimized solution parameters, the necessary conditions that the optimal solution has to satisfy have to be used.

3.3 Constraint Qualification

This section aims to show that under constraint qualification, for the optimal solution of [NLPX], there exists a nonnegative Lagrange multiplier vector that satisfies [LC].

Firstly, we will provide an explanation of the terminology and preliminary concepts required for this section.

Throughout ⟨Definition 3.4⟩–⟨Definition 3.13⟩, $\boldsymbol{X}$ shall be a distance space and $\boldsymbol{A}$ a subset of $\boldsymbol{X}$.

⟨Definition 3.4⟩ $\boldsymbol{A}$ and $\boldsymbol{B}$ are any subsets of $\boldsymbol{X}$. Here, the set of elements contained in $\boldsymbol{A}$ but not in $\boldsymbol{B}$ is called the **difference** between $\boldsymbol{A}$ and $\boldsymbol{B}$ and represents

$$\boldsymbol{A} - \boldsymbol{B} = \{\boldsymbol{x} | \boldsymbol{x} \in \boldsymbol{A} \quad \text{and} \quad \boldsymbol{x} \notin \boldsymbol{B}\}.$$

Moreover, $\boldsymbol{A} - \boldsymbol{B}$ may be considered as $\boldsymbol{A} \backslash \boldsymbol{B}$ in certain cases.

⟨Definition 3.5⟩ $\boldsymbol{A}$ and $\boldsymbol{B}$ are any subsets of $\boldsymbol{X}$ and are considered to be $\boldsymbol{B} \subseteq \boldsymbol{A}$. Here, the difference $\boldsymbol{A} - \boldsymbol{B}$ between $\boldsymbol{A}$ and $\boldsymbol{B}$ is known as a complement of $\boldsymbol{B}$ relative to

$\boldsymbol{A}$ and is expressed as $\boldsymbol{B}^C$. In particular, when $\boldsymbol{A}$ is equivalent to a universal set $\boldsymbol{X}$, $\boldsymbol{B}^C$ is simply referred to as a complementary set.

⟨**Definition 3.6**⟩ Given a point $\boldsymbol{a} \in \boldsymbol{X}$ and a real number $\varepsilon > 0$,

$$\boldsymbol{B}(\boldsymbol{a};\varepsilon) = \{x \in \boldsymbol{X} | d(\boldsymbol{x},\boldsymbol{a}) < \varepsilon\}$$

is stipulated and this $\boldsymbol{B}(\boldsymbol{a};\varepsilon)$ is called $\boldsymbol{a}$'s **$\boldsymbol{\varepsilon}$—open ball** or $\boldsymbol{a}$'s **$\boldsymbol{\varepsilon}$—neighborhood**.

Moreover, ε is called the **radius** of $B(\boldsymbol{a};\varepsilon)$.

⟨**Definition 3.7**⟩ When a point $\boldsymbol{a} \in \boldsymbol{A}$ satisfies $\boldsymbol{B}(\boldsymbol{a};\varepsilon) \subseteq \boldsymbol{A}$ for certain $\varepsilon > 0$, $\boldsymbol{a}$ is called $\boldsymbol{A}$'s **interior point** and the set of interior points of $\boldsymbol{A}$ is represented as ***intA***.

⟨**Definition 3.8**⟩ A point $\boldsymbol{a} \in \boldsymbol{A}^C$ that satisfies $B(\boldsymbol{a};\varepsilon) \subseteq \boldsymbol{A}^C$ for some $\varepsilon > 0$ is called an **exterior point** of $\boldsymbol{A}$ and the set of exterior points of $\boldsymbol{A}$ is denoted as ***extA***.

(Notes on ⟨Definition 3.7⟩ and ⟨Definition 3.8⟩): From ⟨Definition 3.7⟩ and ⟨Definition 3.8⟩,

$$\boldsymbol{a} \in \boldsymbol{intA} \Leftrightarrow \boldsymbol{a} \in \boldsymbol{ext\,A}^C$$

$$\boldsymbol{a} \in \boldsymbol{extA} \Leftrightarrow \boldsymbol{a} \in \boldsymbol{int\,A}^C$$

is obvious.

⟨**Definition 3.9**⟩ When a point $\boldsymbol{a} \in A$ is neither an interior point nor an exterior point of A, i.e., when in respect of any $\varepsilon > 0$,

$$\boldsymbol{B}(\boldsymbol{a};\varepsilon) \cap \boldsymbol{A} \neq \phi \quad \text{and} \quad \boldsymbol{B}(\boldsymbol{a};\varepsilon) \cap \boldsymbol{A}^C \neq \phi$$

holds, $\boldsymbol{a}$ is called a **boundary point** of A. Moreover, the set of boundary points of $\boldsymbol{A}$ is denoted as $\partial \boldsymbol{A}$.

⟨**Definition 3.10**⟩ When all the neighborhoods of points $\boldsymbol{a}$ contained in X contain points in A different from $\boldsymbol{a}$, $\boldsymbol{a}$ is called the **limit point** of A. Furthermore, the set of limit points of A is expressed as follows:

$$\boldsymbol{L}_A = \{\boldsymbol{a} \in \boldsymbol{X} |^{\forall}\varepsilon > 0, (\boldsymbol{B}(\boldsymbol{a};\varepsilon) - \{\boldsymbol{a}\}) \cap \boldsymbol{A} \neq \phi\}$$

However, $\{\boldsymbol{a}\}$ is a **singleton** composed only from 1 point $\boldsymbol{a}$.

The point $\boldsymbol{a}$ satisfying $\boldsymbol{a} \in \boldsymbol{A}$ and $\boldsymbol{a} \notin \boldsymbol{L}_A$ is known as the **isolated point** of $\boldsymbol{A}$.

(Notes on ⟨Definition 3.10⟩): Even if $\boldsymbol{a} \in \boldsymbol{L}_A$, it is not necessarily the case that $\boldsymbol{a} \in \boldsymbol{A}$.

⟨**Definition 3.11**⟩ When all of the points in the set $\boldsymbol{A}$ are internal points of $\boldsymbol{A}$, $\boldsymbol{A}$ is an **open set**.

⟨**Definition 3.12**⟩ When all the sets of limit points of a set A are contained in A, i.e., when $\boldsymbol{A} \supseteq \boldsymbol{L}_A$, $\boldsymbol{A}$ is called a **closed set**.

⟨**Definition 3.13**⟩

$$\overline{A} = \boldsymbol{A} \cup \boldsymbol{L}_A$$

is called the **closure** of $\boldsymbol{A}$.

[Theorem 3.2] The necessary and sufficient condition for $\boldsymbol{x} \in \boldsymbol{L}_A$ is that there is a sequence of points $\{\boldsymbol{x}^\gamma\}$ formed of $\boldsymbol{A}$ points different from $\boldsymbol{x}$ and $\lim_{\gamma \to \infty} \boldsymbol{x}^\gamma = \boldsymbol{x}$.

(Proof)

(Necessity) If $\boldsymbol{x} \in \boldsymbol{L}_A$, for any $\gamma \in \{1, 2, \cdots\}$, $\left(\boldsymbol{B}\left(x; \frac{1}{\gamma}\right) - \{x\}\right) \cap \boldsymbol{A} \neq \phi$. Thus, if a point $\boldsymbol{x}^\gamma$ is selected to be in each set $\left(\boldsymbol{B}\left(x; \frac{1}{\gamma}\right) - \{x\}\right) \cap \boldsymbol{A}$, $\{\boldsymbol{x}^\gamma\} \subseteq \boldsymbol{A}$ and $\lim_{\gamma \to \infty} \boldsymbol{x}^\gamma = \boldsymbol{x}$.

(Sufficiency) Let us assume that there exists a sequence of points $\{\boldsymbol{x}^\gamma\}$, which satisfies $\lim_{\gamma \to \infty} \boldsymbol{x}^\gamma = \boldsymbol{x}$ and $\boldsymbol{x}^\gamma \neq \boldsymbol{x}$. Here, for any $\varepsilon > 0$, there exists certain number n_ε and for all γ that satisfy $\gamma > n_\varepsilon$, $d(\boldsymbol{x}^\gamma, \boldsymbol{x}) < \varepsilon$. Consequently, for all γ such that $\gamma > n_\varepsilon$, $\boldsymbol{x}^\gamma \in \boldsymbol{B}(x; \varepsilon)$ and $\boldsymbol{x}^\gamma \in A$. Considering $\boldsymbol{x}^\gamma \neq \boldsymbol{x}$, $(\boldsymbol{B}(x; \varepsilon) - \{x\}) \cap A \neq \phi$ is obtained for any $\varepsilon > 0$. QED

[Theorem 3.3] The fact that $\boldsymbol{A}$ is a closed set ($\boldsymbol{A} \supseteq \boldsymbol{L}_A$) is equivalent to all of the following (i)–(iii).

(i) $\boldsymbol{A}^C$ is an open set.
(ii) When $\{\boldsymbol{a}^n\}$ is a sequence of points in A, if $\lim_{n \to \infty} \boldsymbol{a}^n = \boldsymbol{a}$, $\boldsymbol{a} \in \boldsymbol{A}$.
(iii) $A \supseteq \partial \boldsymbol{A}$.

(Proof) Show in order (i) ⇒ (ii) ⇒ (iii) ⇒ $\boldsymbol{A} \supseteq \boldsymbol{L}_A$ ⇒ (i).

(i) ⇒ (ii) Contradict the conclusion and let $\{\boldsymbol{a}^n\} \subseteq \boldsymbol{A}$, $\lim_{n \to \infty} \boldsymbol{a}^n = \boldsymbol{a}$ and $\boldsymbol{a} \notin \boldsymbol{A}$. From condition (i), $\boldsymbol{A}^C$ is an open set. Thus, there exists $\varepsilon > 0$ and $B(\boldsymbol{a}; \varepsilon) \subseteq \boldsymbol{A}^C$. When $n \to \infty$, $\boldsymbol{a}^n \to \boldsymbol{a}$, hence, for this $\varepsilon > 0$, some number n_ε is defined and if $n > n_\varepsilon$, $d(\boldsymbol{a}^n, \boldsymbol{a}) < \varepsilon$. Accordingly, for $n > n_\varepsilon$, $\boldsymbol{a}^n \in \boldsymbol{B}(\boldsymbol{a}; \varepsilon) \subseteq \boldsymbol{A}^C$. Although $\{\boldsymbol{a}^n\} \subseteq \boldsymbol{A}$, when $n > n_\varepsilon$, $\boldsymbol{a}^n \in \boldsymbol{A}$ and $\boldsymbol{a}^n \notin \boldsymbol{A}$. This is self-contradictory.

(ii) ⇒ (iii) If $\boldsymbol{A} \supseteq \partial \boldsymbol{A}$, there exists a point $\boldsymbol{a}$ that satisfies $\boldsymbol{a} \notin \boldsymbol{A}$ and $\boldsymbol{a} \in \partial \boldsymbol{A}$. From $\boldsymbol{a} \in \partial \boldsymbol{A}$, for any $\varepsilon > 0$,

$$\boldsymbol{B}(\boldsymbol{a}; \varepsilon) \cap \boldsymbol{A} \neq \phi \text{ and } \boldsymbol{B}(\boldsymbol{a}; \varepsilon) \cap \boldsymbol{A}^C \neq \phi.$$

Furthermore, from $\boldsymbol{a} \notin \boldsymbol{A}$, for any $\varepsilon > 0$,

$$(\boldsymbol{B}(\boldsymbol{a};\varepsilon)-\{\boldsymbol{a}\})\cap \boldsymbol{A}\neq \phi.$$

Thus, $\boldsymbol{a}\in \boldsymbol{L}_A$. Accordingly, from [Theorem 3.2], there are a sequence of points $\{\boldsymbol{a}^n\}$ that converge to $\boldsymbol{a}$ but from condition (ii), $\boldsymbol{a}\in \boldsymbol{A}$. However, this contradicts the assumption $\boldsymbol{a}\notin \boldsymbol{A}$.

(iii) $\Rightarrow$ ($\boldsymbol{A}\supseteq \boldsymbol{L}_A$) If as assuming $\boldsymbol{A}\supseteq \boldsymbol{L}_A$, there is a point $\boldsymbol{a}$ on $\phi\neq(\boldsymbol{L}_A-\boldsymbol{A})$. From $\boldsymbol{a}\in \boldsymbol{L}_A$, for any $\varepsilon>0$,

$$\phi\neq(\boldsymbol{B}(\boldsymbol{a};\varepsilon)-\{\boldsymbol{a}\})\cap \boldsymbol{A}\subseteq \boldsymbol{B}(\boldsymbol{a};\varepsilon)\cap \boldsymbol{A}. \tag{3.29}$$

However, from $\boldsymbol{a}\notin \boldsymbol{A}$ and $\boldsymbol{a}\in \boldsymbol{B}(\boldsymbol{a};\varepsilon)$, for all $\varepsilon>0$,

$$\boldsymbol{B}(\boldsymbol{a};\varepsilon)\cap \boldsymbol{A}^C\neq \phi. \tag{3.30}$$

If you combine (3.29), (3.30), and condition (iii), $\boldsymbol{a}\in \partial \boldsymbol{A}\subseteq \boldsymbol{A}$ is obtained and contradicts the assumption $\boldsymbol{a}\notin \boldsymbol{A}$.

($\boldsymbol{A}\supseteq \boldsymbol{L}_A$) $\Rightarrow$ (i) If $\boldsymbol{A}\supseteq \boldsymbol{L}_A$, for any $\boldsymbol{a}\in \boldsymbol{A}^C$, $\boldsymbol{a}\in(\boldsymbol{L}_A)^C$. Thus, there exists some $\varepsilon_a>0$ and

$$\begin{aligned}(\boldsymbol{B}(\boldsymbol{a};\varepsilon_a)-\{\boldsymbol{a}\})\cap \boldsymbol{A}&=\boldsymbol{B}(\boldsymbol{a};\varepsilon_a)\cap \boldsymbol{A}\\&=\phi.\end{aligned}$$

Thus, for any $\boldsymbol{a}\in \boldsymbol{A}^C$, a $\varepsilon_a>0$ which is $\boldsymbol{B}(\boldsymbol{a};\varepsilon_a)\cap \boldsymbol{A}=\phi$ exists. QED

[Theorem 3.4] Let $\boldsymbol{A}$ and $\boldsymbol{B}$ be any subsets of $\boldsymbol{X}$. In this case, the following holds.

(i) If $\boldsymbol{A}\subseteq \boldsymbol{B}$,

(i-i) $\boldsymbol{L}_A\subseteq \boldsymbol{L}_B$

and

(i-ii) $\overline{\boldsymbol{A}}\subseteq \overline{\boldsymbol{B}}$.

(ii)

(ii-i) $\overline{\boldsymbol{A}}$ is a closed set.

and

(ii-ii) $\overline{\boldsymbol{A}}$ is the smallest closed set containing $\boldsymbol{A}$. In other words, when

$$\boldsymbol{F}_A=\{\boldsymbol{B}\subseteq \boldsymbol{X}|\boldsymbol{A}\subseteq \boldsymbol{B},\boldsymbol{B}\text{ is a closed set}\}$$

for $\overline{\boldsymbol{A}}\subseteq \boldsymbol{F}_A$ and any $\boldsymbol{B}\subseteq \boldsymbol{F}_A$, $\overline{\boldsymbol{A}}\subseteq \boldsymbol{B}$.

(iii) The necessary and sufficient condition for $\boldsymbol{A}$ is $\boldsymbol{A}=\overline{\boldsymbol{A}}$.

(iv) $\overline{\overline{\boldsymbol{A}}}=\overline{\boldsymbol{A}}$, however $\overline{\overline{\boldsymbol{A}}}$ is $\overline{(\overline{\boldsymbol{A}})}$.

(Proof)

(i-i) If $\boldsymbol{x}$ is to be any point in L_A, for all $\varepsilon > 0$,

$$\phi \neq (\boldsymbol{B}(\boldsymbol{x};\varepsilon) - \{\boldsymbol{x}\}) \cap \boldsymbol{A} \subseteq (\boldsymbol{B}(\boldsymbol{x};\varepsilon) - \{\boldsymbol{x}\}) \cap \boldsymbol{B}.$$

Hence, $\boldsymbol{x} \in \boldsymbol{L_B}$ for any $\boldsymbol{x} \in \boldsymbol{L_A}$.
(i-ii) It is clear from $\boldsymbol{A} \subseteq \boldsymbol{B}$, (i-i) and the definition of a closure.
(ii-i) Show $\overline{\boldsymbol{A}} \supseteq \boldsymbol{L}_{\overline{A}}$. If for any $\varepsilon > 0$, $r = \dfrac{2}{3}\varepsilon$ is defined, there exists a point $\boldsymbol{y}$ that belongs to $(\boldsymbol{B}(\boldsymbol{x};r) - \{\boldsymbol{x}\}) \cap \overline{\boldsymbol{A}}$ for all $\boldsymbol{x} \in \boldsymbol{L}_{\overline{A}}$. Thus, there exists a point $\boldsymbol{z} \neq \boldsymbol{x}$, which satisfies $\boldsymbol{z} \in \left(\boldsymbol{B}\left(\boldsymbol{y}; \dfrac{d(\boldsymbol{x},\boldsymbol{y})}{2}\right) - \{\boldsymbol{y}\}\right) \cap \boldsymbol{A}$. Furthermore,

$$\begin{aligned} d(\boldsymbol{x},\boldsymbol{z}) &\leqq d(\boldsymbol{x},\boldsymbol{y}) + d(\boldsymbol{y},\boldsymbol{z}) \\ &< \frac{3}{2} d(\boldsymbol{x},\boldsymbol{y}) \\ &< \frac{3}{2} \cdot \frac{2}{3}\varepsilon = \varepsilon. \end{aligned}$$

hence, for an arbitrary $\boldsymbol{x} \in \boldsymbol{L}_{\overline{A}}$, $\boldsymbol{x} \in \boldsymbol{L_A}$. Thus,

$$\boldsymbol{L}_{\overline{A}} \subseteq \boldsymbol{L_A} \subseteq \overline{\boldsymbol{A}} \tag{3.31}$$

is derived.
(ii-ii) From the definition of $\overline{\boldsymbol{A}}$, $\overline{\boldsymbol{A}} = \boldsymbol{A} \cup \boldsymbol{L_A} \supseteq \boldsymbol{A}$. Furthermore, from (ii-i), $\overline{\boldsymbol{A}} \subseteq \boldsymbol{F_A}$. Therefore, it requires to be shown that $\overline{\boldsymbol{A}} \subseteq \boldsymbol{B}$ for any set of $\boldsymbol{F_A}$, $\boldsymbol{B}$. For all $\boldsymbol{x} \in \overline{\boldsymbol{A}}$, $\boldsymbol{x} \in \boldsymbol{A}$ or $\boldsymbol{x} \in \boldsymbol{L_A}$. If $\boldsymbol{x} \in \boldsymbol{A}$, from $\boldsymbol{A} \subseteq \boldsymbol{B}$, $\boldsymbol{x} \in \boldsymbol{B}$. However, if $\boldsymbol{x} \in \boldsymbol{L_A}$, from (i-i) and $\boldsymbol{B} \supseteq \boldsymbol{L_B}$, $\boldsymbol{L_A} \subseteq \boldsymbol{L_B} \subseteq \boldsymbol{B}$, and therefore $\overline{\boldsymbol{A}} \subseteq \boldsymbol{B}$ is obtained.
(iii)
(Necessity) If $\boldsymbol{A}$ is a closed set, from $\boldsymbol{A} \supseteq \boldsymbol{L_A}$, $\overline{\boldsymbol{A}} = \boldsymbol{A} \cup \boldsymbol{L_A} = \boldsymbol{A}$.
(Sufficiency) From (ii-i), $\overline{\boldsymbol{A}}$ is a closed set. Moreover, because it is assumed that $\boldsymbol{A} = \overline{\boldsymbol{A}}$, $\boldsymbol{A}$ is a closed set.
(iv) By (3.31) in the proof of (ii), $\overline{\boldsymbol{A}} \supseteq \boldsymbol{L_A}$, accordingly, $\overline{\overline{\boldsymbol{A}}} = \overline{\boldsymbol{A}} \cup \boldsymbol{L}_{\overline{A}} = \overline{\boldsymbol{A}}$ is established. QED

[Theorem 3.5] $\boldsymbol{x} \in \overline{\boldsymbol{A}}$ is equivalent to every one of the following (i)–(iii).

(i) $\boldsymbol{x} \in \boldsymbol{A} \cup \partial \boldsymbol{A}$.
(ii) For any $\varepsilon > 0$, $\boldsymbol{B}(\boldsymbol{x};\varepsilon) \cap \boldsymbol{A} \neq \phi$.
(iii) There exists a sequence of points $\{\boldsymbol{x}^\gamma\}$ belonging to $\boldsymbol{A}$ $\lim_{\gamma\to\infty} \boldsymbol{x}^\gamma = \boldsymbol{x}$.

(Proof) Show in the order: $(\boldsymbol{x} \in \overline{\boldsymbol{A}}) \to$ (i) $\to$ (ii) $\to$ (iii) $\to$ $(\boldsymbol{x} \in \overline{\boldsymbol{A}})$.

$\left(x \in \overline{A}\right) \rightarrow$ (i) If $x \in \overline{A}, x \in A$ or $x \in L_A$. If $x \in A$, for any $\varepsilon > 0, x \in B(x; \varepsilon) \cap A$. Thus, $B(x; \varepsilon) \cap A \neq \phi$. If the situation when $x \notin A$ and $x \in L_A$ is considered, for any $\varepsilon > 0$,

$$B(x; \varepsilon) \cap A^C \neq \phi \text{ and } B(x; \varepsilon) \cap A \neq \phi,$$

hence $x \in \partial A$. Therefore, $x \in A \cup \partial A$.

(i) $\rightarrow$ (ii) If $x \in A, x \in B(x; \varepsilon) \cap A \neq \phi$ for any $\varepsilon > 0$. Moreover, if $x \in \partial A$, from the definition of boundary point, we immediately obtain $B(x; \varepsilon) \cap A \neq \phi$ for any $\varepsilon > 0$.

(ii) $\rightarrow$ (iii) For any $\varepsilon > 0$, assume $B(x; \varepsilon) \cap A \neq \phi$. Here, if we select $\varepsilon = \frac{1}{\gamma} > 0$ $(\gamma = 1, 2, \ldots)$, for any $\gamma \in \{1, 2, \ldots\}$, $B\left(x; \frac{1}{\gamma}\right) \cap A \neq \phi$. Accordingly, if $x^\gamma \in B\left(x; \frac{1}{\gamma}\right) \cap A$, $\lim_{\gamma \rightarrow \infty} x^\gamma = x$.

(iii) $\rightarrow$ $\left(x \in \overline{A}\right)$ If there exists a sequence of points $\{x^\gamma\}(\gamma = 1, 2, \ldots)$ that belong to A and $\lim_{\gamma \rightarrow \infty} x^\gamma = x$ holds, from [Theorem 3.4 ; (ii)], $\{x^\gamma\}$ is a converging sequence of points of the closed set $\overline{A}$. Thus, using [Theorem 3.3 ; (ii)], $x \in \overline{A}$. QED

(Notes on [Theorem 3.5]): A point x that satisfies (ii) is called a point of closure of A. Based on this theorem, $\overline{A}$ maybe called the set of points of closure of A.

Until now, the universal set X was assumed to be the metric space; however, henceforth, X will be considered to be R^n.

⟨Definition 3.14⟩ When a subset A of R^n contains closed line segments joining all two points contained in A, i.e., when for any $x, y \in A$ and $\lambda \in [0, 1]$,

$$(1 - \lambda)x + \lambda y \in A$$

holds, A is called a **convex set** (Figs. 3.1 and 3.2).

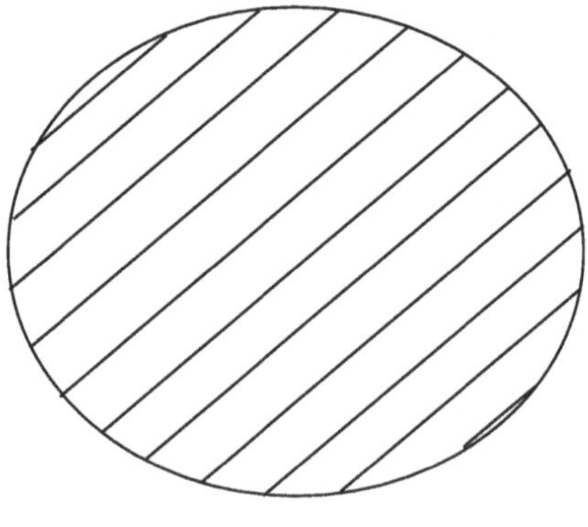

Fig. 3.1 Convex set

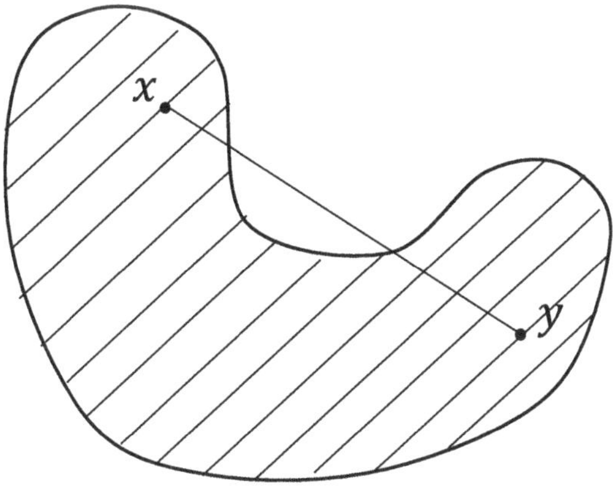

Fig. 3.2 Non-convex set

(Notes on ⟨Definition 3.14⟩): A singleton is a convex set. In this book, for convenience, an empty set is considered a convex set.

⟨Definition 3.15⟩ When the subset $\boldsymbol{A} \neq \phi$ of $\boldsymbol{R}^n$ satisfies $\alpha \boldsymbol{x} \in \boldsymbol{A}$ for any $\boldsymbol{x} \in \boldsymbol{A}$ and $\alpha \geqq 0$, $\boldsymbol{A}$ is called a **cone**.

⟨Definition 3.16⟩ If a subset $\boldsymbol{A} \neq \phi$ of $\boldsymbol{R}^n$ satisfies

$$\text{for any } \boldsymbol{x}, \boldsymbol{y} \in \boldsymbol{A} \text{ and any } \alpha, \beta \geqq 0, \alpha \boldsymbol{x} + \beta \boldsymbol{y} \in \boldsymbol{A}$$

$\boldsymbol{A}$ is called a **convex cone**. Furthermore, when the convex cone $\boldsymbol{A}$ is a closed set, $\boldsymbol{A}$ is known as a **closed convex cone** (Figs. 3.3 and 3.4).

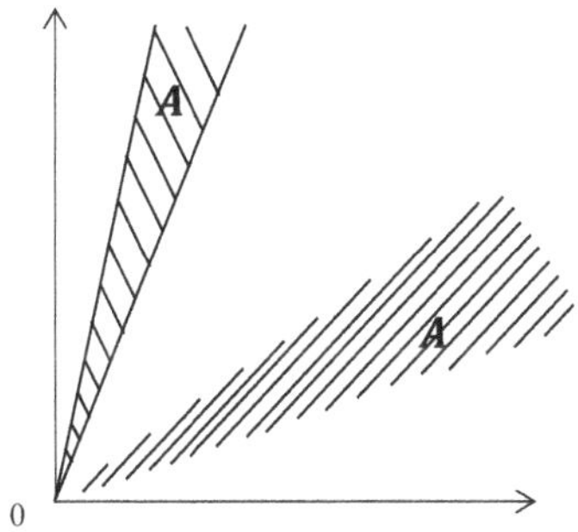

Fig. 3.3 Cone

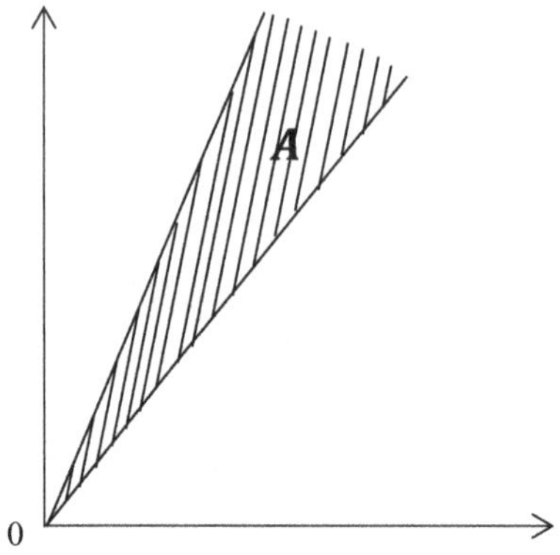

Fig. 3.4 Convex cone

(Notes on ⟨Definition 3.16⟩)

The fact that a convex cone is a convex set and $0 \in \boldsymbol{A}$ is clear from the definitions. The definition of convex cone may be replaced with the following.

$$\text{for any } \boldsymbol{x}, \boldsymbol{y} \in \boldsymbol{A}, \boldsymbol{x} + \boldsymbol{y} \in \boldsymbol{A}$$

and for any $\boldsymbol{x} \subset \boldsymbol{A}$ and $\alpha \geqq 0$, $\alpha \boldsymbol{x} \in \boldsymbol{A}$.

⟨Definition 3.17⟩ Let $\boldsymbol{K}$ be a convex cone of $\boldsymbol{R}^n$. In this case, $\boldsymbol{K}^*$ and $\boldsymbol{K}_*$ are defined, respectively, as follows:

$$\boldsymbol{K}^* = \left\{ \boldsymbol{y} \in \boldsymbol{R}^n \middle| \boldsymbol{y}^t \boldsymbol{x} \geqq 0, {}^{\forall}\boldsymbol{x} \in \boldsymbol{K} \right\}$$

$$\boldsymbol{K}_* = \left\{ \boldsymbol{z} \in \boldsymbol{R}^n \middle| \boldsymbol{z}^t \boldsymbol{x} \leqq 0, {}^{\forall}\boldsymbol{x} \in \boldsymbol{K} \right\}$$

and $\boldsymbol{K}^*$ is called a **nonnegative dual convex cone** of $\boldsymbol{K}$ and $\boldsymbol{K}_*$ is called the $\boldsymbol{K}$ **nonpositive dual convex cone**.

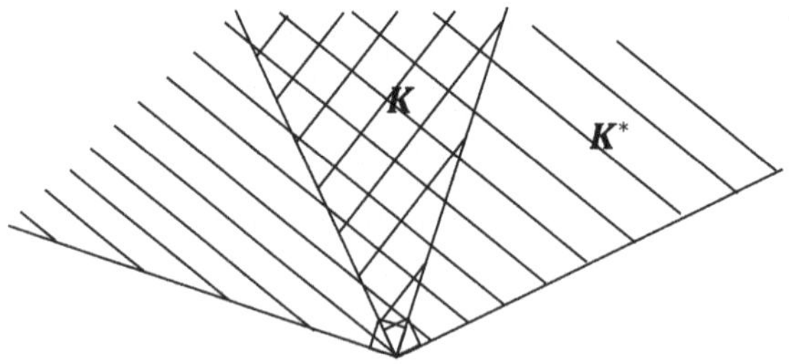

Fig. 3.5 K and K^*

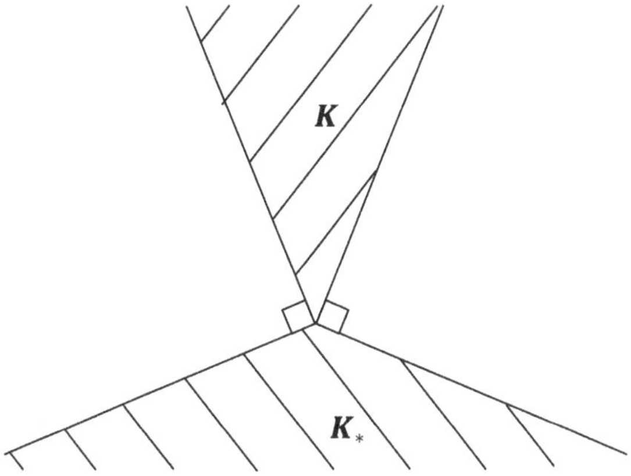

Fig. 3.6 K and K_*

(Notes on ⟨Definition 3.17⟩): If there is no particular requirement to differentiate K^* and K_*, K^* can simply be called a dual convex cone or a polar convex cone.

$K^*(K_*)$ is a union of sets of vectors that have an acute angle (obtuse angle) with all vectors in K. From $0 \in K^*$ and $0 \in K_*$, $K^* \neq \phi$ and $K_* \neq \phi$. Moreover, K is not necessarily a closed set, K^* and K_* are closed sets (Figs. 3.5 and 3.6).

⟨Definition 3.18⟩ For a finite number of vectors of R^n, $x^1, \cdots, x^k$, and any real number $\alpha_1 \geqq 0, \cdots, \alpha_k \geqq 0$, $\sum_{i=1}^k \alpha_i x^i$ is called a **nonnegative linear combination** of $x^1, \cdots, x^k$. In particular, when, for any $i \in \{1, \cdots, k\}$, $\alpha_i \geqq 0$ and $\sum_{i=1}^k \alpha_i = 1$, $\sum_{i=1}^k \alpha_i x^i$ is called $x^i, \cdots, x^k$'s **convex linear combination**.

⟨Definition 3.19⟩ Given a subset $A \neq \phi$ of R^n, all of the nonnegative linear combinations of any finite number of vectors belonging to A is expressed as follows:

$$K(A) = \left\{x \in R^n \middle| x = \sum_{i=1}^k \alpha_i x^i, \alpha_i \geqq 0,\ x^i \in A,\ i = 1, \cdots, k\right\}$$

and $K(A)$ is called a **convex cone spanned** by A. In particular, when A is a finite subset $\{x^1, \cdots, x^k\}$ composed of a finite number of vectors $x^1, \cdots, x^k$, $K(A)$ is called a finite cone and as required $K(A)$ can be expressed as follows:

$$K\left(x^1, \cdots, x^k\right)$$

(Notes on ⟨Definition 3.19⟩): $K(A)$ is a minimum convex cone containing A.

[Theorem 3.6] K and T are convex cones of R^n. In this case, the following holds.

(i) $K^*(K_*)$ is a closed convex cone.
(ii) If $K \subseteq T$, $K^* \supseteq T^*(K_* \supseteq T_*)$.

(iii) $K^{**} \supseteq K(K_{**} \supseteq K)$.
(iv) For K to be a closed convex cone, the necessary and sufficient condition is $K = K^{**}(K = K_{**})$.
(v) If K is a finite cone, $K = K^{**}(K = K_{**})$,
where $(K^*)^* = K^{**}$ and $(K_*)_* = K_{**}$

(Proof) If this can be proved for a nonnegative dual convex cone, it can be proved in a similar manner for a nonpositive dual convex cone.

(i) Let $\{x^p\}$ be any sequence of points of K^* and if we let $x^p \to x$ as $p \to \infty$, for any $y \in K$, $(x^p)^t y \geqq 0$. If the continuity of inner products is considered, this indicates $x^t y \geqq 0$. Thus, $x \in K^*$. Accordingly, from [Theorem 3.3], K^* is a closed set.
Next, if any $x^i \in K^*(i = 1, 2)$ and $\alpha_i \geqq 0$ $(i = 1, 2)$ are selected, for all $y \in K$,

$$(x^1)^t y \geqq 0 \quad \text{and} \quad (x^2)^t y \geqq 0. \tag{3.32}$$

Hence,

$$\left(\alpha_1 (x^1)^t + \alpha_2 (x^2)^t\right) y \geqq 0.$$

Thus, $(\alpha_1 x^1 + \alpha_2 x^2) \in K^*$ and K^* is a convex cone.
(ii) Choosing any $y \in T^*$, for all $x \in T$, $y^t x \geqq 0$. Accordingly, from $K \subseteq T$, for any $x \in K \subseteq T$, $y^t x \geqq 0$. Therefore, $y \in K^*$. Hence, we obtain $T^* \subseteq K^*$.
(iii) If x is any vector in K, for all $y \in K^*$, $y^t x \geqq 0$. Thus, $x \in (K^*)^* = K^{**}$ is shown and $K \subset K^{**}$ holds.
(iv)
(Necessity) From (iii), $K^{**} \supseteq K$. Accordingly, only $K^{**} \subseteq K$ must be demonstrated. Here, we confirm that for the pair of $K^{**} \subseteq K$, i.e., if $y \notin K$, $y \notin K^{**}$.
(i) indicates K is a closed convex cone, and thus a closed convex set. Consequently, if $y \notin K$, from [Theorem 8] (Appendix), there is $p \neq 0$ and α that satisfy

$$p^t y < \alpha < p^t x \tag{3.33}$$

for any $x \in K$. K is a convex cone, hence, for any $\beta > 0$, $\beta x \in K$. Accordingly, from the second inequality in $\alpha < p^t(\beta x) = \beta(p^t x)$, $\frac{\alpha}{\beta} < p^t x$ is established. Here, if we let $\beta \to +\infty, \frac{\alpha}{\beta} \to 0$. Therefore, for any $x \in K, p^t x \geqq 0$, and therefore $p \in K^*$. However, because K is a convex cone, $0 \in K$. If we use the second inequality in (3.33) and let $x = \mathbf{0}$,

$$\alpha < p^t x = 0. \tag{3.34}$$

Accordingly, from (3.34) and the first inequality of (3.33),

$$p^t y < \alpha < 0. \tag{3.35}$$

Using $P \in K^*$, it is evident that (3.35) indicates $y \notin K^{**}$.
(Sufficiency) From $K = K^{**} = (K^*)^*$ and (i), K is a closed convex cone.
(v) K is a convex cone, and hence, from (iii), $K^{**} \supseteq K$. Consequently, $K^{**} \subseteq K$ is shown. If the conclusion is contradicted and it is assumed that there is a $b \in K^{**}$, which satisfies $b \notin K$ and $A^t y = b$ does not have a nonnegative solution for $K = \{x \in R^n | A^t y = x, {}^{\forall} y \geqq 0\}$. Therefore, from [Theorem 1.7], x exists that might be $b^t x > 0$ and $Ax \leqq 0$. Hence, for any $y \geqq 0$,

$$(-x)^t A^t y \geqq 0$$

Accordingly, $-x \in K^*$. Furthermore, from $b \in K^{**}$,

$$b^t(-x) = -b^t x \geqq 0$$

However, this opposes $b^t x > 0$, and thus $K^{**} \subseteq K$. QED

[Theorem 3.7] If A is a convex set, $\overline{A}$ is a convex set.

(Proof) From [Theorem 3.5], there are sequences of points $\{x^\gamma\}$, $\{y^\gamma\}(\gamma = 1, 2, \cdots)$, which satisfy for any $x,\ y \in \overline{A}$,

$$\lim_{\gamma \to \infty} x^\gamma = x \text{ and } \lim_{\gamma \to \infty} y^\gamma = y. \tag{3.36}$$

Therefore, for any $\alpha \in [0, 1]$

$$\lim_{\gamma \to \infty} [\alpha x^\gamma + (1 - \alpha) y^\gamma] = \alpha x + (1 - \alpha) y$$

and

$$\alpha x^\gamma + (1 - \alpha) y^\gamma \in A$$

hold. Thus, $\alpha x + (1 - \alpha) y$ becomes the limit of the sequence of points $\{\alpha x^\gamma + (1 - \alpha) y^\gamma\}$ of A. Hence, $\alpha x + (1 - \alpha) y \in \overline{A}$. QED

Next, notation and terminology used in this chapter will be explained. Unless otherwise specified, (3.A1) is assumed in this section.

(Assumption)

(3.A1) $f(x)$ and $g_1(x), \cdots, g_m(x)$ are continuously differentiable real-valued function defined on open convex sets D of R^n.

If a domain $\boldsymbol{D}$, contains even one boundary point, there is no guarantee that minute fluctuations from that point to all directions would be contained in the region. To exclude this sort of situation, $\boldsymbol{D}$ is assumed to be an open set.

⟨Definition 3.20⟩ When $\bar{\boldsymbol{x}} \in \boldsymbol{C}$ and $\boldsymbol{\xi} \in \boldsymbol{R}^n$ are provided, there are a n dimensional vector function $\varphi(\theta) = [\varphi_1(\theta), \ \cdots, \ \varphi_n(\theta)]^t$ with the closed interval $\left[0, \ \overline{\theta}\right]$ as the domain with some positive number $\overline{\theta}$ as the upper edge and when the following conditions are satisfied:

(a) for any $\theta \in \left[0, \ \overline{\theta}\right]$, $\varphi(\theta) = \bar{\boldsymbol{x}} \in \boldsymbol{C}$ and $\varphi(\theta) \in \boldsymbol{C}$,

(b) for any $j \in \{1, \ \cdots, \ n\}$, $\boldsymbol{\xi}_j = \lim\limits_{h \to +0} \dfrac{\varphi_j(h+0) - \varphi_j(0)}{h} = \varphi_j'(0)$[4]

This $\varphi(\theta)$ is known as a contained path with an **attainable direction vector**.

⟨Definition 3.21⟩ For a given $\bar{\boldsymbol{x}} \in \boldsymbol{C}$, $\boldsymbol{\xi} \in \boldsymbol{R}^n$ which satisfies $\nabla g_E(\bar{\boldsymbol{x}})\boldsymbol{\xi} \geqq \boldsymbol{0}$ is called a locally constrained direction vector for $\bar{\boldsymbol{x}} \in \boldsymbol{C}$[5] and the set of these vectors is as follows:

$$L(\bar{\boldsymbol{x}}) = \{\boldsymbol{\xi} \in \boldsymbol{R}^n | \nabla g_E(\bar{\boldsymbol{x}})\boldsymbol{\xi} \geqq \boldsymbol{0}\},$$

where, $\nabla g_E(\bar{\boldsymbol{x}})\boldsymbol{\xi} \geqq \boldsymbol{0}$ represent for any $i \in \boldsymbol{E}(\bar{\boldsymbol{x}})$, $\nabla g_i(\bar{\boldsymbol{x}})\boldsymbol{\xi} \geqq 0$.

⟨Definition 3.22⟩ $\boldsymbol{W}(\bar{\boldsymbol{x}})$ is a closure of a convex cone spanned by $\boldsymbol{A}(\bar{\boldsymbol{x}})$, i.e.,

$$\boldsymbol{W}(\bar{\boldsymbol{x}}) = \boldsymbol{K}(\boldsymbol{A}(\bar{\boldsymbol{x}})) \cup \boldsymbol{L}_{\boldsymbol{K}(\boldsymbol{A}(\bar{\boldsymbol{x}}))}.$$

However, $\boldsymbol{K}(\boldsymbol{A}(\bar{\boldsymbol{x}}))$ and $\boldsymbol{L}_{\boldsymbol{K}(\boldsymbol{A}(\bar{\boldsymbol{x}}))}$ are the convex cone spanned by $\boldsymbol{A}(\bar{\boldsymbol{x}})$ and the limit point of $\boldsymbol{K}(A(\bar{\boldsymbol{x}}))$. Furthermore, each element of $\boldsymbol{W}(\bar{\boldsymbol{x}})$ is known as **weakly attainable direction vectors**.

⟨Definition 3.23⟩ For a given $\bar{\boldsymbol{x}} \in \boldsymbol{C}$, the closure of a set $\{\lambda(\boldsymbol{x} - \bar{\boldsymbol{x}}) | \lambda \geqq 0, \ \boldsymbol{x} \in \boldsymbol{C}\}$ is expressed as $\boldsymbol{V}(\bar{\boldsymbol{x}})$. To summarize,

$$\{\lambda(\boldsymbol{x} - \bar{\boldsymbol{x}}) | \lambda \geqq 0, \ \boldsymbol{x} \in \boldsymbol{C}\} = \cup_{\boldsymbol{x} \in \boldsymbol{C}} \{\lambda(\boldsymbol{x} - \bar{\boldsymbol{x}}) | \lambda \geqq 0, \ \boldsymbol{x} \in \boldsymbol{C}\}.$$

⟨Definition 3.24⟩ When, for $\bar{\boldsymbol{x}} \in \boldsymbol{C}$, $L(\bar{\boldsymbol{x}}) \subseteq W(\bar{\boldsymbol{x}})$ holds, g is considered to be a **weak constraint qualification** at $\bar{\boldsymbol{x}} \in \boldsymbol{C}$.

Here, it will be shown that when $\bar{\boldsymbol{x}} \in \boldsymbol{C}$ is the optimal solution of [NLPX], if weak constraint qualification is assumed, for $\bar{\boldsymbol{x}}$, a nonnegative Lagrange multiplier vector is defined. Hence, as preparation for it, let us describe a lemma.

---

[4] When expressing the right-hand derivative of $\varphi(\theta)$ at $\theta = 0$, one ought to usually use a symbol that is different from the ordinary derivative symbol, φ', but to avoid complication, φ' will be used. $\boldsymbol{\xi}$ and the set of attainable direction vectors in $\bar{\boldsymbol{x}}$ is represented by $\boldsymbol{A}(\bar{\boldsymbol{x}})$.

[5] $\boldsymbol{L}(\bar{\boldsymbol{x}})$ can be referred to as a **closed linearizing cone** of $\boldsymbol{C}$ on $\bar{\boldsymbol{x}}$.

[Lemma 3.3] $L(\overline{x})$ and $W(\overline{x})$ have the following characteristics.

(i) $\boldsymbol{L}(\overline{\boldsymbol{x}})$ is a closed convex cone.
and
(ii) $\boldsymbol{W}(\overline{\boldsymbol{x}})$ is a minimal closed convex cone containing $\boldsymbol{A}(\overline{\boldsymbol{x}})$.

(Proof)

(i) For any $\alpha_i \geqq 0$ and $\boldsymbol{y}^i \in \boldsymbol{L}(\overline{\boldsymbol{x}})(i=1,2)$,

$$\nabla g_E(\overline{\boldsymbol{x}})\left(\alpha_1\boldsymbol{y}^1+\alpha_2\boldsymbol{y}^2\right)=\alpha_1\nabla g_E(\overline{\boldsymbol{x}})\boldsymbol{y}^1+\ \alpha_2\nabla g_E(\overline{\boldsymbol{x}})\boldsymbol{y}^2$$
$$\geqq \mathbf{0}.$$

Thus, $\alpha_1\boldsymbol{y}^1+\alpha_2\boldsymbol{y}^2 \in \boldsymbol{L}(\overline{\boldsymbol{x}})$. Therefore, $\boldsymbol{L}(\overline{\boldsymbol{x}})$ is a convex cone.

Next, show the fact that $(\boldsymbol{L}(\overline{\boldsymbol{x}}))^C$ is an open set. For any $\boldsymbol{y} \in (\boldsymbol{L}(\overline{\boldsymbol{x}}))^C=\{\boldsymbol{\eta} \in \boldsymbol{R}^n|\boldsymbol{\eta}\notin \boldsymbol{L}(\overline{\boldsymbol{x}})\}$, there exists certain $j \in \{1,\cdots,m\}$ and $\nabla g_j(\overline{\boldsymbol{x}})\boldsymbol{\eta}<0$. Accordingly, there is an appropriate $\delta>0$ and for all $z \in \boldsymbol{B}(\boldsymbol{\eta};\delta)$, $\nabla g_j(\overline{\boldsymbol{x}})z<0$. Therefore $(\boldsymbol{L}(\overline{\boldsymbol{x}}))^C$ is an open set. Consequently, from [Theorem 3.3], $\boldsymbol{L}(\overline{\boldsymbol{x}})$ becomes a closed set.

(ii) From ⟨Definition 3.22⟩, $\boldsymbol{W}(\overline{\boldsymbol{x}})$ is a closure of convex cone $\boldsymbol{K}(\boldsymbol{A}(\overline{\boldsymbol{x}}))$ spanned by $\boldsymbol{A}(\overline{\boldsymbol{x}})$. If the fact that $\boldsymbol{K}(\boldsymbol{A}(\overline{\boldsymbol{x}}))$ is the minimum convex cone containing $\boldsymbol{A}(\overline{\boldsymbol{x}})$, $\boldsymbol{W}(\overline{\boldsymbol{x}})$ being the minimum closed set containing $\boldsymbol{K}(\boldsymbol{A}(\overline{\boldsymbol{x}}))$ ([Theorem 3.4 ; (ii-ii)]) and [Theorem 3.7] are combined, the following conclusion is obtained.

QED

[Lemma 3.4]

$$\boldsymbol{W}(\overline{\boldsymbol{x}}) \subseteq \boldsymbol{L}(\overline{\boldsymbol{x}}).$$

(Proof) Let $\boldsymbol{\xi} \in \boldsymbol{R}^n$ be any vector belonging to $\boldsymbol{A}(\overline{\boldsymbol{x}})$ and $\varphi(\theta)$ be a contained path with an origin $\overline{\boldsymbol{x}}$ and a direction vector $\boldsymbol{\xi}$. Here, from ⟨Definition 3.20⟩, there exists certain $\overline{\theta}>0$ and for any $j \in \boldsymbol{E}(\overline{\boldsymbol{x}})$,

$$g_j[\varphi(0)]=g_j(\overline{\boldsymbol{x}})=0, \tag{3.37}$$

and for any $j \in \boldsymbol{E}(\overline{\boldsymbol{x}})$ and $\theta \in \left[0,\overline{\theta}\right]$,

$$g_j\{\varphi(\theta)\}\geqq 0 \tag{3.38}$$

holds. Thus, at $\theta=0$, for any $j \in \boldsymbol{E}(\overline{\boldsymbol{x}})$,

$$\frac{dg_j[\varphi(\theta)]}{d\theta}=\nabla g_j(\overline{\boldsymbol{x}})\varphi'(0)$$
$$=\nabla g_j(\overline{\boldsymbol{x}})\boldsymbol{\xi}$$
$$\geqq 0.$$

Therefore, for any $\boldsymbol{\xi} \in \boldsymbol{A}(\overline{\boldsymbol{x}})$, $\boldsymbol{\xi} \in \boldsymbol{L}(\overline{\boldsymbol{x}})$, i.e., $\boldsymbol{A}(\overline{\boldsymbol{x}}) \subseteq \boldsymbol{L}(\overline{\boldsymbol{x}})$. From [Lemma 3.3], $\boldsymbol{L}(\overline{\boldsymbol{x}})$ is a convex cone and $\boldsymbol{K}(\boldsymbol{A}(\overline{\boldsymbol{x}}))$ is the minimum convex cone containing $\boldsymbol{A}(\overline{\boldsymbol{x}})$. From this and $\boldsymbol{A}(\overline{\boldsymbol{x}}) \subseteq \boldsymbol{L}(\overline{\boldsymbol{x}})$, $\boldsymbol{K}(\boldsymbol{A}(\overline{\boldsymbol{x}})) \subseteq \boldsymbol{L}(\overline{\boldsymbol{x}})$. Moreover, $\boldsymbol{L}(\overline{\boldsymbol{x}})$ is a closed set and so $\boldsymbol{L}(\overline{\boldsymbol{x}}) \supseteq \boldsymbol{L}_{\boldsymbol{L}(\overline{\boldsymbol{x}})}$. Hence, if you note that $\boldsymbol{L}(\overline{\boldsymbol{x}}) \supseteq \boldsymbol{K}(\boldsymbol{A}(\overline{\boldsymbol{x}}))$, $\boldsymbol{L}(\overline{\boldsymbol{x}}) \supseteq \boldsymbol{L}_{\boldsymbol{K}(\boldsymbol{A}(\overline{\boldsymbol{x}}))}$. Therefore, $\boldsymbol{W}(\overline{\boldsymbol{x}}) = \left(\boldsymbol{K}(\boldsymbol{A}(\overline{\boldsymbol{x}})) \cup \boldsymbol{L}_{\boldsymbol{K}(\boldsymbol{A}(\overline{\boldsymbol{x}}))}\right) \subseteq \boldsymbol{L}(\overline{\boldsymbol{x}})$. QED

[Lemma 3.5] If a constraint set $\boldsymbol{C}$ is a convex set,

(i) $\boldsymbol{V}(\overline{\boldsymbol{x}})$ is a closed convex cone

and

(ii) $\boldsymbol{V}(\overline{\boldsymbol{x}}) \subseteq \boldsymbol{W}(\overline{\boldsymbol{x}})$.

(Proof)

(i) $\boldsymbol{V}(\overline{\boldsymbol{x}})$ is closed so from [Theorem 3.4; (ii)], it is a closed set; moreover, from the definition, $\boldsymbol{V}(\overline{\boldsymbol{x}})$ is a cone. Therefore, it must be demonstrated that $\boldsymbol{V}(\overline{\boldsymbol{x}})$ is a convex set.

For any $\boldsymbol{x}^i \in \boldsymbol{C}(i = 1, 2)$ and $\lambda \geqq 0$, $\lambda(\boldsymbol{x}^i - \overline{\boldsymbol{x}})$. From the assumption, $\boldsymbol{C}$ is a convex set, thus, for any $\mu \in [0, 1]$, $(1 - \mu)\boldsymbol{x}^1 + \mu\boldsymbol{x}^2 \in \boldsymbol{C}$. Accordingly, for any $\mu \in [0, 1]$ and $\lambda \geqq 0$,

$$\begin{aligned}(1-\mu)\lambda(\boldsymbol{x}^1 - \overline{\boldsymbol{x}}) &+ \mu\lambda(\boldsymbol{x}^2 - \overline{\boldsymbol{x}}) \\ &= \lambda\left[(1-\mu)(\boldsymbol{x}^1 - \overline{\boldsymbol{x}}) + \mu(\boldsymbol{x}^2 - \overline{\boldsymbol{x}})\right] \\ &= \lambda\left[(1-\mu)\boldsymbol{x}^1 + \mu\boldsymbol{x}^2 - \overline{\boldsymbol{x}}\right] \in \boldsymbol{V}(\overline{\boldsymbol{x}}).\end{aligned}$$

(ii) For any $\boldsymbol{y} \in \boldsymbol{V}(\overline{\boldsymbol{x}})$, there is $\boldsymbol{x}^y \in \boldsymbol{C}$ and $\boldsymbol{y} \in \{\lambda(\boldsymbol{x}^y - \overline{\boldsymbol{x}}) | \lambda \geqq \boldsymbol{0}\}$. Thus, there is $\lambda_y \geqq 0$, i.e., $\boldsymbol{y} = \lambda_y(\boldsymbol{x}^y - \overline{\boldsymbol{x}})$. Here, if for a $\theta \in [0, 1]$, $\varphi(\theta) = \overline{\boldsymbol{x}} + \theta(\boldsymbol{x}^y - \overline{\boldsymbol{x}})$ is defined, $\varphi(0) = \overline{\boldsymbol{x}}$. Moreover, from the convexity of $\boldsymbol{C}$, for any $\theta \in [0, 1]$, $\varphi(\theta) = (1 - \theta)\overline{\boldsymbol{x}} + \theta\boldsymbol{x}^y \in \boldsymbol{C}$. Furthermore, $\varphi'(0) = \boldsymbol{x}^y - \overline{\boldsymbol{x}}$ is evident and hence $\boldsymbol{x}^y - \overline{\boldsymbol{x}} \in \boldsymbol{A}(\overline{\boldsymbol{x}})$. From this fact and from the fact that $\boldsymbol{A}(\overline{\boldsymbol{x}}) \subseteq \boldsymbol{W}(\overline{\boldsymbol{x}})$ and $\boldsymbol{A}(\overline{\boldsymbol{x}})$ as well as $\boldsymbol{W}(\overline{\boldsymbol{x}})$ being a cone, for any $\lambda \geqq 0$, $\lambda(\boldsymbol{x}^y - \overline{\boldsymbol{x}}) \in \boldsymbol{A}(\overline{\boldsymbol{x}}) \subseteq \boldsymbol{W}(\overline{\boldsymbol{x}})$. Therefore, $\boldsymbol{y} = \lambda_y(\boldsymbol{x}^y - \overline{\boldsymbol{x}}) \in \boldsymbol{W}(\overline{\boldsymbol{x}})$. QED

[Lemma 3.6] If $f(\boldsymbol{x})$ has a maximum value at $\boldsymbol{x} = \overline{\boldsymbol{x}} \in \boldsymbol{C}$, $\nabla f(\overline{\boldsymbol{x}}) \in (\boldsymbol{W}(\overline{\boldsymbol{x}}))_*$.

(Proof) Let $\boldsymbol{\xi} \in \boldsymbol{R}^n$ be a vector belonging to $\boldsymbol{A}(\overline{\boldsymbol{x}})$ and $\varphi(\theta)$ a contained path with origin $\overline{\boldsymbol{x}}$ and direction vector $\boldsymbol{\xi}$. Here, there are some $\overline{\theta} > 0$ and for any $\theta \in [0, \overline{\theta}]$,

$$f[\varphi(\theta)] \leqq f[\varphi(\theta)] = f(\overline{\boldsymbol{x}})$$

holds. Hence,

$$\begin{aligned}\nabla f(\overline{\boldsymbol{x}})\boldsymbol{\xi} &= \nabla f(\overline{\boldsymbol{x}})\varphi'(0) \\ &\leqq 0.\end{aligned} \tag{3.39}$$

Moreover, for any $\boldsymbol{\xi} \in \boldsymbol{A}(\overline{\boldsymbol{x}})$, $\boldsymbol{\xi} \in \boldsymbol{W}(\overline{\boldsymbol{x}})$. Here, to demonstrate that (3.39) holds for any $\boldsymbol{\xi} \in \boldsymbol{W}(\overline{\boldsymbol{x}})$, let us assume that $\boldsymbol{\eta} \in \boldsymbol{W}(\overline{\boldsymbol{x}})$ exists such that

$$\nabla f(\overline{\boldsymbol{x}})\boldsymbol{\eta} > 0. \tag{3.40}$$

From ⟨Definition 3.22⟩, $\boldsymbol{W}(\overline{\boldsymbol{x}}) = \boldsymbol{K}(\boldsymbol{A}(\overline{\boldsymbol{x}})) \cup \boldsymbol{L}_{\boldsymbol{K}(\boldsymbol{A}(\overline{\boldsymbol{x}}))}$. If $\boldsymbol{\eta} \in \boldsymbol{K}(\boldsymbol{A}(\overline{\boldsymbol{x}}))$, $\boldsymbol{\eta}$ can be expressed as a nonnegative linear combination of a finite number (=k) vectors of $\boldsymbol{A}(\overline{\boldsymbol{x}})$, i.e., it can be expressed as follows:

$$\boldsymbol{\eta} = \sum_{i=1}^{k} \lambda_i \boldsymbol{q}^i,\ 1 \leqq k,\ \ \boldsymbol{q}^i \in \boldsymbol{A}(\overline{\boldsymbol{x}}),\ \ \ \lambda_i \geqq 0 \tag{3.41}$$

hence

$$\begin{aligned}\nabla f(\overline{\boldsymbol{x}})\boldsymbol{\eta} &= \nabla f(\overline{\boldsymbol{x}}) \sum_{i=1}^{k} \lambda_i \boldsymbol{q}^i \\ &\leqq 0.\end{aligned} \tag{3.42}$$

However, (3.42) contradicts (3.40).

Hence, the situation $\boldsymbol{\eta} \in \boldsymbol{L}_{\boldsymbol{K}(\boldsymbol{A}(\overline{\boldsymbol{x}}))}$ is considered. From the definition of a limit point, for any $\varepsilon > 0$, there exists $\boldsymbol{v} \in \boldsymbol{K}(\boldsymbol{A}(\overline{\boldsymbol{x}}))$, which is $\boldsymbol{v} \in (\boldsymbol{B}(\boldsymbol{\eta}; \varepsilon) - \{\boldsymbol{\eta}\})$. Accordingly, even for sufficiently small $\tilde{\varepsilon} > 0$, there exists $\widehat{\boldsymbol{v}}$ that satisfies $\widehat{\boldsymbol{v}} \in \boldsymbol{B}(\boldsymbol{\eta}; \tilde{\varepsilon}) - \{\boldsymbol{\eta}\}\} \cap \boldsymbol{K}(\boldsymbol{A}(\overline{\boldsymbol{x}}))$. $\widehat{\boldsymbol{v}} \in (\boldsymbol{B}(\boldsymbol{\eta}; \tilde{\varepsilon}) - \{\boldsymbol{\eta}\})$, from the continuity of $\nabla f(\overline{\boldsymbol{x}})\boldsymbol{v}$ about $\boldsymbol{v}$, indicates

$$\nabla f(\overline{\boldsymbol{x}})\widehat{\boldsymbol{v}} > 0 \tag{3.43}$$

However, by $\widehat{\boldsymbol{v}} \in \boldsymbol{K}(\boldsymbol{A}(\overline{\boldsymbol{x}}))$, $\widehat{\boldsymbol{v}}$ can be expressed as a nonnegative linear combination of an appropriate finite number of vectors of $\boldsymbol{A}(\overline{\boldsymbol{x}})$, and hence

$$\nabla f(\overline{\boldsymbol{x}})\widehat{\boldsymbol{v}} \leqq 0$$

but this contradicts (3.43).

Therefore, it is shown that for any $\boldsymbol{\xi} \in \boldsymbol{W}(\overline{\boldsymbol{x}})$, $\nabla f(\overline{\boldsymbol{x}})\boldsymbol{\xi} \leqq 0$. QED

The important theorem in this is introduced next.

[Theorem 3.8] Let $\overline{\boldsymbol{x}} \in \boldsymbol{C}$ be the optimum solution of [NLPX]. Here, if g satisfies weak constraint qualification at $\overline{\boldsymbol{x}}$, g is Lagrange regular at $\overline{\boldsymbol{x}}$.

(Proof) If it is assumed that $f(\boldsymbol{x})$ achieves its maximum at $\boldsymbol{x} = \overline{\boldsymbol{x}} \in \boldsymbol{C}$,

$$\nabla f(\overline{\boldsymbol{x}}) \in (\boldsymbol{W}(\overline{\boldsymbol{x}}))_*. \tag{3.44}$$

Furthermore, from the condition, g satisfies weak constraint qualification on $\overline{\boldsymbol{x}}$ and hence $\boldsymbol{L}(\overline{\boldsymbol{x}}) \subseteq \boldsymbol{W}(\overline{\boldsymbol{x}})$. Accordingly,

$$(\boldsymbol{L}(\overline{\boldsymbol{x}}))_{*} \supseteq (\boldsymbol{W}(\overline{\boldsymbol{x}}))_{*}. \tag{3.45}$$

Thus,

$$\nabla f(\overline{\boldsymbol{x}}) \in (\boldsymbol{L}(\overline{\boldsymbol{x}}))_{*} \tag{3.46}$$

is obtained. Here, define $\boldsymbol{B}(\overline{\boldsymbol{x}}) = \{\boldsymbol{y}_E(-\nabla g_E(\overline{\boldsymbol{x}}))|\boldsymbol{y}_E \geqq \boldsymbol{0}\}$ and show $\boldsymbol{L}(\overline{\boldsymbol{x}}) = (\boldsymbol{B}(\overline{\boldsymbol{x}}))_{*}$. For any $\boldsymbol{\xi} \in \boldsymbol{L}(\overline{\boldsymbol{x}})$ and $\boldsymbol{y}_E(-\nabla g_E(\overline{\boldsymbol{x}})) \in (\boldsymbol{B}(\overline{\boldsymbol{x}}))$,

$$\begin{aligned} \boldsymbol{y}_E(-\nabla g_E(\overline{\boldsymbol{x}}))\boldsymbol{\xi} &= -\boldsymbol{y}_E \nabla g_E(\overline{\boldsymbol{x}})\boldsymbol{\xi} \\ &\leqq 0 \end{aligned}$$

holds hence $\boldsymbol{\xi} \in (\boldsymbol{B}(\overline{\boldsymbol{x}}))_{*}$. Therefore $\boldsymbol{L}(\overline{\boldsymbol{x}}) \subseteq (\boldsymbol{B}(\overline{\boldsymbol{x}}))_{*}$; however, from the definition of $\boldsymbol{B}(\overline{\boldsymbol{x}})$, if $\boldsymbol{y}_E \geqq \boldsymbol{0}$ is considered to be a unit vector, each row of $[-\nabla g_E(\overline{\boldsymbol{x}})] \in \boldsymbol{B}(\overline{\boldsymbol{x}})$. Consequently, for all $\boldsymbol{u} \in (\boldsymbol{B}(\overline{\boldsymbol{x}}))_{*}$, each row of $[-\nabla g_E(\overline{\boldsymbol{x}})]\boldsymbol{u} \leqq 0$. Therefore, $\nabla g_E(\overline{\boldsymbol{x}})\boldsymbol{u} \geqq \boldsymbol{0}$. Hence, in respect of any $\boldsymbol{u} \in (\boldsymbol{B}(\overline{\boldsymbol{x}}))_{*}$, $\boldsymbol{u} \in \boldsymbol{L}(\overline{\boldsymbol{x}})$, i.e., $(\boldsymbol{B}(\overline{\boldsymbol{x}}))_{*} \subseteq \boldsymbol{L}(\overline{\boldsymbol{x}})$. From the above, it is certainly the case that $\boldsymbol{L}(\overline{\boldsymbol{x}}) = (\boldsymbol{B}(\overline{\boldsymbol{x}}))_{*}$. Accordingly, $(\boldsymbol{L}(\overline{\boldsymbol{x}}))_{*} = \left((\boldsymbol{B}(\overline{\boldsymbol{x}}))_{*}\right)_{*} = (\boldsymbol{B}(\overline{\boldsymbol{x}}))_{**}$. Combining this with (3.46) yields the following:

$$\nabla f(\overline{\boldsymbol{x}}) \in (\boldsymbol{L}(\overline{\boldsymbol{x}}))_{*} = (\boldsymbol{B}(\overline{\boldsymbol{x}}))_{**}. \tag{3.47}$$

Furthermore, $\boldsymbol{B}(\overline{\boldsymbol{x}})$ is a closed convex cone, and hence from [Theorem 3.6; (iv)],

$$\boldsymbol{B}(\overline{\boldsymbol{x}}) = (\boldsymbol{B}(\overline{\boldsymbol{x}}))_{**}. \tag{3.48}$$

(3.47) and (3.48) indicate $\nabla f(\overline{\boldsymbol{x}}) \in \boldsymbol{B}(\overline{\boldsymbol{x}})$; thus, from the definition of $\boldsymbol{B}(\overline{\boldsymbol{x}})$, there exists

$$-\nabla f(\overline{\boldsymbol{x}}) = \overline{\boldsymbol{y}}_E(\nabla g_E(\overline{\boldsymbol{x}})) \tag{3.49}$$

that satisfies $\overline{\boldsymbol{y}}_E \geqq \boldsymbol{0}$. Hence, if we define

$$\begin{aligned} \overline{\boldsymbol{y}} = (\overline{\boldsymbol{y}}_E, \overline{\boldsymbol{y}}_F) &\text{ and } \overline{\boldsymbol{y}}_F = \boldsymbol{0}, \\ \boldsymbol{y} &\geqq \boldsymbol{0}, \\ \nabla f(\overline{\boldsymbol{x}}) + \overline{\boldsymbol{y}}\nabla g(\overline{\boldsymbol{x}}) &= \nabla f(\overline{\boldsymbol{x}}) + \overline{\boldsymbol{y}}_E \nabla g_E(\overline{\boldsymbol{x}}) \\ &= \boldsymbol{0} \end{aligned} \tag{3.50}$$

and

$$\begin{aligned} \overline{\boldsymbol{y}} g(\overline{\boldsymbol{x}}) &= \overline{\boldsymbol{y}}_E g_E(\overline{\boldsymbol{x}}) \\ &= 0. \end{aligned}$$

Therefore, g is a Lagrange regular on $\overline{\boldsymbol{x}} \in \boldsymbol{C}$. QED

Multiple constraint qualifications that bring about similar conclusions as with [Theorem 3.8], in addition to weak constraint qualifications, have been proposed. Here, before we move on, let us consider the mutual relationships between all constraint qualifications, including weak constraint qualifications.

The definition of constraint qualification is presented below. However, for only ⟨Definition 3.25⟩–⟨Definition 3.28⟩, the differentiability of g is unnecessary.

⟨**Definition 3.25**⟩ Let g be a concave function on the convex set $\boldsymbol{D}$. If there exists $\overline{\boldsymbol{x}} \in \boldsymbol{D}$, which is $g(\overline{\boldsymbol{x}}) > \boldsymbol{0}$, g is then said to satisfy **Slater's constraint qualification.**[6]

⟨**Definition 3.26**⟩ Let g be a concave function on the convex set $\boldsymbol{D}$. For any $\boldsymbol{p} \geq \boldsymbol{0}$, should some $\overline{\boldsymbol{x}} \in \boldsymbol{D}$ exist and $\boldsymbol{p}^t g(\overline{\boldsymbol{x}}) > 0$, g is said to satisfy **Karlin's constraint qualification.**[7]

⟨**Definition 3.27**⟩ Let g be a concave function on the convex set $\boldsymbol{D}$. If two points $\boldsymbol{x}^1$, $\boldsymbol{x}^2$, which are at least different with g as a narrow concave function on $\boldsymbol{x}^1$ are contained in $\boldsymbol{C}$, g is said to satisfy **a strict constraint qualification**.

⟨**Definition 3.28**⟩ For any $i \in \boldsymbol{E}(\overline{\boldsymbol{x}})$, if any of the following holds:

(i) g_i is a convex function on $\overline{\boldsymbol{x}}$

or

(ii) g_i is a linear function on $\boldsymbol{R}^n$[8]

Then, g is said to satisfy **reverse convex constraint qualification** on $\overline{\boldsymbol{x}} \in \boldsymbol{C}$.

⟨**Definition 3.29**⟩ When $\nabla g_i(\overline{\boldsymbol{x}})(i \in \boldsymbol{E}(\overline{\boldsymbol{x}}))$ is linearly independent, i.e., when

$$rank(\nabla g_{\boldsymbol{E}}(\overline{\boldsymbol{x}})) = \#\boldsymbol{E}(\overline{\boldsymbol{x}})$$

holds, g is said to satisfy **regularity condition** on $\overline{\boldsymbol{x}} \in \boldsymbol{C}$.

(**Notes on** ⟨**Definition 3.29**⟩): Arrow et al. (1961, p.179) consider ⟨Definition 3.29⟩ to be the **nondegeneracy condition** but recently it is more mainstream to refer to it as the regularity condition. Thus, we shall refer to it as regularity condition in this book.

⟨**Definition 3.30**⟩ The following three conditions:

(a) Constraint set $\boldsymbol{C}$ is a convex set;
(b) $\boldsymbol{C}$ has an inner point
(c) $\nabla g_{\boldsymbol{E}}(\overline{\boldsymbol{x}}) \neq \boldsymbol{0}$
are satisfied.

---

[6] See Slater (1950).

[7] See Karlin (1959).

[8] A linear function usually does not contain a constant term but it is fine to consider it containing one. See notes on ⟨Definition 2.9⟩.

⟨**Definition 3.31**⟩ When $\boldsymbol{L}(\overline{\boldsymbol{x}}) \subseteq \boldsymbol{A}(\overline{\boldsymbol{x}})$ holds for $\overline{\boldsymbol{x}} \in \boldsymbol{C}$, g is said to satisfy **Kuhn–Tucker's constraint qualification**[9] at $\overline{\boldsymbol{x}} \in \boldsymbol{C}$.

⟨**Definition 3.32**⟩ When there is $\boldsymbol{\xi}^* \in \boldsymbol{R}^n$, which is $\nabla g_{E'}(\overline{\boldsymbol{x}})\boldsymbol{\xi}^* \geqq \boldsymbol{0}$ and $\nabla g_{E''}(\overline{\boldsymbol{x}})\boldsymbol{\xi}^* > \boldsymbol{0}$, g is said to satisfy the **Arrow–Hurwicz–Uzawa constraint qualification**[10] on $\overline{\boldsymbol{x}} \in \boldsymbol{C}$.

However, $\boldsymbol{E}'(\overline{\boldsymbol{x}}) = \{i \in \boldsymbol{E}(\overline{\boldsymbol{x}}) | g_i \text{ is a convex function on } \overline{\boldsymbol{x}}\}$.

$$\boldsymbol{E}''(\overline{\boldsymbol{x}}) = \{i \in \boldsymbol{E}(\overline{\boldsymbol{x}}) | g_i \text{ is not a convex function on } \overline{\boldsymbol{x}}\}.$$

Furthermore, ⟨Definition 3.33⟩–⟨Definition 3.35⟩ are the constraint qualifications that have smoothed the concaveness or convexity of g in ⟨Definition 3.25⟩, ⟨Definition 3.28⟩ and ⟨Definition 3.32⟩ to be quasi-concaveness or quasi-convexity of g.

⟨**Definition 3.33**⟩ Let $g_E(\overline{\boldsymbol{x}})$ be a pseudo convex function. Here, if there exists $\widehat{\boldsymbol{x}} \in \boldsymbol{C}$ that is $g_E(\widehat{\boldsymbol{x}}) > \boldsymbol{0}$, g is said to satisfy **Slater's weak constraint qualification** on $\overline{\boldsymbol{x}} \in \boldsymbol{C}$.

⟨**Definition 3.34**⟩ When for any $i \in \boldsymbol{E}(\overline{\boldsymbol{x}})$,

(i) g_i is a pseudo convex function on $\overline{\boldsymbol{x}}$

or

(ii) g_i is a linear function on $\boldsymbol{R}^n$,

g is said to satisfy **weakly inverse convex constraint** on $\overline{\boldsymbol{x}} \in \boldsymbol{C}$.

⟨**Definition 3.35**⟩ When there exists $z \in \boldsymbol{R}^n$ that is $\nabla g_{Q'}(\overline{\boldsymbol{x}})z \geqq \boldsymbol{0}$ and $\nabla g_{Q''}(\overline{\boldsymbol{x}})z > \boldsymbol{0}$, g is said to satisfy **Arrow–Hurwicz–Uzawa's weak constraint qualification** on $\overline{\boldsymbol{x}} \in \boldsymbol{C}$.

However, $\boldsymbol{Q}'(\overline{\boldsymbol{x}}) = \{i \in \boldsymbol{E}(\overline{\boldsymbol{x}}) | g_i \text{ is a quasiconvex function on } \overline{\boldsymbol{x}}\}$,

$$\boldsymbol{Q}''(\overline{\boldsymbol{x}}) = \{i \in \boldsymbol{E}(\overline{\boldsymbol{x}}) | g_i \text{ is not a quasiconvex function on } \overline{\boldsymbol{x}}\}.$$

Firstly, explain the relationship of Kuhn–Tucker constraint qualification and weak constraint qualification.

[Theorem 3.9] If g satisfies the Kuhn–Tucker constraint qualification on $\overline{\boldsymbol{x}} \in \boldsymbol{C}$, g satisfies weak constraint qualification on $\overline{\boldsymbol{x}} \in \boldsymbol{C}$.

(Proof) It is immediately derived from ⟨Definition 3.31⟩ and ⟨Definition 3.22⟩.

QED

Next, a small lemma is demonstrated to show the mutual relationship between all the constraint qualifications other than Kuhn–Tucker's constraint qualifications and Arrow–Hurwicz–Uzawa's weak constraint qualifications.

---

[9] See Kuhn and Tucker (1951).

[10] See Arrow et al. (1961).

[Lemma 3.7] The following relationships hold between ⟨Definition 3.25⟩–⟨Definition 3.27⟩.

(i) The constraint qualifications of Slater and Karlin are mutually equivalent.
(ii) Strong constraint qualification indicates Slater's constraint qualification and Karlin's constraint qualification.

(Proof)
(i) If Slater's constraint qualification is satisfied, there exists a $\bar{\boldsymbol{x}} \in \boldsymbol{D}$, which becomes $g(\bar{\boldsymbol{x}}) > 0$. Hence, for any $\boldsymbol{p} \geq \boldsymbol{0}$, $\boldsymbol{p}^t g(\bar{\boldsymbol{x}}) > 0$ holds and Karlin's constraint qualification is satisfied.
Next, the opposite predicate to this shall be shown by forming its contrapositive. If it is assumed that Slater's constraint qualification, for any $\boldsymbol{x} \in \boldsymbol{D}$, there exists certain number $i \in \{1, \cdots, m\}$ and $g_i(\bar{\boldsymbol{x}}) \leqq 0$. In this case, if Karlin's constraint qualification is satisfied, for any $\boldsymbol{p} \geq \boldsymbol{0}$, there exists some $\bar{\boldsymbol{x}} \in \boldsymbol{D}$ and $\boldsymbol{p}^t g(\bar{\boldsymbol{x}}) > 0$. Accordingly, if the ith unit column vector $\boldsymbol{e}^i$ is selected as $\boldsymbol{p}$, $\boldsymbol{e}^i \geq 0$ and $(\boldsymbol{e}^i)^t g(\bar{\boldsymbol{x}}) = g_i(\bar{\boldsymbol{x}}) > 0$, which contradicts the assumed $g_i(\bar{\boldsymbol{x}}) \leqq 0$. Hence, Karlin's constraint qualification is not satisfied.
(ii) If Slater's constraint qualification is derived from strong constraint qualification, (i) concludes the proof. From these assumptions, $\boldsymbol{D}$ is a convex set and hence for any $\boldsymbol{x}^1, \boldsymbol{x}^2 \in \boldsymbol{C} \subseteq \boldsymbol{D}$ and $\lambda \in [0, 1]$, $(1 - \lambda)\boldsymbol{x}^1 + \lambda \boldsymbol{x}^2 \in \boldsymbol{D}$. Moreover, g is a strictly concave function on $\boldsymbol{x}^1$, and thus the following holds for a different $\boldsymbol{x}^1$, $\boldsymbol{x}^2 \in \boldsymbol{C}$ and any $\lambda \in (0, 1)$.

$$\begin{aligned} g\left[(1-\lambda)\boldsymbol{x}^1 + \lambda \boldsymbol{x}^2\right] &> (1-\lambda) g\left(\boldsymbol{x}^1\right) + \lambda g\left(\boldsymbol{x}^2\right) \\ &\geqq \boldsymbol{0}. \end{aligned}$$

Hence, by letting $\bar{\boldsymbol{x}} = (1-\lambda)\boldsymbol{x}^1 + \lambda \boldsymbol{x}^2$, for $\bar{\boldsymbol{x}} \in \boldsymbol{C} \subseteq \boldsymbol{D}$, $g(\bar{\boldsymbol{x}}) > \boldsymbol{0}$. Accordingly, Slater's constraint qualification is satisfied. QED

The following is an alternative representation of [Lemma 3.7; (i)].

[Theorem 3.10][11] Let g be a concave function on the convex set $\boldsymbol{D}$ of $\boldsymbol{R}^n$. Here,

I. [there exists certain $\bar{\boldsymbol{x}} \in \boldsymbol{D}$ and $g(\bar{\boldsymbol{x}}) > \boldsymbol{0}$],

or

II. [for any $\boldsymbol{x} \in \boldsymbol{D}$, there exists a $\boldsymbol{p} \geq \boldsymbol{0}$ that satisfies $\boldsymbol{p}^t g(\boldsymbol{x}) \leqq 0$].

However, I or II indicate that only one or the other of I, II always holds.

(Notes on [Theorem 3.10]): If in [Theorem 3.10], we let $f(\boldsymbol{x}) = \boldsymbol{A}\boldsymbol{x}$ (however $\boldsymbol{A}$ is a $m \times n$ matrix), Gordan's theorem ([Theorem 1.10]) is derived. Thus, [Theorem 3.10] is an expansion of Gordan's theorem[12] in respect of linear inequality systems to nonlinear inequality systems and is known as **a generalization of Gordan's**

[11] See Fan et al. (1957).

[12] See Gordan (1873).

theorem. For details on theorem using [Theorem 3.10], please see Mangasarian (1969).

[Lemma 3.8] When one of the following conditions hold,

(i) g satisfies Slater's constraint qualification, Karlin's constraint qualification or strong constraint qualification,
(ii) g satisfies the regularity condition on $\overline{\boldsymbol{x}} \in \boldsymbol{C}$,
(iii) g satisfies ⟨Definition 3.30⟩ on $\overline{\boldsymbol{x}} \in \boldsymbol{C}$,

g satisfies the Arrow–Hurwicz–Uzawa's constraint qualification on $\overline{\boldsymbol{x}} \in \boldsymbol{C}$.

(Proof)

(i) From [Lemma 3.7], Slater's constraint qualification and Karlin's constraint qualification are equivalent, and a strong constraint qualification indicates Slater's constraint qualification, thus, to show (i), we only require to demonstrate that Slater's constraint qualification can be used to derive Arrow–Hurwicz–Uzawa's constraint qualification. From Slater's constraint qualification, there exists $\widehat{\boldsymbol{x}} \in \boldsymbol{C} \subseteq \boldsymbol{D}$ that is $g(\widehat{\boldsymbol{x}}) > \boldsymbol{0}$. Furthermore, because g is a concave function and differentiable on $\boldsymbol{D}$, for $\overline{\boldsymbol{x}} \in \boldsymbol{C} \subseteq \boldsymbol{D}$,

$$\begin{aligned}\nabla g_E(\overline{\boldsymbol{x}})(\widehat{\boldsymbol{x}} - \overline{\boldsymbol{x}}) &\geqq g_E(\widehat{\boldsymbol{x}}) - g_E(\overline{\boldsymbol{x}}) \\ &= g_E(\widehat{\boldsymbol{x}}) \\ &> \boldsymbol{0}\end{aligned}$$

hold. Accordingly, if we let $\boldsymbol{\xi}^* = \widehat{\boldsymbol{x}} - \overline{\boldsymbol{x}}$, for this $\boldsymbol{\xi}^* \in \boldsymbol{R}^n$, $\nabla g_E(\overline{\boldsymbol{x}})\boldsymbol{\xi}^* > \boldsymbol{0}$.
(ii) From $rank(\nabla g_E(\overline{\boldsymbol{x}})) = \#\boldsymbol{E}(\overline{\boldsymbol{x}})$, for any $\boldsymbol{u} > \boldsymbol{0}$, there exists some $\boldsymbol{y} \in \boldsymbol{R}^n$ and $\nabla g_E(\overline{\boldsymbol{x}})\boldsymbol{y} = \boldsymbol{u} > \boldsymbol{0}$ holds.[13]
(iii) The constraint set $\boldsymbol{C}$ is a convex set hence from [Lemma 3.5], for any $\boldsymbol{x} \in \boldsymbol{C}$, $\boldsymbol{x} - \overline{\boldsymbol{x}} \in \boldsymbol{W}(\overline{\boldsymbol{x}})$. However, $\overline{\boldsymbol{x}}$ is a vector belonging to an arbitrary $\boldsymbol{C}$. Furthermore, from [Lemma 3.4], $\boldsymbol{W}(\overline{\boldsymbol{x}}) \subseteq \boldsymbol{L}(\overline{\boldsymbol{x}})$. Thus, for any $\boldsymbol{x} \in \boldsymbol{C}$,

$$\boldsymbol{x} - \overline{\boldsymbol{x}} \in \boldsymbol{W}(\overline{\boldsymbol{x}}) \subseteq \boldsymbol{L}(\overline{\boldsymbol{x}}). \tag{3.51}$$

Now, let us assume that there exists some number $j_0 \in \boldsymbol{E}(\overline{\boldsymbol{x}})$ and that for any $\boldsymbol{\xi} \in \boldsymbol{L}(\overline{\boldsymbol{x}})$, $\nabla g_{j_0}(\overline{\boldsymbol{x}})\boldsymbol{\xi}^* = \boldsymbol{0}$. $\boldsymbol{C}$ has an internal point; so, if we let $\boldsymbol{x}^0 \in$ ***int C***, there exists $\boldsymbol{\delta} > 0$ that satisfies $\boldsymbol{B}(\boldsymbol{x}^0; \delta) \subseteq \boldsymbol{C}$. Hence, if we define $\boldsymbol{x} = \boldsymbol{x}^0 + \frac{\delta}{2}\boldsymbol{e}^k$ $(k = 1, \cdots, n)$, $\rho = \frac{\delta}{2} > 0$ and $d(\boldsymbol{x}, \boldsymbol{x}^0) = \|\boldsymbol{x} - \boldsymbol{x}^0\| = \max_{i \in N} |x_i - x_i^0|$,[14]

[13] Let $\boldsymbol{A}$ be a $m \times n$ matrix. Here, the necessary and sufficient conditions for $\boldsymbol{Ax} = \boldsymbol{b}$ to have a solution is $rank(\boldsymbol{A}) = rank\ [\boldsymbol{A} \vdots \boldsymbol{b}]$. ([Theorem 19; (iv)] (Appendix)).

[14] The argument still holds even whatever other vectors and norms are used. For norms, see ⟨Definition 9⟩ and (Appendix).

$$d\left(\boldsymbol{x},\boldsymbol{x}^0\right)=\frac{\delta}{2}<\delta$$

and

$$\boldsymbol{x}=\left(\boldsymbol{x}^0+\rho\boldsymbol{e}^k\right)\in\boldsymbol{B}\left(\boldsymbol{x}^0;\delta\right)\subseteq\boldsymbol{C}, \tag{3.52}$$

where $\boldsymbol{N}=\{1,\cdots,n\}$. Therefore, for any $\boldsymbol{x}\in\boldsymbol{C}$,

$$\begin{aligned}\boldsymbol{L}(\overline{\boldsymbol{x}})\ni\ \boldsymbol{\xi}&=\boldsymbol{x}-\overline{\boldsymbol{x}}\\&=\boldsymbol{x}^0+\rho\boldsymbol{e}^k-\overline{\boldsymbol{x}}\\&=\left(\boldsymbol{x}^0-\overline{\boldsymbol{x}}\right)+\rho\boldsymbol{e}^k.\end{aligned}$$

Thus, the following self-contradiction is derived.

$$\begin{aligned}0&=\nabla g_{j_0}(\overline{\boldsymbol{x}})\boldsymbol{\xi}\\&=\rho(\nabla g(\overline{\boldsymbol{x}}))_{j_0k}\\&\neq 0,\end{aligned}$$

where $(\nabla g(\overline{\boldsymbol{x}}))_{j_0k}$ is the kth element of $\nabla g_{j_0}(\overline{\boldsymbol{x}})$. Accordingly, for any $j\subset\boldsymbol{E}(\overline{\boldsymbol{x}})$, there exists $\boldsymbol{\xi}^j\in\boldsymbol{L}(\overline{\boldsymbol{x}})$ and $\nabla g_j(\overline{\boldsymbol{x}})\boldsymbol{\xi}^j\neq 0$. Therefore from (3.51) and the definition of $\boldsymbol{L}(\overline{\boldsymbol{x}})$,

$$\begin{gathered}\text{for any } j\in\boldsymbol{E}(\overline{\boldsymbol{x}}),\ \nabla g_j(\overline{\boldsymbol{x}})\boldsymbol{\xi}^j>0\\ \text{and}\\ \text{for any } k\in\boldsymbol{E}(\overline{\boldsymbol{x}}),\ \nabla g_j(\overline{\boldsymbol{x}})\boldsymbol{\xi}^k\geqq 0.\end{gathered} \tag{3.53}$$

Thus, if we set $\boldsymbol{\xi}^*=\sum_{j\in\boldsymbol{E}(x)}\boldsymbol{\xi}^j$, from (3.53), we obtain $\nabla g_{\boldsymbol{E}}(\overline{\boldsymbol{x}})\boldsymbol{\xi}^*>\boldsymbol{0}$. QED

The fact that from ⟨Definition 3.25⟩, ⟨Definition 3.28⟩ and ⟨Definition 3.32⟩, ⟨Definition 3.33⟩, ⟨Definition 3.34⟩ and ⟨Definition 3.35⟩ are derived and the relationships between ⟨Definition 3.33⟩–⟨Definition 3.35⟩ are shown below.

[Lemma 3.9] The following relationships hold mutually between each of the constraint qualifications.

(i) Slater's constraint qualification satisfies Slater's weak constraint qualification.
(ii) Inverse convex constraint qualification satisfies the weak inverse convex constraint qualification.
(iii) Arrow–Hurwicz–Uzawa constraint qualification satisfies Arrow–Hurwicz–Uzawa weak constraint qualification.

(iv) If g satisfies Slater's weak constraint qualification or weak inverse convex constraint qualification on $\bar{\boldsymbol{x}} \in \boldsymbol{C}$, then g satisfies Arrow–Hurwicz–Uzawa's weak constraint qualification on $\bar{\boldsymbol{x}} \in \boldsymbol{C}$.

(Proof)

(i)~(iii) If on $\bar{\boldsymbol{x}} \in \boldsymbol{C}$, g_i is a concave (convex) function, from [Theorem 2.4; (ii)], g_i is a pseudo concave (pseudo convex) function. Therefore, the conclusion is evident.
(iv) If g satisfies Slater's weak constraint qualification on $\bar{\boldsymbol{x}} \in \boldsymbol{C}$, there exists $\bar{\boldsymbol{x}} \in \boldsymbol{C}$ such that

$$\begin{aligned} \mathbf{0} &< g_E(\bar{\boldsymbol{x}}) \\ &= g_E(\widehat{\boldsymbol{x}}) - g_E(\bar{\boldsymbol{x}}) \end{aligned}$$

$g_E(\bar{\boldsymbol{x}})$ is a pseudoconcave function hence $g_E(\widehat{\boldsymbol{x}}) > g_E(\bar{\boldsymbol{x}})$ indicates $\nabla\, g_E(\bar{\boldsymbol{x}})(\widehat{\boldsymbol{x}} - \bar{\boldsymbol{x}}) > \mathbf{0}$. Thus, if we let $z = \widehat{\boldsymbol{x}} - \bar{\boldsymbol{x}}$, $\nabla\, g_E(\bar{\boldsymbol{x}})z > \mathbf{0}$. Consequently, g satisfies Arrow–Hurwicz–Uzawa's weak constraint qualification on $\bar{\boldsymbol{x}} \in \boldsymbol{C}$.

Next, if g is $\bar{\boldsymbol{x}} \in \boldsymbol{C}$ and satisfies weak inverse convex constraint qualification, it is the set defined in ⟨Definition 3.35⟩, $\boldsymbol{Q}''(\bar{\boldsymbol{x}}) = \phi$. Therefore, if we let $z = \mathbf{0}$, $\nabla g_{Q'}(\bar{\boldsymbol{x}})z = \nabla g(\bar{\boldsymbol{x}})z = \mathbf{0}$. Accordingly, g satisfies Arrow–Hurwicz–Uzawa's **weak constraint** qualification on $\bar{\boldsymbol{x}} \in \boldsymbol{C}$. QED

Utilizing [Lemma 3.8] and [Lemma 3.9], the fact that Kuhn–Tucker's constraint qualification and any constraint qualification other than weak constraint qualification indicates Arrow–Hurwicz–Uzawa's weak constraint qualification is shown. Thus, if it is shown that Arrow–Hurwicz–Uzawa's weak constraint qualification derives weak constraint qualification, it is evident that each of the constraint qualification other than Kuhn–Tucker's constraint qualification is a sufficient condition for weak constraint qualification.

[Theorem 3.11] If g satisfies Arrow–Hurwicz–Uzawa's weak constraint qualification on $\bar{\boldsymbol{x}} \in \boldsymbol{C}$, g satisfies the weak constraint qualification on $\bar{\boldsymbol{x}} \in \boldsymbol{C}$.

(Proof) Define $\boldsymbol{\xi}$ as any vector on $L(\bar{\boldsymbol{x}})$, α as any positive number and for any $\theta \geqq 0$,

$$\varphi(\theta) = \bar{\boldsymbol{x}} + (\boldsymbol{\xi} + \alpha\boldsymbol{\xi}^*)\theta. \tag{3.54}$$

However, $\boldsymbol{\xi}^*$ is the same as z in Arrow–Hurwicz–Uzawa's constraint qualification (⟨Definition 3.35⟩).

First, show that $\varphi(\theta)$ is a contained path with origin $\bar{\boldsymbol{x}}$ and direction vector $\boldsymbol{\xi} + \alpha\boldsymbol{\xi}^*$. $\varphi(\theta)$ is defined on the nonnegative domain of θ and in fact, $\varphi(\theta) = \bar{\boldsymbol{x}} \in \boldsymbol{C}$ is evident. Moreover, for any $h > 0$,

$$\varphi(h+0)-\varphi(0)-(\boldsymbol{\xi}+\alpha\boldsymbol{\xi}^*)h=0.$$

Therefore,

$$\lim_{h\to+\infty}\left\|\frac{\varphi(h+0)-\varphi(0)-(\boldsymbol{\xi}+\alpha\boldsymbol{\xi}^*)h}{h}\right\|=0.$$

Thus, from the definition of a derivative, $\varphi'(0) = \boldsymbol{\xi} + \alpha\boldsymbol{\xi}^*$. Furthermore, it only requires to be shown that if θ is considered to be sufficiently small, $\varphi(\theta) \in \boldsymbol{C}$.

For any $j \in \boldsymbol{E}(\overline{\boldsymbol{x}})$,

$$\begin{aligned}\left.\frac{dg_j[\varphi(\theta)]}{d\theta}\right|_{\theta=0} &= \nabla g_i(\overline{\boldsymbol{x}})(\boldsymbol{\xi}+\alpha\boldsymbol{\xi}^*).\\ &= g_j(\overline{\boldsymbol{x}})\boldsymbol{\xi}+\alpha\nabla g_i(\overline{\boldsymbol{x}})\boldsymbol{\xi}^*\\ &\geqq \alpha\nabla g_i(\overline{\boldsymbol{x}})\boldsymbol{\xi}^*.\end{aligned} \tag{3.55}$$

Consider $\boldsymbol{E}(\overline{\boldsymbol{x}})=\boldsymbol{Q}'(\overline{\boldsymbol{x}})\cup\boldsymbol{Q}''(\overline{\boldsymbol{x}})$ and consider each of the cases $j\in\boldsymbol{Q}'(\overline{\boldsymbol{x}})$ or $j\in\boldsymbol{Q}''(\overline{\boldsymbol{x}})$. $j\in\boldsymbol{Q}'(\overline{\boldsymbol{x}})$ Then, for any $\theta \geqq 0$,

$$\begin{aligned}\nabla g_j[\varphi(\theta)][\varphi(\theta)-\varphi(0)] &= \nabla g_j[\varphi(0)][\overline{\boldsymbol{x}}+(\boldsymbol{\xi}+\alpha\boldsymbol{\xi}^*)\theta-\overline{\boldsymbol{x}}]\\ &= \nabla g_j[\varphi(\theta)]\boldsymbol{\xi}\theta+\alpha\nabla g_j[\varphi(\theta)]\boldsymbol{\xi}^*\theta\\ &= \nabla g_j(\overline{\boldsymbol{x}})\boldsymbol{\xi}\theta+\alpha\nabla g_j(\overline{\boldsymbol{x}})\boldsymbol{\xi}^*\theta\\ &\geqq 0.\end{aligned} \tag{3.56}$$

Furthermore, $g_j(\overline{\boldsymbol{x}})$ is a pseudo convex function, and hence (3.56) indicates

$$g_j[\varphi(\theta)]-g_j[\varphi(\theta)]\geqq 0. \tag{3.57}$$

Moreover, for any $j\in\boldsymbol{Q}'(\overline{\boldsymbol{x}})$ and $\theta \geqq 0$,

$$\begin{aligned}g_j[\varphi(\theta)] &\geqq g_j[\varphi(0)]\\ &= g_j(\overline{\boldsymbol{x}})\\ &= 0.\end{aligned} \tag{3.58}$$

However, if $j\in\boldsymbol{Q}''(\overline{\boldsymbol{x}})$,

$$\begin{aligned}\left.\frac{dg_j[\varphi(\theta)]}{d\theta}\right|_{\theta=0} &\geqq \alpha\nabla g_j(\overline{\boldsymbol{x}})\boldsymbol{\xi}^*\\ &> 0.\end{aligned} \tag{3.59}$$

Therefore, if a positive number θ is considered to be sufficiently small,

$$\begin{aligned} g_E[\varphi(\theta)] &\geqq g_E[\varphi(0)] \\ &= g_E(\bar{x}) \\ &= \mathbf{0}. \end{aligned} \tag{3.60}$$

In respect of $\boldsymbol{F}(\bar{\boldsymbol{x}}) = \{1, \cdots, m\} - \boldsymbol{E}(\bar{\boldsymbol{x}})$, when $\theta = 0$

$$g_F[\varphi(0)] = g_F(\bar{x}) > \mathbf{0}. \tag{3.61}$$

If $\theta > 0$ is considered to be sufficiently small,

$$g_F[\varphi(\theta)] \geqq \mathbf{0}. \tag{3.62}$$

Accordingly, from (3.60) to (3.62), for a sufficiently small $\theta > 0$, $\varphi(\theta) \in \boldsymbol{C}$.

Thus, for any $\boldsymbol{\xi} \in \boldsymbol{L}(\bar{\boldsymbol{x}})$ and $\alpha > 0$, $\boldsymbol{\xi} + \alpha\boldsymbol{\xi}^* \in \boldsymbol{A}(\bar{\boldsymbol{x}})$. From this and $\boldsymbol{A}(\bar{\boldsymbol{x}}) \subseteq \boldsymbol{W}(\bar{\boldsymbol{x}})$, $\boldsymbol{\xi} + \alpha\boldsymbol{\xi}^* \in \boldsymbol{W}(\bar{\boldsymbol{x}})$ is derived for any $\boldsymbol{\xi} \in \boldsymbol{L}(\bar{\boldsymbol{x}})$ and $\alpha > 0$. Hence, if we define $\alpha = \frac{1}{\gamma}(\gamma = 1, 2, \cdots)$ and let $z^\gamma = \boldsymbol{\xi} + \left(\frac{1}{\gamma}\right)\boldsymbol{\xi}^*$ corresponding to this α, $\{z^\gamma\}$ becomes a converging sequence of points within the closed set $\boldsymbol{W}(\bar{\boldsymbol{x}})$; hence, when $\gamma \to \infty$, its limit $\boldsymbol{\xi}$ belongs to $\boldsymbol{W}(\bar{\boldsymbol{x}})$ (See [Theorem 3.3]). This shows that any $\boldsymbol{\xi} \in \boldsymbol{L}(\bar{\boldsymbol{x}})$ is contained in $\boldsymbol{W}(\bar{\boldsymbol{x}})$, i.e., $\boldsymbol{L}(\bar{\boldsymbol{x}}) \subseteq \boldsymbol{W}(\bar{\boldsymbol{x}})$. QED

Because of the aforementioned theorem, [Theorem 3.8], [Theorem 3.9], [Lemma 3.7], [Lemma 3.8] and [Lemma 3.9], the following can be obtained, which is the most comprehensive theorem in respect of a Lagrange regular.

[Theorem 3.12] If at [NLPX]'s optimal solution $\bar{\boldsymbol{x}} \in \boldsymbol{C}$, one of the constraint qualifications in ⟨Definition 3.24⟩–⟨Definition 3.35⟩ is satisfied, g is a Lagrange regular.

(Proof) The fact that weak constraint qualification is obtained from Kuhn–Tucker's constraint qualification has been shown by [Theorem 3.9]. Moreover, from [Lemma 3.7], [Lemma 3.8] and [Lemma 3.9], any constraint qualification other than Kuhn–Tucker's constraint qualification becomes the sufficient condition for Arrow–Hurwicz–Uzawa's weak constraint qualification and from [Theorem 3.11], the fact that Arrow–Hurwicz–Uzawa's weak constraint qualification is clear. Thus, combining this and [Theorem 3.8], the proof is complete. QED

Figure 3.7 shows the results above in a list form.

Until now, only inequality constraints, i.e., $g_i(\boldsymbol{x}) \geqq 0$ $(i = 1, \cdots, m)$, had been used as constraint equations; however, if there is an equality constraint $h_j(\boldsymbol{x}) = 0$ $(j = 1, \cdots, l)$ in the constraint, there is no other approach than to remove the variables from equality constraints and apply in to a standard nonlinear programming problems.

Furthermore, the Lagrange function when there is a mix of equality constraints is defined as follows:

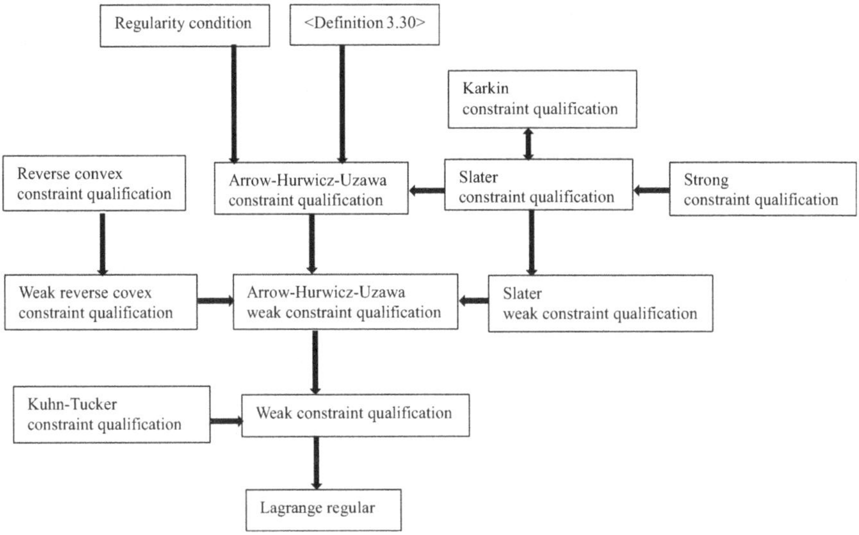

Fig. 3.7 Interrelationship of constraint qualifications

$$L(\boldsymbol{x},\boldsymbol{\lambda},\boldsymbol{\mu}) = f(\boldsymbol{x}) + \sum_{i=1}^{m} \lambda_i g_i(\boldsymbol{x}) + \sum_{j=1}^{l} \mu_j h_j(\boldsymbol{x})$$

and the Lagrange condition is represented as the fact that for $\boldsymbol{x} \in \boldsymbol{C}$, there is a $(\boldsymbol{\lambda},\boldsymbol{\mu}) \in \boldsymbol{R}^{m+l}$ that satisfies

$$\nabla f(\boldsymbol{x}) + \sum_{i=1}^{m} \lambda_i \nabla g_i(\boldsymbol{x}) + \sum_{j=1}^{l} \mu_j \nabla h_j(\boldsymbol{x}) = \boldsymbol{0},$$
$$\lambda_i \geqq 0,\ \ g_i(\boldsymbol{x}) \geqq 0,\ \ \lambda_i g_i(\boldsymbol{x}) = 0,\ \ (i = 1, \cdots, m)$$

and

$$h_j(\boldsymbol{x}) = 0, \quad (j = 1, \cdots, l).$$

Moreover, for optimization problems with no constraints, which are the simplest of nonlinear programming problems, the necessary and sufficient condition for something to be the optimal solution can be determined by [Theorem 17] (Appendix).

3.4 Linear Programming Problems

Generally, problems involving maximizing or minimizing the value of certain linear function under the constraint conditions given in the form of a linear inequality equation system or an equality equation system are referred to as **linear programming problems** (hereinafter abbreviated to be [LP]) and they can be defined as shown below.

⟨**Definition 3.36**⟩ Let $\boldsymbol{A}$ be a $m \times n$ matrix, $\boldsymbol{c}$ and $\boldsymbol{x}$ n-dimensional vectors and $\boldsymbol{b}$ and $\boldsymbol{u}$ m-dimensional vectors. Here, considering the following two [LP],

$$\text{(I)}\quad \max_{\boldsymbol{x}} \boldsymbol{c}^t\boldsymbol{x} \qquad\qquad \text{(II)}\quad \min_{\boldsymbol{u}} \boldsymbol{b}^t\boldsymbol{u}$$

$$\text{subject to } \boldsymbol{Ax} \leqq \boldsymbol{b} \qquad\qquad \text{subject to } \boldsymbol{A}^t\boldsymbol{u} \geqq \boldsymbol{c},$$

$$\boldsymbol{x} \geqq \boldsymbol{0} \qquad\qquad \boldsymbol{u} \geqq \boldsymbol{0}$$

these two contrasting problems are referred to as a **primal problem** (primal problem), whereas the other is referred to as a **dual problem**.

The set of feasible solutions for problem (I) and problem (II) are given as follows:

$$F_{\text{I}} = \{\boldsymbol{x} | \boldsymbol{Ax} \leqq \boldsymbol{b}, \boldsymbol{x} \geqq \boldsymbol{0}\},$$

$$F_{\text{II}} = \{\boldsymbol{u} | \boldsymbol{A}^t\boldsymbol{u} \geqq \boldsymbol{c}, \boldsymbol{u} \geqq \boldsymbol{0}\}$$

respectively. Moreover the set of optimal solutions of problem (I) and problem (II) can be expressed as follows:

$$Q_{\text{I}} = \left\{\boldsymbol{x} \in F_{\text{I}} \middle| {}^{\forall}\boldsymbol{\xi} \in F_{\text{I}}, \boldsymbol{c}^t\boldsymbol{x} \geqq \boldsymbol{c}^t\boldsymbol{\xi}\right\}.$$

$$Q_{\text{II}} = \left\{\boldsymbol{u} \in F_{\text{II}} \middle| {}^{\forall}\boldsymbol{\eta} \in F_{\text{II}}, \boldsymbol{b}^t\boldsymbol{u} \leqq \boldsymbol{b}^t\boldsymbol{\eta}\right\}$$

(Notes on ⟨Definition 3.36⟩): Lagrange function corresponding to problems **(I)** and **(II)** are defined as follows:

$$\begin{aligned} L_{\text{I}}(\boldsymbol{x}, \boldsymbol{u}) &= \boldsymbol{c}^t\boldsymbol{x} + \boldsymbol{u}^t(\boldsymbol{b} - \boldsymbol{Ax}) \\ &= -(-\boldsymbol{u}^t\boldsymbol{b} + \boldsymbol{u}^t\boldsymbol{Ax} - \boldsymbol{c}^t\boldsymbol{x}) \\ &= -\{-\boldsymbol{u}^t\boldsymbol{b} + \boldsymbol{x}^t(\boldsymbol{A}^t\boldsymbol{u} - \boldsymbol{c})\} \end{aligned}$$

and

$$L_{\text{II}}(\boldsymbol{u}, \boldsymbol{x}) = -\boldsymbol{b}^t\boldsymbol{u} + \boldsymbol{x}^t(\boldsymbol{A}^t\boldsymbol{u} - \boldsymbol{c}).$$

Therefore it is evident that the Lagrange function for one of the problems is equal to the Lagrange function of the remaining other problem multiplied by (-1).

For the discussion of [LP], firstly let us rewrite Farkas's theorem[15] ([Theorem 1.7]) as below and show several Lemmas.

[Lemma 3.10] For a given $p \times q$ matrix $\boldsymbol{B}$ and $\boldsymbol{d} \in \boldsymbol{R}^q$,

$$\text{I} \quad [\boldsymbol{B}^t\boldsymbol{\eta} \geqq \boldsymbol{d}] \text{ has a solution } \boldsymbol{\eta} \geqq \boldsymbol{0},$$

or

$$\text{II} \quad [\boldsymbol{B}\boldsymbol{\xi} \leqq \boldsymbol{0},\ \boldsymbol{d}^t\boldsymbol{\xi} > 0] \text{ has a solution } \boldsymbol{\xi} \geq \boldsymbol{0}.$$

However, the description in this Lemma indicates that, as per Chap. 1, either I or II always holds and that I and II cannot simultaneously hold. This applies in the following proof too.

(Proof) From Farkas' theorem ([Theorem 1.7]), in respect of a given $p \times q$ matrix $\boldsymbol{B}$ and $\boldsymbol{d} \in \boldsymbol{R}^q$,

$$\text{(I)} \quad \left[[\boldsymbol{B}^t \ -\boldsymbol{I}_q] \begin{bmatrix} \boldsymbol{\eta} \\ \boldsymbol{\gamma} \end{bmatrix} = \boldsymbol{d} \right] \text{ has a solution } \begin{bmatrix} \boldsymbol{\eta} \\ \boldsymbol{\gamma} \end{bmatrix} \geqq \boldsymbol{0},$$

or

$$\text{(II)} \quad \left[\begin{bmatrix} \boldsymbol{B} \\ -\boldsymbol{I}_q \end{bmatrix} \boldsymbol{\xi} \leqq \boldsymbol{0}, \boldsymbol{d}^t\boldsymbol{\xi} > 0 \right] \text{ has a solution } \boldsymbol{\xi}.$$

Clearly, (I) and these Lemmas I and (II) and this Lemma II are equivalent. QED

[Lemma 3.11] In respect of $m \times n$ matrix $\boldsymbol{A}$, $\boldsymbol{b} \in \boldsymbol{R}^m$ and $\boldsymbol{c} \in \boldsymbol{R}^n$, the following holds.

$$\textbf{[I]} \quad \begin{bmatrix} -\boldsymbol{A} & \boldsymbol{0} \\ \boldsymbol{0} & \boldsymbol{A}^t \\ \boldsymbol{c}^t & -\boldsymbol{b}^t \end{bmatrix} \begin{bmatrix} \boldsymbol{x} \\ \boldsymbol{u} \end{bmatrix} \geqq \begin{bmatrix} -\boldsymbol{b} \\ \boldsymbol{c} \\ 0 \end{bmatrix} \text{ has a solution } \begin{bmatrix} \boldsymbol{x} \\ \boldsymbol{u} \end{bmatrix} \geqq \boldsymbol{0},$$

or

$$\textbf{[II]} \quad \begin{bmatrix} -\boldsymbol{A}^t & \boldsymbol{0} & \boldsymbol{c} \\ \boldsymbol{0} & \boldsymbol{A} & -\boldsymbol{b} \end{bmatrix} \begin{bmatrix} \boldsymbol{v} \\ \boldsymbol{w} \\ t \end{bmatrix} \leqq \boldsymbol{0} \text{ has a solution } \begin{bmatrix} \boldsymbol{v} \\ \boldsymbol{w} \\ t \end{bmatrix} \geq \boldsymbol{0}.$$

$$\begin{bmatrix} -\boldsymbol{b}^t & \boldsymbol{c}^t & \boldsymbol{0} \end{bmatrix} \begin{bmatrix} \boldsymbol{v} \\ \boldsymbol{w} \\ t \end{bmatrix} > 0$$

[15] See Farkas (1902).

Furthermore, the set of solutions of [**I**] is denoted as $\boldsymbol{S}_{[\text{I}]}$, and the set of solutions of [**II**] is denoted as $\boldsymbol{S}_{[\text{II}]}$. Using $\boldsymbol{S}_{[\text{I}]}$ and $\boldsymbol{S}_{[\text{II}]}$, [Lemma 3.10] can be rephrased to be that only one of $\boldsymbol{S}_{[\text{I}]}$ or $\boldsymbol{S}_{[\text{II}]}$ must be nonempty.

(Proof) By letting [Lemma 3.10]'s $\boldsymbol{B}$, $\boldsymbol{\eta}$, $\boldsymbol{d}$ and $\boldsymbol{\xi}$ to be

$$\boldsymbol{B}=\begin{bmatrix}-\boldsymbol{A} & \boldsymbol{0}\\ \boldsymbol{0} & \boldsymbol{A}^t\\ \boldsymbol{c}^t & -\boldsymbol{b}^t\end{bmatrix},\boldsymbol{\eta}=\begin{bmatrix}\boldsymbol{x}\\ \boldsymbol{u}\end{bmatrix},\boldsymbol{d}=\begin{bmatrix}-\boldsymbol{b}\\ \boldsymbol{c}\\ 0\end{bmatrix},\boldsymbol{\xi}=\begin{bmatrix}\boldsymbol{v}\\ \boldsymbol{w}\\ t\end{bmatrix}$$

respectively, it is immediately evident. QED

[Lemma 3.12] For any $\boldsymbol{x} \in \boldsymbol{F}_{\text{I}}$ and $\boldsymbol{u} \in \boldsymbol{F}_{\text{II}}$, $\boldsymbol{c}^t\boldsymbol{x} \leqq \boldsymbol{b}^t\boldsymbol{u}$.

(Proof) For any $\boldsymbol{x} \in \boldsymbol{F}_{\text{I}}$ and $\boldsymbol{u} \in \boldsymbol{F}_{\text{II}}$,

$$\boldsymbol{u}^t\boldsymbol{A}\boldsymbol{x} \leqq \boldsymbol{u}^t\boldsymbol{b}$$

and

$$\boldsymbol{u}^t\boldsymbol{A}\boldsymbol{x} \geqq \boldsymbol{c}^t\boldsymbol{x}.$$

Therefore, $\boldsymbol{c}^t\boldsymbol{x} \leqq \boldsymbol{b}^t\boldsymbol{u}$. QED

[Lemma 3.13] For the following four propositions,

(a) $\boldsymbol{Q}_{\text{I}} \neq \phi \Rightarrow \boldsymbol{Q}_{\text{II}} \neq \phi$ (or $\boldsymbol{Q}_{\text{I}} \neq \phi \Leftarrow \boldsymbol{Q}_{\text{II}} \neq \phi$)
(b) $\boldsymbol{Q}_{\text{I}} \neq \phi$ and $\boldsymbol{Q}_{\text{II}} \neq \phi$
(c) $\boldsymbol{F}_{\text{I}} \neq \phi$ and $\boldsymbol{F}_{\text{II}} \neq \phi$
(d) $\boldsymbol{S}_{[\text{I}]} \neq \phi$

the following hold.

(i) (b), (c), (d) are mutually equivalent.
(ii) (b) indicates (a).

(Proof)
(i) Show in the order (b) $\Rightarrow$ (c) $\Rightarrow$ (d) $\Rightarrow$ (b).
(b) $\Rightarrow$ (c) $\Rightarrow$(c) $\phi \neq \boldsymbol{Q}_{\text{I}} \subseteq \boldsymbol{F}_{\text{I}}$ and $\phi \neq \boldsymbol{Q}_{\text{II}} \subseteq \boldsymbol{F}_{\text{II}}$. Hence, (c) is obtained.
(c)$\Rightarrow$(d) Contradicting the conclusion and letting $\boldsymbol{S}_{[\text{I}]} \neq \phi$, from [Lemma 3.11], $\boldsymbol{S}_{[\text{II}]} \neq \phi$. Accordingly, $\boldsymbol{A}^t\boldsymbol{v} \geqq \boldsymbol{c}t$, $\boldsymbol{A}\boldsymbol{w} \leqq \boldsymbol{b}t$ and $\boldsymbol{b}^t\boldsymbol{v} < \boldsymbol{c}^t\boldsymbol{w}$ have the solution $(\boldsymbol{v}, \boldsymbol{w}, t)^t \geq \boldsymbol{0}$. If $t > 0$ $\left(\frac{\boldsymbol{w}}{t}, \frac{\boldsymbol{v}}{t}\right)^t \geqq \boldsymbol{0}$ becomes a solution of [**I**], and hence contradicts $\boldsymbol{S}_{[\text{I}]} \neq \phi$. Thus, $t = 0$. Hence $\boldsymbol{A}^t\boldsymbol{v} \geqq \boldsymbol{0}$, $\boldsymbol{A}\boldsymbol{w} \leqq \boldsymbol{0}$ and $\boldsymbol{b}^t\boldsymbol{v} < \boldsymbol{c}^t\boldsymbol{w}$ hold but from the condition (c), $\boldsymbol{F}_{I} \neq \phi$ and $\boldsymbol{F}_{\text{II}} \neq \phi$. Therefore, for any $\boldsymbol{x} \in \boldsymbol{F}_{\text{I}}$ and $\boldsymbol{u} \in \boldsymbol{F}_{\text{II}}$,

$$\begin{aligned}0 \leqq x^t(A^t v) &= (x^t A^t)v \\ &\leqq b^t v \\ &< c^t w \\ &\leqq (u^t A)w \\ &= u^t(Aw) \leqq 0\end{aligned}$$

are derived but these are self-contradicting.
(d) $\Rightarrow$ (b) From assumption (d), **[I]** has a nonnegative solution $(\bar{x}^t, \bar{u}^t)$. Therefore, from [Lemma 3.12], for any $x \in F_{\mathrm{I}}$,

$$c^t\bar{x} \geqq b^t\bar{u} \geqq c^t x, \tag{3.63}$$

and

$$c^t\bar{x} \leqq b^t\bar{u}. \tag{3.64}$$

$\bar{x}$ and $\bar{u}$ are feasible and hence (3.63) and (3.64) indicates $\bar{x} \in Q_{\mathrm{I}}$ and $c^t\bar{x} = b^t\bar{u}$. Here, if the equalities shown above, [Lemma 3.12] and $\bar{u} \in F_{\mathrm{II}}$ are combined, for any $u \in F_{\mathrm{II}}$,

$$b^t\bar{u} = c^t\bar{x} \leqq b^t u$$

holds. Thus, $\bar{u} \in Q_{\mathrm{II}}$ is shown.
(ii) If (a) does not hold, $Q_{\mathrm{I}} \neq \phi$ and $Q_{\mathrm{II}} = \phi$ (or $Q_{\mathrm{I}} = \phi$ and $Q_{\mathrm{II}} \neq \phi$). However, this contradicts condition (b). QED

Use the aforementioned lemma and explain the two most fundamental theorems among [LP].

[Theorem 3.13] Existence Theorem The necessary and sufficient condition for problems **(I)** and **(II)** to have an optimal solution is that F_{I} and F_{II} are not empty sets.

(Proof) [Lemma 3.13; (i)] shows this.

[Theorem 3.14] Duality Theorem

(i) For $x^0 \in F_{\mathrm{I}}$ to be optimal, the necessary and sufficient condition is for there to exist $u^0 \in F_{\mathrm{II}}$ that satisfies $c^t x^0 = b^t u^0$.
(ii) For $u^0 \in F_{\mathrm{II}}$ to be optimal, the necessary and sufficient condition is for there to exist $x^0 \in F_{\mathrm{I}}$ that satisfies $c^t x^0 = b^t u^0$.

(Proof)

(i) (Necessity) From the conditions, $\boldsymbol{Q}_{\mathrm{I}} \neq \phi$. Furthermore, with [LP], ⟨Definition 3.28 ; (ii)⟩ is automatically satisfied such that there exists $\boldsymbol{u}^0 \geqq \boldsymbol{0}$ that satisfies $(\boldsymbol{c}^t - (\boldsymbol{u}^0)^t\boldsymbol{A}) \leqq \boldsymbol{0}$ for $\boldsymbol{x}^0 \in \boldsymbol{Q}_{\mathrm{I}}$. Consequently, $\boldsymbol{F}_{\mathrm{II}} \neq \phi$. Moreover, from $\boldsymbol{x}^0 \in \boldsymbol{Q}_{\mathrm{I}} \subseteq \boldsymbol{F}_{\mathrm{I}}$, $\boldsymbol{F}_{\mathrm{I}} \neq \phi$. Therefore, $\boldsymbol{F}_{\mathrm{I}} \neq \phi$ and $\boldsymbol{F}_{\mathrm{II}} \neq \phi$ so if [Lemma 3.13: (i); (c) ⇒ (d)] is considered, $\boldsymbol{S}_{\mathrm{I}} \neq \phi$. Accordingly, as shown in the proof of [Lemma 3.13: (i); (d) ⇒ (b)], there exists $\overline{\boldsymbol{x}} \in \boldsymbol{Q}_{\mathrm{I}}$ and $\overline{\boldsymbol{u}} \in \boldsymbol{Q}_{\mathrm{II}}$ that are $\boldsymbol{c}^t\overline{\boldsymbol{x}} = \boldsymbol{b}^t\overline{\boldsymbol{u}}$. Moreover, the optimality of $\boldsymbol{x}^0$ guarantees $\boldsymbol{c}^t\boldsymbol{x}^0 = \boldsymbol{c}^t\overline{\boldsymbol{x}}$ and $\overline{\boldsymbol{u}} \in \boldsymbol{Q}_{\mathrm{II}} \subseteq \boldsymbol{F}_{\mathrm{II}}$. Hence, if $\overline{\boldsymbol{u}}$ is rewritten as $\boldsymbol{u}^0$, correspondingly to $\boldsymbol{x}^0 \in \boldsymbol{Q}_{\mathrm{I}}$, there exists $\boldsymbol{u}^0 \in \boldsymbol{F}_{\mathrm{II}}$ that satisfies $\boldsymbol{c}^t\boldsymbol{x}^0 = \boldsymbol{b}^t\boldsymbol{u}^0$.

(Sufficiency) From [Lemma 3.12], for any $\boldsymbol{x} \in \boldsymbol{F}_{\mathrm{I}}$,

$$\boldsymbol{c}^t\boldsymbol{x}^0 = \boldsymbol{b}^t\boldsymbol{u}^0 \geqq \boldsymbol{c}^t\boldsymbol{x}$$

holds. Thus, for any $\boldsymbol{x} \in \boldsymbol{F}_{\mathrm{I}}$, $\boldsymbol{c}^t\boldsymbol{x}^0 \geqq \boldsymbol{c}^t\boldsymbol{x}$. Therefore, $\boldsymbol{x}^0 \in \boldsymbol{Q}_{\mathrm{I}}$.

(ii) can be shown in a similar manner to (i). QED

References

Arrow, K. J., & Enthoven, A. C. (1961). Quasi-concave programming. *Econometrica, 29*, 779–800.

Arrow, K. J., Hurwicz, L., & Uzawa, H. (1961). Constraint qualifications in maximization problems. *Naval Research Logistics Quarterly, 8*, 175–191.

Fan, K., Glicksberg, I., & Hoffman, A. J. (1957). Systems of unequalities involving convex functions. *American Mathematical Society Proceedings, 8*, 617–622.

Farkas, J. (1902). Über die Theorie der einfachen Ungleichungen. *Journal für Reine und Angewandte Mathematik, 124*, 1–27.

Gordan, P. (1873). Über die Auflösungen linearer Gleichungen mit reelen Coefficienten. *Mathematische Annalen, 6*(1), 23–28.

Karlin, S. (1959). *Mathematical methods and theory in games, programming, and economics* (Vol. I). Addison Wesley. No ISBN. Reprinted by Dover Publications, 1992. ISBN-13: 9780486670201.

Kuhn, H. W., & Tucker, A. W. (1951). Nonlinear programming. In J. Neyman (Ed.), *Proceedings of the Second Berkeley Symposium on Mathematical Statics and Probability* (pp. 481–492). University of California press.

Mangasarian, O. L. (1969). *Nonlinear programming*. McGraw-Hill. ISBN-13: 978-0898713412.

Slater, M. (1950). Lagrange multiplier revisited: A contribution to nonlinear programming. Cowles Commission Discussion Paper, Mathematics, 403.

Chapter 4
Frobenius' Theorem

4.1 Introduction

In 1908, the theorem relating to nonnegative matrices established by Frobenius was called Frobenius' theorem and it holds a critical place in the study of economics. Generally, the expansion of this theorem to nonlinear transformations $f: \boldsymbol{R}^n{}_+ \to \boldsymbol{R}$ are not clearly described in this chapter. This chapter will primarily describe **Frobenius' theorem**[1] for nonlinear transformations f.

4.2 Notation and Assumptions

In this Chapter, the notations and assumptions used are stipulated below.

(Notation)

$\boldsymbol{x} = (x_1, \cdots, x_n)^t$

$\boldsymbol{y} = (y_1, \cdots, y_n)^t$

$\boldsymbol{l} = (1, \cdots, 1) \in \boldsymbol{R}^n_{++}$

$f(\boldsymbol{x}) = (f_1(\boldsymbol{x}), \cdots, f_n(\boldsymbol{x}))^t \left(\boldsymbol{R}^n_+ \to \boldsymbol{R}^n\right)$

$\boldsymbol{S}_n = \left\{\boldsymbol{x} \in \boldsymbol{R}^n_+ | \boldsymbol{lx} = 1\right\}$: Dimensional n-dimensional single body

$\boldsymbol{L}_f = \{\mu \in \boldsymbol{R} |^\exists \boldsymbol{x} \in \boldsymbol{S}_n | f(\boldsymbol{x}) \geqq \mu \boldsymbol{x}\}$

μ_f: Maximum element of $\boldsymbol{L}_f$

$\boldsymbol{\Lambda}_f = \{\lambda \in \boldsymbol{R} |^\exists \boldsymbol{y} \in \boldsymbol{S}_n | f(\boldsymbol{y}) \geqq \lambda \boldsymbol{y}\}$

λ_f: Maximum element of $\boldsymbol{\Lambda}_f$

$\boldsymbol{U}$: Order n-order square matrix with all elements $=1$

$\boldsymbol{x}^{(m)} = \left(x_1{}^m, \cdots, x_n^m\right)^t \in \boldsymbol{R}^n$

[1] See Frobenius (1908, 1909). Takayama (2010) describes this thorem in this chapter.

© Springer Nature Singapore Pte Ltd. 2024
K. Nakayama, *Eigenvalue Problem and Nonlinear Programming Problem*,
New Frontiers in Regional Science: Asian Perspectives 70,
https://doi.org/10.1007/978-981-99-2943-6_4

$N = \{1, \cdots, n\}$: A set of numbers

$/A/$: A matrix with $|a_{ij}|$ as the (i,j)th element. However, a_{ij} represents the (i,j)th element of A and $|a_{ij}|$ is the absolute value of a_{ij}. The same applies for vectors.

(Assumption)

(4.A1) (NN)

Nonnegativity:For all $x \geqq 0, f(x) \geqq 0$.

(4.A2) (C)

Continuity: $f(x)$ is continuous on R^n_+.

(4.A3) (H)

Homogenuity: There exists a constant number $m \in [0, 1]$ and for any x and $t \geqq 0, f(tx) = t^m f(x)$.

(4.A4) (WM)

Weak monotonicity: When $x \geqq y$, for all $i \in N$ that are $x_i = y_i, f_i(x) = f_i(y)$.

(4.A5) (M)

Monotonicity: If $x \geqq y, f(x) \geqq f(y)$.

(4.A6) (SM)

Strict monotonicity: If $x \geq y, f(x) > f(y)$.

(4.A7) (I)

Indecomposability: Let J be any nonempty proper subsets of N and when vectors x and y[2] satisfy $x_J = y_J$ and $x_{J^c} > y_{J^c}$,[3] there exists a $k \in J$ that becomes $f_k(x) \neq f_k(y)$.

A lemma that is useful for the derivation of the theorem in this chapter is described below.

[Lemma 4.1] n-Order simple $S_n = \{x \geqq 0 | lx = 1\}$ is compact.

(Proof) Since $S_n \subseteq R^n$, it only must be shown that S_n is a (a) bounded and (b) closed set ([Theorem 6; (ii)] (Appendix)).

(a) If we stipulate the norm of a vector x with $\max_{j \in N} |x_j|$, for any $x, y \in S_n$,

$$\begin{aligned} d(x,y) &\equiv \max_{j \in N} |x_j - y_j| \\ &\equiv |\ x_k - y_k\ | \\ &\leqq |x_k| + |y_k| \\ &= x_k + y_k \end{aligned} \tag{4.1}$$

[2]See Sect. 4.5 for more information on indecomposability.

[3]In respect of any $J(\neq\phi) \subset N$, $S_J = \{(x,y) | x_J = y_J, x_{J^c} > y_{J^c}\}$ can always be defined.

holds. If there exists a number $j_0 \in \boldsymbol{N}$ that satisfies $x_{j_0} > 1$, $1 < x_{j_0} \leqq \sum_{j=1}^{n} x_j$ and it goes against the definition of $\boldsymbol{S}_n$. Accordingly, for any $\boldsymbol{x} \in \boldsymbol{S}_n$ each $x_j \leqq 1 (j = 1, \cdots, n)$. Therefore,

$$d(\boldsymbol{x}, \boldsymbol{y}) \leqq x_k + y_k \leqq 2. \tag{4.2}$$

This shows the boundedness of $\boldsymbol{S}_n$.

(b) From the definition of $\boldsymbol{S}_n$, it can be expressed as $\boldsymbol{S}_n{}^c = \{\boldsymbol{x} \in \boldsymbol{R}^n | \ \boldsymbol{x} \ngeqq \boldsymbol{0} \text{ or } \boldsymbol{lx} \neq 1\}$. Accordingly, for any $\boldsymbol{x} \in \boldsymbol{S}_n{}^c$, we consider the case (b-1) $\boldsymbol{x} \ngeqq \boldsymbol{0}$ or (b-2) $\boldsymbol{lx} \neq 1$.

(b-1) If we define $\boldsymbol{I}_x = \{j \in N | x_j < 0\}$, from $\boldsymbol{x} \ngeqq 0$, $\boldsymbol{I}_x \neq \phi$. Here, if we let

$$\delta = -\frac{1}{2} \max_{j \in \boldsymbol{I}_x} x_j = -\frac{1}{2} x_i$$

for any $\boldsymbol{y} \in \boldsymbol{B}(\boldsymbol{x}; \delta)$,

$$d(\boldsymbol{x}, \boldsymbol{y}) = \max_{j \in \boldsymbol{I}_x} |x_j - y_j| < -\frac{1}{2} x_i.$$

Thus,

$$|x_i - y_i| \leqq \max_{j \in N} |x_j - y_j| < -\frac{1}{2} x_i.$$

If $x_i \geqq y_i$, from $x_i < 0$, $y_i < 0$ is evident. Therefore if we let $x_i < y_i$,

$$|x_i - y_i| = y_i - x_i < -\frac{1}{2} x_i.$$

Therefore,

$$y_i < x_i - \frac{1}{2} x_i < 0$$

is obtained hence for any $\boldsymbol{y} \in \boldsymbol{B}(\boldsymbol{x}; \delta)$, $\boldsymbol{y} \ngeqq \boldsymbol{0}$.

(b-2) If $\boldsymbol{lx} \neq 1$, $\boldsymbol{lx} > 1$ or $\boldsymbol{lx} < 1$. Therefore, consider the case when $\boldsymbol{lx} > 1$. $\boldsymbol{lx}$ is a continuous function of $\boldsymbol{x}$ hence there exists a suitable $\rho > 0$ and $\boldsymbol{ly} > 1$ for any $\boldsymbol{y} \in \boldsymbol{B}(\boldsymbol{x}; \delta)$. It is the same when $\boldsymbol{lx} < 1$. QED

(Notes on [Lemma 4.1])

① If $f : \boldsymbol{D} \subseteq \boldsymbol{R}^n \to \boldsymbol{R}$ is continuous on a point $\boldsymbol{x}$ on $\boldsymbol{D}$, and $f(\boldsymbol{x}) > c$ for some constant c, there exists some $\delta > 0$ and for any $\boldsymbol{y} \in \boldsymbol{B}(\boldsymbol{x}; \delta)$, $f(\boldsymbol{y}) > c$ holds.

(Proof) From the definition of continuity, for $\varepsilon = \frac{1}{2}[f(\boldsymbol{x}) - c] > 0$, there exists some $\delta > 0$ and the following holds for any $\boldsymbol{y} \in \boldsymbol{B}(\boldsymbol{x}; \delta)$.

$$
\begin{aligned}
|f(\boldsymbol{y})-f(\boldsymbol{x})| &= |[f(\boldsymbol{y})-c]-[f(\boldsymbol{x})-c]| \\
&< \varepsilon \\
&= \frac{1}{2}[f(\boldsymbol{x})-c].
\end{aligned} \tag{4.3}
$$

If $f(\boldsymbol{y}) > c$, any further proof is unnecessary. Hence, if we let $f(\boldsymbol{y}) \leqq c$,

$$
\begin{aligned}
|[f(\boldsymbol{y})-c]-[f(\boldsymbol{x})-c]| &= f(\boldsymbol{x})-c-[f(\boldsymbol{y})-c] \\
&< \frac{1}{2}[f(\boldsymbol{x})-c].
\end{aligned} \tag{4.4}
$$

Therefore

$$
0 < \frac{1}{2}[f(\boldsymbol{x})-c] < f(\boldsymbol{y})-c.
$$

However, this goes against the assumption ($f(\boldsymbol{y}) \leqq c$). QED

② $\max_{j \in N} |x_j|$ was used as the vector norm but this lemma also holds for any vector norms. Moreover, the proof of note ① does not depend on the type of vector norm.

[Lemma 4.2] Let $f(x) = \sum_{j=0}^{n} a_j x^{n-j}$ be a n-order polynomial of real coefficient a_j $(j = 0, 1, \cdots, n)$ and express the maximum real root of $f(x)=0$ as λ^*. Here,

$$
\text{for any } x > \lambda^*,\ sgn f(x) = sgn\, a_0.
$$

(Proof) Let $x_j (j = 1, \cdots, n)$ be a root of $f(x)=0$. Of $x_j (j = 1, \cdots, n)$ if for example, we assume x_i to be a complex number, $f(\overline{x}_i) = 0$ in respect of its conjugate complex number $\overline{x}_i$ too. Thus, without losing generality, the following rearrangement can be made with the first two p of x_j as complex roots and the rest as real roots.

$$
\begin{aligned}
x_{2j-1} &= \overline{x_{2J}} \quad (j = 1, \cdots, p), \\
x_k &= \overline{x_k} \quad (k = 2p+1, \cdots, n).
\end{aligned} \tag{4.5}
$$

Using (4.5), (4.6) can be derived.

$$
f(x) = a_0 \left[\prod_{j=1}^{p} (x - x_{2j-1})(x - x_{2j}) \right] \left[\prod_{k=2p+1}^{n} (x - x_k) \right]. \tag{4.6}
$$

Additionally, if x is a real number, for any $j \in \{1, \cdots, p\}$,

$$\left(x-x_{2j}\right)=\left(\bar{x}-\bar{x}_{2j-1}\right)=\overline{x-x_{2j-1}}. \tag{4.7}$$

Accordingly for any $x \in \boldsymbol{R}$,

$$f(x)=a_0\prod_{j=1}^{p}\left|x-x_{2j-1}\right|^2\prod_{k=2p+1}^{n}(x-x_k). \tag{4.8}$$

Furthermore, for any $j \in \{1,\cdots,p\}$,

$$\left|x-x_{2j-1}\right|^2=\left[x-Re\left(x_{2j-1}\right)\right]^2+\left[Im\left(x_{2j-1}\right)\right]^2$$

and

$$Im\left(x_{2j-1}\right)\neq 0.$$

Therefore, $\prod_{j=1}^{p}\left|x-x_{2j-1}\right|^2>0$. Hence, if $x>\lambda^*$,

$$\prod_{j=1}^{p}\left|x-x_{2j-1}\right|^2\prod_{k=2p+1}^{n}(x-x_k)>0. \tag{4.9}$$

Thus, for all $x>\lambda^*$ from (4.8) to (4.9), the signs of $f(x)$ match the signs of the coefficients a_0 of the highest order number of x. QED

4.3 Frobenius' Theorem in Respect of Nonlinear Transformations

As preparation for deriving Frobenius' theorem in respect of nonlinear transformation f, the following theorem is demonstrated. Accordingly, the proofs of [Theorem 4.1]–[Theorem 4.3] are on the whole based on Uekawa and Kimura (1987).

[Theorem 4.1]

(i) If (NN) and (C) hold, the real number set $\boldsymbol{L}_f$ has a nonnegative maximum element ($\equiv\mu_f$).

(ii) If the nonlinear transformation $F:\boldsymbol{R}_+^n\rightarrow\boldsymbol{R}^n$ satisfies (NN) and (C) and $F(\boldsymbol{x})\geqq f(\boldsymbol{x})$ for any $\boldsymbol{x}\in\boldsymbol{S}_n$, it is $\left[\boldsymbol{L}_F'\text{s maximum element}\right]\geqq\left[\boldsymbol{L}_f'\text{s maximum element}\right]$.

(iii) Let a semipositive vector with nonzero elements be $\boldsymbol{x}$ and if f satisfies (WM), (I) and $f(\boldsymbol{0})=\boldsymbol{0}$, there exists $k\in\boldsymbol{N}$ such that $x_k=0$ and $f_k(\boldsymbol{x})>0$.

(Proof)

(i) From [Lemma 4.1], $\boldsymbol{S}_n$ is a compact set. In addition, from (C), each $f_i(\boldsymbol{x})(i = 1, \cdots, n)$ is continuous on $\boldsymbol{S}_n$. Therefore each $f_i(\boldsymbol{x})$ achieves its maximum value on $\bar{\boldsymbol{y}}^i \in \boldsymbol{S}_n$.[4] Additionally from (NN), for any $\boldsymbol{x} \in \boldsymbol{S}_n$, $f(\boldsymbol{x}) \geqq \boldsymbol{0} = 0 \cdot \boldsymbol{x}$, therefore $0 \in \boldsymbol{L}_f \neq \phi$.

Consequently, for any $\mu \in \boldsymbol{L}_f$, $\boldsymbol{x}^\mu \leqq \boldsymbol{S}_n$ satisfying $\mu\boldsymbol{x}^\mu \leqq f(\boldsymbol{x}^\mu)$ is defined and for all $i \in \boldsymbol{N}$,

$$\begin{aligned}\mu x_i^\mu &\leqq f_i(\boldsymbol{x}^\mu)\\ &\leqq f_i(\bar{\boldsymbol{y}}^i)\\ &\leqq \max_{i\in \boldsymbol{N}} f_i(\bar{\boldsymbol{y}}^i)\end{aligned}$$

holds. However, x_i^μ is the ith element of $\boldsymbol{x}^\mu$. Here, if we define $\max_{i\in\boldsymbol{N}} f_i(\bar{\boldsymbol{y}}^i) = K$ and express the equation above as a matrix, $\mu\boldsymbol{x}^\mu \leqq K\boldsymbol{l}^t$. In addition, by multiplying both sides of this by $\boldsymbol{l}$ on the left, $\mu\boldsymbol{l}\boldsymbol{x}^\mu \leqq \boldsymbol{l}\boldsymbol{l}^tK$ is obtained but if $\boldsymbol{l}\boldsymbol{x}^\mu = 1$ and $\boldsymbol{l}\boldsymbol{l}^t = n$ are considered, it becomes $\mu \leqq nK$ anyway. Thus, from [Theorem 2] (Appendix), there exists [$\boldsymbol{L}_f$'s upper bound] $=$ $sup\ \boldsymbol{L}_f = \mu_f$ and a sequence of numbers $\{\mu_\gamma\}$ that converges to μ_f can be created. Moreover, from the definition of $0 \in \boldsymbol{L}_f$ and an upper limit, $\mu_f \geqq 0$ is also evident. From the definition of $\boldsymbol{L}_f$, $\boldsymbol{x}^\gamma \in \boldsymbol{S}_n$ satisfying $\mu_\gamma\boldsymbol{x}^\gamma \leqq f(\boldsymbol{x}^\gamma)$ corresponding to each $\mu_\gamma \in \boldsymbol{L}_f$ can be defined but $\{\boldsymbol{x}^\gamma\}$ is a sequence of points of the compact set $\boldsymbol{S}_n$. Therefore $\{\boldsymbol{x}^\gamma\}$ has a sub-sequence of points $\{\boldsymbol{x}^{\gamma_i}\}$ that converges to $\boldsymbol{x}^f \in \boldsymbol{S}_n$. Then, the subsequence $\{\mu_{\gamma_i}\}$ corresponding to this $\{\boldsymbol{x}^{\gamma_i}\}$ converges to μ_f and $\mu_{\gamma_i}\boldsymbol{x}^{\gamma_i} \leqq f(\boldsymbol{x}^{\gamma_i})$ holds for any $i \in \boldsymbol{N}$. Therefore, from (C),

$$\begin{aligned}\mu_f\boldsymbol{x}^f &= \lim_{\gamma_i\to\infty} \mu_{\gamma_i}\boldsymbol{x}^{\gamma_i}\\ &\leqq \lim_{\gamma_i\to\infty} f(\boldsymbol{x}^{\gamma_i})\\ &= f(\boldsymbol{x}^f).\end{aligned}$$

Therefore, it is shown that $0 \leqq \mu_f \in \boldsymbol{L}_f$ is the maximum element of $\boldsymbol{L}_f$.

(ii) From (NN) and (C), (i) becomes applicable hence $\boldsymbol{L}_F = \{\mu \in \boldsymbol{R} | {}^\exists\boldsymbol{x} \in \boldsymbol{S}_n | F(\boldsymbol{x}) \geqq \mu\boldsymbol{x}\}$ also has a maximum element $=\mu_F$. Thence for μ_f and $\boldsymbol{x}^f$ in the proof of (i),

[4] $\boldsymbol{L}_f$ is only stipulated for $\boldsymbol{S}_n$ hence it cannot be used instead even if the domain of f is restricted to $\boldsymbol{S}_n$. Generally, when $\boldsymbol{X}$ is a compact set, the continuous functions $\boldsymbol{g} : \boldsymbol{X} \to \boldsymbol{R}$ has a smallest value or a largest value in the range $\boldsymbol{X}$ ([Theorem 10] (Appendix)).

$$\begin{aligned}\mu_f &\leqq f(\boldsymbol{x}^f) \\ &\leqq F(\boldsymbol{x}^f).\end{aligned}$$

Hence, $\mu_f \in \boldsymbol{L}_F$ is obtained. If we consider the fact that μ_F is the maximum element of $\boldsymbol{L}_F$, $\mu_f \leqq \mu_F$ is evident.

(iii) $\boldsymbol{x} \geq \boldsymbol{0}$ and $\boldsymbol{x} \ngtr \boldsymbol{0}$. Therefore if we define $\boldsymbol{J} = \{j \in N | x_i = 0\}$, $\boldsymbol{J} \neq \phi$ and $\boldsymbol{J} \subset \boldsymbol{N}$. Thus, for any $j \in \boldsymbol{J}$,

$$\begin{aligned}f_j(\boldsymbol{x}) &\geqq f_j(\boldsymbol{0}) \\ &= 0\end{aligned} \tag{4.10}$$

holds. However, from the definition of $\boldsymbol{J}$, $\boldsymbol{x}_J = \boldsymbol{0}$ and $\boldsymbol{x}_{J^C} > \boldsymbol{0}$. Therefore, from (I) and (10), there exists $k \in \boldsymbol{J} \subset N$ which satisfies $f_k(\boldsymbol{x}) > f_k(\boldsymbol{0}) = 0$. QED

Henceforth, the nonnegative maximum element $=\mu_f$ of the set of real numbers $\boldsymbol{L}_f$ and the vector $\boldsymbol{x}^f \in \boldsymbol{S}_n$ corresponding to μ_f are referred to as the nonlinear transformation f's **Frobenius root** and **Frobenius vector**, respectively. Additionally, [Theorem 4.2]–[Theorem 4.4] described below are referred to altogether as Frobenius' theorem in respect of nonlinear transformation f. Firstly, let us show the Frobenius theorem in respect of an indecomposable nonlinear transformation f from [Theorem 4.1].

[Theorem 4.2] When the nonlinear transformation f satisfies (NN), (C), (H), (WM) and (I), the following hold in respect of [Theorem 4.1]'s μ_f and $\boldsymbol{x}^f \in \boldsymbol{S}_n$.

(i) $\mu_f \boldsymbol{x}^f = f(\boldsymbol{x}^f) > \boldsymbol{0}$.
(ii) If $\mu \neq \mu_f$, there is no $\boldsymbol{x} \in \boldsymbol{S}_n$ that satisfies $\mu\boldsymbol{x} = f(\boldsymbol{x})$.[5]
(iii)
(iii-i) If it is the same order number $m \in [0, 1)$ of homogeneity, $\boldsymbol{x}^f$ has one meaning.
(iii-ii) If $m = 1$, $\boldsymbol{x}^f$ has one meaning if the scalar multiplier is excluded.
(iv) If $m = 1$ and $\mu \neq \mu_f$, there does not exist $\boldsymbol{x} \geq \boldsymbol{0}$ which satisfies $\mu\boldsymbol{x} = f(\boldsymbol{x})$.

(Proof)

(i) To show $\mu_f \boldsymbol{x}^f = f(\boldsymbol{x}^f)$, assume $\mu_f \boldsymbol{x}^f \neq f(\boldsymbol{x}^f)$. However, $\mu_f \in \boldsymbol{L}_f$, thus, considering $\mu_f\ \boldsymbol{x}^f \leqq f(\boldsymbol{x}^f)$, the assumption becomes $\mu_f\ \boldsymbol{x}^f \leq f(\boldsymbol{x}^f)$. Hence, if we set $\boldsymbol{J} = \left\{j \in \boldsymbol{N} \middle| \mu_f x_j^f \leqq f_j(\boldsymbol{x}^f)\right\}$, $\boldsymbol{J} \neq \phi$.

[5] In other words, when $\mu \neq \mu_f$, $\mu\boldsymbol{x} \leq f(\boldsymbol{x})$ in respect of any $\boldsymbol{x} \in \boldsymbol{S}_n$. To be more precise, $\mu\boldsymbol{x} \leq f(\boldsymbol{x})$ in respect of any $\boldsymbol{x} \in \boldsymbol{S}_n$ can be derived from $\mu\boldsymbol{x} \leqq f(\boldsymbol{x})$ and $\mu\boldsymbol{x} \neq f(\boldsymbol{x})$.

If we let $\boldsymbol{J} = \boldsymbol{N}$, $\mu_f \boldsymbol{x}^f < f(\boldsymbol{x}^f)$. Therefore, $\tilde{\mu}_f > \mu_f$ can be chosen to satisfy $\tilde{\mu}_f \boldsymbol{x}^f \leqq f(\boldsymbol{x}^f)$, which contradicts [Theorem 4.1; (i)] hence $\boldsymbol{J} \subset \boldsymbol{N}$. Additionally, for any $j \in \boldsymbol{J}$, if we define a function $g_j(\boldsymbol{x})$ as

$$g_j(\boldsymbol{x}) = f_j(\boldsymbol{x}) - \mu_f x_j$$

from the definition of $\boldsymbol{J}$, $g_j(\boldsymbol{x}^f) > 0$. In addition, from (C), each g_j is continuous. Therefore, if $\boldsymbol{y} = (y_1, \cdots, y_n)^t$ that would be $y_j > x_j^f$, $y_j = x_j^f$ corresponding to $j \in \boldsymbol{J}$ or $j \notin \boldsymbol{J}$ is chosen to be sufficiently close to $\boldsymbol{x}^f$, for any $j \in \boldsymbol{J}$, $\mu_f y_j < f_j(\boldsymbol{y})$. Moreover, for any $j \notin \boldsymbol{J}$,

$$\begin{aligned} \mu_f y_j &= \mu_f x_j^f \\ &= f_j(\boldsymbol{x}^f) \\ &\leqq f_j(\boldsymbol{y}) \end{aligned}$$

holds and particularly, for some $k \notin \boldsymbol{J}$,

$$f_k(\boldsymbol{x}^f) \neq f_k(\boldsymbol{y})$$

Hence for this $k \notin \boldsymbol{J}$, $f_k(\boldsymbol{x}^f) < f_k(\boldsymbol{y})$ holds. Therefore, if the number of elements of $\boldsymbol{J} - \left\{ j \in \boldsymbol{N} \middle| \mu_f x_j^f < f_j(\boldsymbol{x}^f) \right\}$ and $\tilde{\boldsymbol{J}} - \left\{ j \in \boldsymbol{N} \middle| \mu_f y_j < f_j(\boldsymbol{y}) \right\}$ are compared, $\#\boldsymbol{J} < \#\tilde{\boldsymbol{J}}$.

Furthermore, if we consider $1 = \boldsymbol{l}\boldsymbol{x}^f < \boldsymbol{l}\boldsymbol{y}$ and let $\boldsymbol{w} = (\boldsymbol{l}\boldsymbol{y})^{-1}\boldsymbol{y}$,

$$\boldsymbol{w} \in \boldsymbol{S}_n$$

and

$$\begin{aligned} f(\boldsymbol{w}) - \mu_f \boldsymbol{w} &= f\left[(\boldsymbol{l}\boldsymbol{y})^{-1}\boldsymbol{y}\right] - \mu_f (\boldsymbol{l}\boldsymbol{y})^{-1}\boldsymbol{y} \\ &= (\boldsymbol{l}\boldsymbol{y})^{-m} f(\boldsymbol{y}) - (\boldsymbol{l}\boldsymbol{y})^{-1}\mu_f \boldsymbol{y} \\ &= (\boldsymbol{l}\boldsymbol{y})^{-1}\left[(\boldsymbol{l}\boldsymbol{y})^{1-m} f(\boldsymbol{y}) - \mu_f \boldsymbol{y}\right] \\ &\geqq (\boldsymbol{l}\boldsymbol{y})^{-1}\left[f(\boldsymbol{y}) - \mu_f \boldsymbol{y}\right] \\ &\geq \boldsymbol{0}. \end{aligned}$$

Therefore, if $\tilde{\tilde{\boldsymbol{J}}} = \left\{ j \in \boldsymbol{N} \middle| \mu_f w_j < f_j(\boldsymbol{w}) \right\}$, $\#\boldsymbol{J} < \#\tilde{\boldsymbol{J}} \leqq \#\tilde{\tilde{\boldsymbol{J}}}$. However, w_j is the jth element of $\boldsymbol{w}$. If $\mu_f \boldsymbol{w} < f(\boldsymbol{w})$ does not hold, let $\boldsymbol{w}$ be viewed as $\boldsymbol{x}^f$ and if the same process as above is repeated, $\#\tilde{\tilde{\boldsymbol{J}}}$ increases every time. Thus, we cannot take it as $\mu_f \boldsymbol{w} < f(\boldsymbol{w})$

without losing generality, however, this contradicts [Theorem 4.1; (i)]. Therefore, $\mu_f \boldsymbol{x}^f = f(\boldsymbol{x}^f)$ has been shown.

Next, let us suppose there exist $\boldsymbol{x} \geq \boldsymbol{0}$ and μ that satisfy $\mu_f \boldsymbol{x} = f(\boldsymbol{x})$. Then, from (NN), $\mu \boldsymbol{x} = f(\boldsymbol{x}) \geqq \boldsymbol{0}$ holds and $\boldsymbol{x} \geq \boldsymbol{0}$ hence, $\mu \geqq 0$. In addition, to show $\boldsymbol{x} > \boldsymbol{0}$, if we assume that $\boldsymbol{x} \ngtr \boldsymbol{0}$, from [Theorem 4.1; (iii)], there exist $i \in \boldsymbol{N}$, which become $x_i = 0$ and $f_i(\boldsymbol{x}) > 0$. Therefore, the next self-contradiction

$$0 = \mu \boldsymbol{x}_i = f_i(\boldsymbol{x}) > 0$$

is derived. Therefore, $\boldsymbol{x} > \boldsymbol{0}$. If $\boldsymbol{x}^f \in \boldsymbol{S}_n$ and $\mu \boldsymbol{x}^f = f(\boldsymbol{x}^f)$ are considered, of course, $\boldsymbol{x}^f > \boldsymbol{0}$. Additionally, in respect of the $j (j = 1, \cdots, n)$ unit vector $\boldsymbol{e}^j \in \boldsymbol{R}^n$,

$$f\left(x_j^f \boldsymbol{e}^j\right) \geqq 0 \tag{4.11}$$

is shown. Hence, for any $j \in \boldsymbol{N}$,

$$\begin{aligned} f_j(\boldsymbol{x}^f) &> f_j\left[x_j^f \boldsymbol{e}^j\right] \\ &\geqq 0. \end{aligned} \tag{4.12}$$

By (4.12),

$$\mu_f \boldsymbol{x}^f = f(\boldsymbol{x}^f) > \boldsymbol{0}.$$

Combining this result and $\boldsymbol{x}^f > \boldsymbol{0}$, $\mu_f > 0$ is also evident.

(ii) If the conclusion is denied and assume that $\mu \boldsymbol{x} = f(\boldsymbol{x})$ for $\mu \neq \mu_f$ and some $\boldsymbol{x} \in \boldsymbol{S}_n$, as shown in the proof of (i), $\boldsymbol{x} > \boldsymbol{0}$. Here, let us define $a = \min_{j \in N} \left(\frac{x_i}{x_i^f} \right) = \frac{x_k}{x_k^f}$ and $z = \boldsymbol{x} - a\boldsymbol{x}^f$. From $\boldsymbol{x} > \boldsymbol{0}$ and $\boldsymbol{x}^f > \boldsymbol{0}$, $a > 0$. Moreover, from $\boldsymbol{x}^f \in \boldsymbol{S}_n$, $1 = \boldsymbol{l}\boldsymbol{x} \geqq a(\boldsymbol{l}\boldsymbol{x}^f) = a$ is also clear. Accordingly, $a \in (0, 1]$ and $\boldsymbol{x} \geqq a\boldsymbol{x}^f$, thus, $z \geqq \boldsymbol{0}$. If for all $j \in \boldsymbol{N}$, $z_j > 0$, $x_j > a x_j^f$ holds and by substituting in $a = \frac{x_k}{x_k^f}$, $x_j x_k^f = x_k x_j^f$ is obtained. Here, $j \in \boldsymbol{N}$ is arbitrary hence by choosing $j{=}k$, $x_k x_k^f = x_k x_k^f$. This contradicts the aforementioned inequality. Hence, there exists $i \in \boldsymbol{N}$ which satisfies $z_i = 0$. Therefore, for this $i \in \boldsymbol{N}$,

$$\begin{aligned}0 &= \mu_f z_i \\ &= \mu_f\left(x_i - a x_i^f\right) \\ &= \mu_f x_i - a\mu_f x_i^f \\ &> \mu x_i - a\mu_f x_i^f \\ &= f_i(\boldsymbol{x}) - a^{1-m} a^m \mu_f x_i^f \\ &= f_i(\boldsymbol{x}) - a^{1-m} f_i\left(a\boldsymbol{x}^f\right) \\ &\geqq f_i(\boldsymbol{x}) - f_i\left(a\boldsymbol{x}^f\right) \\ &\geqq 0.\end{aligned}$$

This is self-contradictory.

(iii)

(iii-i) If it is assumed that there exists a $\boldsymbol{x} \geq \boldsymbol{0}$ which satisfies $m \in [0,\ 1)$ and $\mu_f \boldsymbol{x} = f(\boldsymbol{x})$ which is different from $\boldsymbol{x}^f$, as described in the proof of (i), $\boldsymbol{x} > \boldsymbol{0}$. Therefore, using the a and z defined in the proof of (ii), we obtain the below.

$$\begin{aligned}\mu_f z &= \mu_f\left(\boldsymbol{x} - a\boldsymbol{x}^f\right) \\ &= f(\boldsymbol{x}) - a\left(\mu_f \boldsymbol{x}^f\right) \\ &= f(\boldsymbol{x}) - a^{1-m} f\left(a\boldsymbol{x}^f\right) \\ &= f\left(z + a\boldsymbol{x}^f\right) - a^{1-m} f\left(a\boldsymbol{x}^f\right).\end{aligned} \tag{4.13}$$

Here, if $\boldsymbol{x}$ is proportional to $\boldsymbol{x}^f$, in other words, $\boldsymbol{x} = b\boldsymbol{x}^f$, $0 < a = \frac{x_k}{x_k^f} = b$. Additionally, from $\boldsymbol{x} \neq \boldsymbol{x}^f$, $a \neq 1$. Thus, $\boldsymbol{x} = a\boldsymbol{x}^f$, hence $z = \boldsymbol{x} - a\boldsymbol{x}^f = \boldsymbol{0}$. Considering (4.13),

$$\begin{aligned}\boldsymbol{0} &= \mu_f z \\ &= f\left(a\boldsymbol{x}^f\right) - a^{1-m} f\left(a\boldsymbol{x}^f\right) \\ &= \left(1 - a^{1-m}\right) f\left(a\boldsymbol{x}^f\right) \\ &\neq \boldsymbol{0}\end{aligned}$$

are derived but these are self-contradicting. Accordingly, when $m \in [0,\ 1)$, $\boldsymbol{x}$ is not proportional to $\boldsymbol{x}^f$. If $m = 1$, we just need to show that $\boldsymbol{x}^f$ has one meaning, when $\boldsymbol{x}$ proportional to $\boldsymbol{x}^f$ are removed. Thus, henceforth, we will consider $\boldsymbol{x}$ that is not proportional to $\boldsymbol{x}^f$ and let it be $m \in [0,\ 1]$ to prove the conclusion remaining in (iii-i) and (iii-ii).

If $\max_{i \in N} \frac{x_i}{x_i^f} = a$, $\boldsymbol{x}$ is proportional to $\boldsymbol{x}^f$ hence, $\max_{i \in N} \frac{x_i}{x_i^f} > a$. Thus, if we let $\boldsymbol{Q} = \left\{ i \in \boldsymbol{N} \middle| z_i = x_i - a x_i^f = 0 \right\}$, $\boldsymbol{Q} \neq \phi$ and $\boldsymbol{Q} \subset \boldsymbol{N}$. In addition, to show $a > 1$, if we let $a \leqq 1$, for any $i \in \boldsymbol{Q}$,

$$\begin{aligned}0 &= \mu_f z_i \\ &= f_i(z + ax^f) - a^{1-m} f_i(ax^f) \\ &\geqq f_i(z + ax^f) - f_i(ax^f) \\ &\geqq 0.\end{aligned} \tag{4.14}$$

Therefore, for any $i \in \boldsymbol{Q}$, $f_i(\boldsymbol{z} + a\boldsymbol{x}^f) = f_i(a\boldsymbol{x}^f)$. However, for any $i \notin \boldsymbol{Q}$, from the definition of $\boldsymbol{Q}$ and $\boldsymbol{z} \geqq \boldsymbol{0}$, $z_i > 0$. Thus, $z_i + ax_i^f > ax_i^f$. At this time, (I) means $f_j(\boldsymbol{z} + a\boldsymbol{x}^f) \neq f_j(a\boldsymbol{x}^f)$ for some $j \in \boldsymbol{Q}$ and contradicts (4.14). Therefore $a > 1$.

Additionally, if $\tilde{a} = \max_{i \in N} \left(\frac{x_i}{x_i^f} \right)$, $\tilde{\boldsymbol{z}} = \tilde{a}\boldsymbol{x}^f - \boldsymbol{x}$ and $\tilde{\boldsymbol{Q}} = \{i \in \boldsymbol{N} | \tilde{z}_i = 0\}$ are defined, $\tilde{\boldsymbol{Q}} \neq \phi$ and $\tilde{\boldsymbol{Q}} \subset \boldsymbol{N}$. Here, to show $\tilde{a} \leqq 1$, if $\tilde{a} > 1$ is set, the argument becomes the same as in the proof from $a \leqq 1$, leading to a contradiction, thus, $\tilde{a} \leqq 1$.
Summarizing the above, $1 \geqq \tilde{a} > a > 1$ is obtained, however, this is again a self-contradiction. Therefore, (iii-i) and (iii-ii) are shown.
(iv) When $m = 1$ and $\mu \neq \mu_f$, for some $\boldsymbol{x} \geq \boldsymbol{0}$, if $\mu\boldsymbol{x} = f(\boldsymbol{x})$ holds,

$$\boldsymbol{y} = (\boldsymbol{lx})^{-1}\boldsymbol{x} \in \boldsymbol{S}_n \tag{4.15}$$

and

$$\begin{aligned}\mu(\boldsymbol{lx})\boldsymbol{y} &= \mu\boldsymbol{x} \\ &= f(\boldsymbol{x}) \\ &= f[(\boldsymbol{lx})\boldsymbol{y}] \\ &= \boldsymbol{lx} f(\boldsymbol{y}).\end{aligned} \tag{4.16}$$

Thus,

$$\mu\boldsymbol{y} = f(\boldsymbol{y}). \tag{4.17}$$

It is evident that (4.17) goes against (ii). QED

[Theorem 4.3] If a nonlinear transformation f satisfies (NN), (C), (H), and (WM), a $\boldsymbol{y}^f \in \boldsymbol{S}_n$, which becomes $=\mu_f$ corresponding to the maximum element $=\mu_f$ of $\boldsymbol{L}_f$ can be defined.

(Proof) If we define a function $G^\gamma(\boldsymbol{x})$ as

$$G^{\gamma}(\boldsymbol{x})=f(\boldsymbol{x})+\left(\frac{1}{\gamma}\right)\boldsymbol{U}\boldsymbol{x}^{(m)},\quad (\gamma=1,2,\ \cdots) \tag{4.18}$$

Each $G^{\gamma}(\boldsymbol{x})(\gamma = 1, 2, \cdots)$ satisfies all of the conditions of [Theorem 4.2]. In addition, for any $\boldsymbol{x} \in \boldsymbol{S}_n$,

$$G^{\gamma}(\boldsymbol{x})\geqq G^{\gamma+1}(\boldsymbol{x})\geqq f(\boldsymbol{x}),\quad (\gamma=1,2,\cdots) \tag{4.19}$$

holds. Here, if we let $max\,\boldsymbol{L}_{G^{\gamma}}=\overline{\mu}_{\gamma}$, from [Theorem 4.2; (i)], there exists some $\overline{\boldsymbol{y}}^{\gamma} \in \boldsymbol{S}_n$ and

$$0<\overline{\mu}_{\gamma}\overline{\boldsymbol{y}}^{\gamma}=G^{\gamma}(\overline{\boldsymbol{y}}^{\gamma}). \tag{4.20}$$

Moreover, for all γ, by [Theorem 4.1; (ii)] and (4.19),

$$\overline{\mu}_{\gamma}\geqq\overline{\mu}_{\gamma+1}\geqq\mu_f.$$

Thus, $\{\overline{\mu}_{\gamma}\}$ is a monotonically nonincreasing sequence of numbers with a lower bound. Consequently, if the set with each term of $\{\overline{\mu}_{\gamma}\}$ as elements are expressed as $\boldsymbol{S}\{\overline{\mu}_{\gamma}\}$ and $inf\ \boldsymbol{S}\{\overline{\mu}_{\gamma}\}$ as $\overline{\mu}$,[6]

$$\lim_{\gamma\to\infty}\mu_{\gamma}=inf\,\boldsymbol{S}\{\mu_{\gamma}\}=\mu. \tag{4.21}$$

In Addition, the sequence of points $\{\overline{\boldsymbol{y}}^{\gamma}\}$ of the compact set $\boldsymbol{S}_n$ has a subsequence of points $\{\overline{\boldsymbol{y}}^{\gamma_i}\}$ that converges to $\boldsymbol{y}^f \in \boldsymbol{S}_n$ and

$$\lim_{\gamma_i\to\infty}\overline{\mu}_{\gamma_i}=\overline{\mu} \tag{4.22}$$

is shown. Here, if we note that each G^{γ} and f satisfy (C) as well as on (4.18),

$$\begin{aligned}\overline{\mu}\boldsymbol{y}^f &= \lim_{\gamma_i\to\infty}\overline{\mu}_{\gamma_i}\overline{\boldsymbol{y}}^{\gamma_i}\\ &=\lim_{\gamma_i\to\infty}\left\{f(\overline{\boldsymbol{y}}^{\gamma_i})+\left(\frac{1}{\gamma_i}\right)\boldsymbol{U}\overline{\boldsymbol{y}}^{\gamma_i(m)}\right\}\\ &=f(\boldsymbol{y}^f).\end{aligned} \tag{4.23}$$

[6]If there exists some $\delta > 0$ and in respect of any $\gamma \in \{1, 2, \cdots\}$, $\overline{\mu}+\delta\leqq\overline{\mu}_{\gamma}$ holds, $\overline{\mu}<\overline{\mu}+\delta\leqq\overline{\mu}_{\gamma}$. Therefore, if we take $inf\,\boldsymbol{S}(\overline{\mu}_{\gamma})=\overline{\mu}$, there exists $\overline{\mu}_{\gamma(\varepsilon)}\in\boldsymbol{S}\{\overline{\mu}_{\gamma}\}$ which satisfies $\overline{\mu}_{\gamma(\varepsilon)}<\overline{\mu}+\varepsilon$ in respect of any $\varepsilon > 0$. Therefore, when $\gamma \geqq \gamma(\varepsilon)$, $|\overline{\mu}-\overline{\mu}_{\gamma}|=\overline{\mu}_{\gamma}-\overline{\mu}\leqq\overline{\mu}_{\gamma(\varepsilon)}-\overline{\mu}<\varepsilon$. Therefore $\lim_{\gamma\to\infty}\overline{\mu}_{\gamma}=\overline{\mu}$.

(4.23) means $\overline{\mu} \in \boldsymbol{L}_f$, thus, $\overline{\mu} \leqq \mu_f$. However, μ_f is a lower bound of $\boldsymbol{S}\{\overline{\mu}_\gamma\}$, hence, $\mu_f \leqq inf\ \boldsymbol{S}\{\overline{\mu}_\gamma\} = \overline{\mu}$. Therefore, $\mu_f = \overline{\mu}$ is obtained. QED

By the various theorems above, the following fact can be obtained.

[Theorem 4.4] When a nonlinear transformation f satisfies (NN), (C), (H), and (WM), the following holds.

(i) $\mu_f = \lambda_f \geqq 0$
(ii)

$$\begin{bmatrix} \text{There exists some positive integer } p \text{ and for all } \boldsymbol{x} \in \boldsymbol{S}_n, \\ f^p(\boldsymbol{x}) \equiv \underbrace{f(\cdots f(f(\boldsymbol{x}))}_{\leftarrow\ p\ \rightarrow} = 0 \end{bmatrix} \Rightarrow \lambda_f = 0.$$

(iii) If (WM) is strengthened to be (M), [Theorem 4.4; (ii)] becomes the necessary and sufficient condition for $\lambda_f = 0$.

(Proof)

(i) When (NN) and (C) hold, from [Theorem 4.1; (i)], $\boldsymbol{L}_f$ has a maximum element $\mu_f \geqq 0$ and if (H) and (WM) is further assumed, from [Theorem 4.3], in response to this μ_f, there exists $\boldsymbol{y} \in \boldsymbol{S}_n$ that satisfies $f(\boldsymbol{y}) = \mu_f \boldsymbol{y}$. Thus, $\mu_f \in \boldsymbol{\Lambda}_f$. Additionally, from the definitions of $\boldsymbol{L}_f$ and $\boldsymbol{\Lambda}_f$, $\boldsymbol{\Lambda}_f \subseteq \boldsymbol{L}_f$ is evident. Therefore, $\mu_f = \lambda_f$.
(ii) From (i), $\boldsymbol{y} \in \boldsymbol{S}_n$ can be defined that is $f(\boldsymbol{y}) = \lambda_f \boldsymbol{y}$ for λ_f, thus,

$$f^p(\boldsymbol{y}) = \lambda_f^s \boldsymbol{y} \quad \left(s = \sum_{t=0}^{p-1} m^t\right)$$

holds. From the above equation and [Theorem 4.4] (ii),

$$\boldsymbol{0} = f^p(\boldsymbol{y}) = \lambda_f^s \boldsymbol{y}$$

is derived but from $\boldsymbol{y} \geq \boldsymbol{0}$, there exists $k \in \boldsymbol{N}$, which is $y_k > 0$ and

$$0 = f_k^p(\boldsymbol{y}) = \lambda_f^s y_k,$$

however, $f_k^p(\boldsymbol{y})$ is the kth element of $f^p(\boldsymbol{y})$. Therefore, $\lambda_f = 0$ is obtained.
(iii)
(Sufficiency) (M) implies (WM) hence, sufficiency is self-evident.
(Neccessity) Firstly, select any $\boldsymbol{y} > \boldsymbol{0}$ from $\boldsymbol{S}_n$ and a $\theta > 0$, which satisfies

$$f(y) \leqq \theta y \tag{4.24}$$

and define $f^{\theta}(y) = y > \mathbf{0}$. If the definitions of (NN), (H), (M), and f^{θ} are applied to (4.24),

$$\mathbf{0} \leqq f^{s+1}(y) \leqq \theta^{m^S} f^{s}(y) \quad (s = 0, 1, 2, \cdots). \tag{4.25}$$

In Addition, for simplification, the $i \in \boldsymbol{N}$th element of $f^s(y)$ is expressed as $f_i^s(y)$ and if we define $\boldsymbol{N}_s = \{i \in \boldsymbol{N} | f_i^s(y) > 0\} \quad (s = 0, 1, 2, \cdots)$,

$$\boldsymbol{N}_s \supseteq \boldsymbol{N}_{s+1} \quad (s = 0, 1, 2, \cdots). \tag{4.26}$$

From definitions, $\boldsymbol{N} = \boldsymbol{N}_0 \supseteq \boldsymbol{N}_1$ is obvious. Here, to show that (4.26) holds for all $s \geqq 1$, if the conclusion is defined, there exists $k \in \boldsymbol{N}$, which becomes $k \in \boldsymbol{N}_{\bar{s}+1}$ and $k \notin \boldsymbol{N}_{\bar{s}}$ for some integer $\bar{s} \geqq 1$. Accordingly, from the definition of N_s, $f_k^{\bar{s}}(y) = 0$ holds and from the fact that this, (4.25) and k are elements of $\boldsymbol{N}_{\bar{s}+1}$, the following self-contradiction can be obtained.

$$0 < f_k^{\bar{s}+1}(y) \leqq \theta^{m^{\bar{s}}} f_k^{\bar{s}}(y) = 0$$

Next, to derive $\boldsymbol{N}_n = \phi$, let us assume $\boldsymbol{N}_n \neq \phi$. (4.26) means

$$\boldsymbol{N} = \boldsymbol{N}_0 \supseteq \boldsymbol{N}_1 \supseteq \boldsymbol{N}_2 \supseteq \cdots N_{n-1} \supseteq \boldsymbol{N}_n \tag{4.27}$$

and also $\boldsymbol{N}_n \neq \phi$, hence, as long as $s \leqq n - 1$, $\boldsymbol{N}_s \neq \phi$. Thus, if we define

$$\eta = \max_{i \in N_s} \frac{f_i^{s+1}(\mathbf{y})}{f_i^s(\mathbf{y})} = \frac{f_j^{s+1}(\mathbf{y})}{f_j^s(\mathbf{y})} \qquad (s = 0, 1, 2, \cdots, n-1)$$

from (H),

$$\begin{aligned} \mu y &\leqq f(y), \\ y &= \frac{f^s(y)}{\sum_{i=1}^n f_i^s(y)} \in \boldsymbol{S}_n, \\ \mu &= \eta \left(\sum_{i=1}^n f_i^s(y)\right)^{1-m} \geqq 0. \end{aligned}$$

$\mu \in \boldsymbol{L}_f$. From this, theorem (i) and condition $\lambda_f = 0$,

$$0 = \lambda_f = \mu_f \geqq \mu \geqq 0.$$

Thus, $\mu = 0$ and $\eta = 0 = \dfrac{f_j^{s+1}(y)}{f_j^s(y)}$ $(s = 0, 1, 2, \cdots, n-1)$. Consequently, $j \in \boldsymbol{N}_s$ and $j \notin \boldsymbol{N}_{s+1} (s = 0, 1, 2, \cdots, n - 1)$. Therefore,

$$N = N_0 \supset N_1 \supset N_2 \supset \cdots \supset N_{n-1} \supset N_n$$

or

$$(\#N_s - 1) \geqq \#N_{s+1} \quad (s = 0, 1, 2, \cdots, n-1).$$

Yet $\#N_0 = \# N = n$, Thus, $\#N_n = 0$, in other words, $N_n = \phi$ and contradicts the assumption of proof by contradiction. Therefore, $N_n = \phi$ immediately derives $f^n(y) = \mathbf{0}$. Accordingly, for any $x \geq \mathbf{0}$, if we define $\varpi = \max_{j \in N} \frac{x_i}{y_i}$, $\varpi > 0$ satisfies $x \leqq \varpi y$. Thus,

$$\begin{aligned} \mathbf{0} &\leqq f^n(x) \\ &\leqq \varpi^{m^n} f^n(y) \\ &= \mathbf{0}. \end{aligned}$$

Therefore, for any $x \geq \mathbf{0}$, $f^n(x) = \mathbf{0}$ is obtained. Additionally, from (4.25), for any positive integer $q \geqq n$ too, $f^q(x) = \mathbf{0}$ can be derived. In addition, if (4.27) and $N_n = \phi$ are considered,

$$n = \#N_0 \geqq \#N_1 \geqq \#N_2 \geqq \cdots \geqq \#N_{n-1} \geqq \#N_n = 0.$$

Thus, depending on the reduction of $\#N_s (s = 0, 1, 2, \cdots, n-1)$, for some integer $p \leqq n-1$, $\#N_p = 0$. Consequently, for any $x \geq \mathbf{0}$, $f^p(x) = \mathbf{0}$ may be the case and in this case too, it is obvious that $f^q(x) = \mathbf{0}$ for any integer $q \geqq p$. QED

(Notes on [Theorem 4.4])

① [Theorem 4.4] (ii) means [nonlinear transformation $f : R^n_+ \to R^n$'s Frobenius root] = [Maximum element of the set of images $f(x)$ of a point x belonging to n-order element by f, which are more than x multiplied by some real number]. If f is a linear transformation, [Theorem 4.4] (ii) is well known, however, is not written or proven for the nonlinear case, thus, we have tried to clarify that it holds for the nonlinear transformation too.

② The necessary and sufficient condition [Theorem 4.4] (ii) for it to be a Frobenius root $\lambda_f = 0$ can be strengthened to the following [Theorem 4.4] (ii)' as evident from the proof.

$$\text{[Theorem 4.4] (ii)'} \begin{bmatrix} \text{There exists some positive integer } p \leqq n \text{ and} \\ \text{for any } x \geq \mathbf{0},\ f^p(x) = \mathbf{0} \\ \text{and} \\ \text{for any positive integer } q \geqq p,\ f^q(x) = \mathbf{0}. \end{bmatrix}$$

③ Nikaido (1968)'s Theorem 10.3 (iii) (pp. 152–155) demonstrates the necessary and sufficient condition for (4.1) to be $\mu_f = 0$ under the four conditions (NN), (C), (H), and (M) and next $\mu_f = \lambda_f$ is derived. However, at some stage in the proof of

$\mu_f = 0$ (homogeneity order $=1$), (SM) is used in place of the assumed (M). However, in this theorem, we have shown that Theorem 10.3 (iii) can be proven fully under (NN), (C), (H), and (M) with no dependency on (SM). Thus, it would be fine to call [Theorem 4.4; (iii)], particularly as the strengthened version of Nikaido's Theorem 10.3.

We had been using nonlinear transformations $f : \boldsymbol{R}_+^n \to \boldsymbol{R}^n$ so far but here we add an addendum as to why the domain of f was chosen to be $\boldsymbol{R}_+^n = \{\boldsymbol{x} \geqq \boldsymbol{0} | \boldsymbol{x} \in \boldsymbol{R}^n\}$, not $\boldsymbol{R}^n$.

In [Theorem 4.3], it is assumed that f satisfies (H) but at that time the order of homogeneity is $m \in [0, 1]$. Accordingly, if the domain is $\boldsymbol{R}^n$, when $m \in (0, 1)$, there is a possibility that the $\boldsymbol{x}^{(m)} = \left(x_1^m, \cdots, x_n^m\right)^t$ that appears within the proof may not be within the domain.[7] This is the main reason that the domain is taken to be $\boldsymbol{R}_+^n$. It does not need saying that if it is limited to $m = 1$ (including linear transformations too), even if the domain is expanded to $\boldsymbol{R}^n$, the theorem is not impacted in any way.

4.4 Frobenius' Theorem in Respect of Linear Transformations

Up to the last section, f had been treated as a nonlinear transformation. However, if f is a linear transformation, using a n-order square matrix $\boldsymbol{A}$ with the image $f(\boldsymbol{e}^j) \equiv \boldsymbol{a}^j$ of the $j(j \in \boldsymbol{N})$ unit vector $\boldsymbol{e}^j$ by f as the jth column, the image of any $\boldsymbol{x} \in \boldsymbol{R}_+^n$ by $f=f(\boldsymbol{x})$ can be expressed as $\boldsymbol{Ax}$. It is not hard to imagine that the various theories relating to the nonlinear transformations f so far are closely linked to eigenvalue problems of matrix $\boldsymbol{A}$.

However, if n-order square matrix $\boldsymbol{A}$ is taken to be a nonnegative matrix, four conditions—(NN), (C), (H), (WM)—are satisfied hence here below, $\boldsymbol{A}$ shall be described as **a gross substitution matrix**.[8]

[Theorem 4.5] When $\boldsymbol{A} = [a_{ij}]$ is a n-order gross substitution matrix and μ is any eigenvalue of $\boldsymbol{A}$, the following holds.

(i) For some real eigenvalue $\mu(\boldsymbol{A})$ of $\boldsymbol{A}$, a semipositive eigenvector corresponds and

$$\mu(\boldsymbol{A}) \geqq Re(\mu) \text{ and } \mu(\boldsymbol{A}) + \max_{j \in \boldsymbol{N}} \{|a_{ii}| - a_{ii}\} \geqq |\mu|.$$

However, $Re(\mu)$ shows the real part of μ.

(ii) For any subset $\boldsymbol{J}$ of $\boldsymbol{N}$, $\mu(\boldsymbol{A}) \geqq \mu(\boldsymbol{A}_{JJ})$

[7] For example, if $x_i < 0$ $(i \in \boldsymbol{N})$, x_i^m may be a complex number.

[8] A square matrix with each non-diagonal element $\geqq 0$. It is also called a **Metzler matrix**. (C), (H) and (WM) are satisfied but (NN) is not necessarily satisfied. As for Metzler matrix, see Metzler (1945), Kemp and Kimura (1978), Farina and Rinaldi (2000) and Narendra and Shorten (2009).

(iii) For any $\mu \geqq \mu(\boldsymbol{A}), f_{ij}(\mu) \geqq 0 \ (i, j = 1, \cdots, n)$
and
for any $\mu > \mu(\boldsymbol{A}), f_{ii}(\mu) > 0 \ (i = 1, \cdots, n)$,
where $f(\mu) = det \ (\mu \boldsymbol{I}_n - \boldsymbol{A})$, $f_{ij}(\mu)$ is a cofactor of $(\mu \boldsymbol{I}_n - \boldsymbol{A})_{ij}$, $(\mu \boldsymbol{I}_n - \boldsymbol{A})_{ij}$ is a (i, j) element of $(\mu \boldsymbol{I}_n - \boldsymbol{A})$.
(iv) $(\rho \boldsymbol{I}_n - \boldsymbol{A}^{-1}) \geq [\boldsymbol{0}]$ and $\rho > \mu(\boldsymbol{A})$ are mutually equivalent.
(v) $\mu(\boldsymbol{A})$ is the maximum element of a set of real numbers $= \tilde{\boldsymbol{L}}_A$ such that $\mu \boldsymbol{x} \leqq \boldsymbol{A}\boldsymbol{x}$ has a solution $\boldsymbol{x} \geq \boldsymbol{0}$, while, $\tilde{\boldsymbol{L}}_A = \left\{\mu \in \boldsymbol{R} \middle| {}^{\exists}\boldsymbol{x} \geq \boldsymbol{0} | \boldsymbol{A}\boldsymbol{x} \geqq \mu \boldsymbol{x}\right\}$.
(vi) When $\boldsymbol{A}\boldsymbol{x} \geqq \boldsymbol{0}$ has a solution $\boldsymbol{x} \geq \boldsymbol{0}$ and only then, $\mu(\boldsymbol{A}) \geqq 0$.

(Proof)

(i) If a $d \geqq 0$ which satisfies $(d\boldsymbol{I}_n + \boldsymbol{A}) \geqq [\boldsymbol{0}]$ is chosen, from [Theorem 4.3], a $\overline{\boldsymbol{x}} \geq \boldsymbol{0}$ which is $\overline{\mu}\overline{\boldsymbol{x}} = (d\boldsymbol{I}_n + \boldsymbol{A})\overline{\boldsymbol{x}}$ can be defined depending on $max \, \boldsymbol{L}_{(d\boldsymbol{I}_n + \boldsymbol{A})} \equiv \overline{\mu}$. Thus, if we let $\mu(\boldsymbol{A}) = (\overline{\mu} - d)$, $\mu(\boldsymbol{A})$ is a real eigenvalue of $\boldsymbol{A}$ and $\overline{\boldsymbol{x}} \geq \boldsymbol{0}$ is an eigenvector corresponding to $\mu(\boldsymbol{A})$. Additionally, if we let $\boldsymbol{x}$ be an eigen column vector of $\boldsymbol{A}$ for μ,

$$(\mu + d)\boldsymbol{x} = (d\boldsymbol{I}_n + \boldsymbol{A})\boldsymbol{x}.$$

Hence,

$$\boldsymbol{y} = (\boldsymbol{l}/\boldsymbol{x}/)^{-1}/\boldsymbol{x}/ \in \boldsymbol{S}_n$$

and

$$|\mu + d|\boldsymbol{y} \leqq (d\boldsymbol{I}_n + \boldsymbol{A})\boldsymbol{y}.$$

Therefore, the following is obtained.

$$\begin{aligned} \mu(\boldsymbol{A}) &= (\overline{\mu} - d) \\ &\geqq |\mu + d| - d \\ &\geqq Re(\mu + d) - d \\ &= Re(\mu). \end{aligned} \tag{4.28}$$

Next, let us focus on the sign of the diagonal elements $= a_{ii}(i \in \boldsymbol{N})$ of $\boldsymbol{A}$, define $\boldsymbol{J} = \{j \in \boldsymbol{N} | a_{jj} < 0\}$ and show the remaining conclusions depending on whether it is $\boldsymbol{J} = \phi$ or not.
(A) $\boldsymbol{J} = \phi$ $\boldsymbol{A} \geqq [\boldsymbol{0}]$ hence $(0 \cdot \boldsymbol{I}_n + \boldsymbol{A}) \geqq [\boldsymbol{0}]$. Thus, by letting it to be $d = 0$ in (4.28),

$$\begin{aligned}\mu(\boldsymbol{A}) + \max_{i\in \boldsymbol{N}}\{|a_{ii}| - a_{ii}\} &= \mu(\boldsymbol{A}) \\ &\geqq |\mu|\end{aligned} \tag{4.29}$$

(B) $\boldsymbol{J} \neq \phi$ For any $j \in \boldsymbol{J}$, $\{|a_{jj}| - a_{jj}\} = 2|a_{jj}| > 0$. Hence if we define $= \max_{i\in \boldsymbol{N}} |a_{jj}| = |a_{kk}|$, $(d\boldsymbol{I}_n + \boldsymbol{A}) \geqq [\boldsymbol{0}]$. Therefore

$$\begin{aligned}\mu(\boldsymbol{A}) &\geqq |\mu + d| - d \\ &\geqq |\mu| - |d| - d \\ &= |\mu| - 2|a_{kk}|.\end{aligned} \tag{4.30}$$

Additionally, from the direct calculation,

$$\begin{aligned}\max_{i\in \boldsymbol{N}}\{|a_{ii}| - a_{ii}\} &= \max\left\{2\max_{j\in \boldsymbol{J}} |a_{jj}|, 0\right\} \\ &= 2|a_{kk}|.\end{aligned} \tag{4.31}$$

Thus, from (4.30) and (4.31),

$$\begin{aligned}\mu(\boldsymbol{A}) + \max_{i\in \boldsymbol{N}}\{|a_{ii}| - a_{ii}\} &= \mu(\boldsymbol{A}) + 2|a_{kk}| \\ &\geqq |\mu| - 2|a_{kk}| + 2|a_{kk}| \\ &= |\mu|.\end{aligned} \tag{4.32}$$

Combining (4.29) and (4.32), the required conclusion is obtained.
(ii) When $\boldsymbol{J} = \boldsymbol{N}$, it is self-evident so let $\boldsymbol{J} \subset \boldsymbol{N}$. If we define a suitable $d \geqq 0$ which satisfies $(d\boldsymbol{I}_n + \boldsymbol{A}) \geqq [\boldsymbol{0}]$ and define a n-order square matrix $\boldsymbol{B}$ as

$$\boldsymbol{B} = \begin{bmatrix} d\boldsymbol{I}_J + \boldsymbol{A}_{JJ} & \boldsymbol{0} \\ \boldsymbol{0} & \boldsymbol{0} \end{bmatrix}$$

from [Theorem 4.3], there exists $\boldsymbol{x_J} \in \boldsymbol{S_{\#J}}$ in response to $\max \boldsymbol{L}_{(d\boldsymbol{I_J}+\boldsymbol{A_{JJ}})} = \mu_{\boldsymbol{J}}$ and $\mu_{\boldsymbol{J}}\boldsymbol{x_J} = (d\boldsymbol{I_J} + \boldsymbol{A_{JJ}})\boldsymbol{x_J}$ holds. In addition, if we define $\boldsymbol{x}^t = \left(\boldsymbol{x}_{\boldsymbol{J}}^t\ \boldsymbol{0}\right)$, $\boldsymbol{x} \in \boldsymbol{S}_n$ and $\mu_{\boldsymbol{J}}\boldsymbol{x} = \boldsymbol{B}\boldsymbol{x}$. Therefore

$$\mu_{\boldsymbol{J}} \leqq \max L_{\boldsymbol{B}}. \tag{4.33}$$

On the other hand, from the definitions of $\boldsymbol{B}$ and $(d\boldsymbol{I}_n + \boldsymbol{A}) \geqq [\boldsymbol{0}]$, for any $\boldsymbol{y} \geqq \boldsymbol{0}$, $(d\boldsymbol{I}_n + \boldsymbol{A})\boldsymbol{y} \geqq \boldsymbol{B}\boldsymbol{y}$, hence from [Theorem 4.1: (ii)],

$$\max \boldsymbol{L_B} \leqq \max \boldsymbol{L}_{(d\boldsymbol{I}_n+\boldsymbol{A})} \tag{4.34}$$

is obtained. Therefore, if we let $\boldsymbol{L}_{(d\boldsymbol{I}_n+\boldsymbol{A})}=\overline{\mu}$, from (4.33) and (4.34), we immediately obtain

$$\mu_{\boldsymbol{J}} \leqq \overline{\mu}. \tag{4.35}$$

Additionally, from the definition of $\mu_{\boldsymbol{J}}$, $(\mu_{\boldsymbol{J}} - d)\boldsymbol{x}_{\boldsymbol{J}} = \boldsymbol{A}_{\boldsymbol{JJ}}\boldsymbol{x}_{\boldsymbol{J}}$ and $\boldsymbol{x}_{\boldsymbol{J}} \geq \boldsymbol{0}$. In other words, $(\mu_{\boldsymbol{J}} - d)$ is one of the eigenvalues of $\boldsymbol{A}_{\boldsymbol{JJ}}$ hence,

$$\mu_{\boldsymbol{J}} - d \leqq \overline{\mu} - d.$$

Thus, if we let $\mu(\boldsymbol{A}) = \overline{\mu} - d$, it is shown that $\mu(\boldsymbol{A}) \geqq \mu(\boldsymbol{A}_{\boldsymbol{JJ}})$ for any proper subset $\boldsymbol{J} \subset \boldsymbol{N}$ of $\boldsymbol{N}$.

(iii) Prove using the induction approach in respect of $\boldsymbol{A}$'s number of orders n.

Firstly, let us show the case when $n = 2$. If we define $\mu(\boldsymbol{A}) = \frac{1}{2} \times \left[(a_{11} + a_{22}) + \sqrt{(a_{11} - a_{22})^2 + 4a_{12}a_{21}}\right]$ it is certainly the case that $det(\mu(\boldsymbol{A})\boldsymbol{I}_2 - \boldsymbol{A}) = 0$. From the definition of $\mu(\boldsymbol{A})$, for any $i \in (1, 2)$,

$$\begin{aligned} f_{ii}[\mu(\boldsymbol{A})] &= \mu(\boldsymbol{A}) - a_{jj} \\ &= \frac{1}{2}\left[(-1)^j(a_{11} - a_{22}) + \sqrt{(a_{11} - a_{22})^2 + 4a_{12}a_{21}}\right] \\ &\geqq 0 \quad (j = 1, 2). \end{aligned} \tag{4.36}$$

Moreover, for any $\mu > \mu(\boldsymbol{A})$,

$$f_{ii}(\mu) - f_{ii}[\mu(\boldsymbol{A})] = \mu - \mu(\boldsymbol{A}) > 0. \tag{4.37}$$

Accordingly, for any $\mu > \mu(\boldsymbol{A})$,

$$\begin{aligned} f_{ii}(\mu) &> f_{ii}[\mu(\boldsymbol{A})] \\ &\geqq 0 \quad (i = 1, 2) \end{aligned}$$

is obtained. Furthermore, for any different i, $j \in \{1, 2\}, f_{ij}(\mu) = a_{ji} \geqq 0$ is also evident so (iii) is shown.

Next, let us assume that up to $(n - 1)$-order, (iii) holds. Moreover, $M^i_{kj}(\mu)$ represents the cofactor of $f_{ii}(\mu)$ in respect of $(\delta_{kj} - a_{kj})$. However, if δ_{kj} is a Kronecker's delta, in other words, $k \neq j (k = j)$, $\delta_{kj} = 0(\delta_{kj} = 1)$. In this case, from the cofactor expansion of $f_{ij}(\mu)$,

$$f_{ij}(\mu) = \sum\nolimits_{k \neq i} a_{kj} M^i_{kj}(\mu) \quad (i \neq j; i = 1, \cdots, n). \tag{4.38}$$

However, if $f(\mu)$ is expanded for the ith row,

$$\begin{aligned} f(\mu) &= det(\mu \boldsymbol{I}_n - \boldsymbol{A}) \\ &= \sum_{j=1}^{n} (\mu \boldsymbol{I}_n - \boldsymbol{A})_{ij} f_{ij}(\mu) \\ &= (\mu - a_{ii}) f_{ii}(\mu) - \sum_{j \neq i} a_{ij} f_{ij}(\mu). \end{aligned} \tag{4.39}$$

Therefore, (4.38) can be substituted into (4.39) which gives (4.40).

$$f(\mu) = (\mu - a_{ii}) f_{ii}(\mu) - \sum_{k,j \neq i} a_{ij} a_{ki} M^i_{kj}(\mu). \tag{4.40}$$

Additionally, from the assumption of the induction approach, $f_{ii}(\mu) = 0$ has a real root $\tilde{\mu}_i$ which satisfies (iii) hence

$$\begin{aligned} f(\tilde{\mu}_i) &= -\sum_{k,j \neq i} a_{ij} a_{ki} M^i_{kj}(\tilde{\mu}_i) \\ &\leqq 0. \end{aligned} \tag{4.41}$$

To show that $\boldsymbol{A}$ has an eigenvalue if we suppose that $f(\mu) = 0$ does not have a real root, for any $\mu \in \boldsymbol{R}$,

$$f(\mu) = \prod_j |\mu - \mu_{2j-1}|^2 > 0$$

holds. However, this contradicts (4.41), thus, $\boldsymbol{A}$ has an eigenvalue. Hence if we let $\mu(\boldsymbol{A})$ be the maximum eigenvalue of $\boldsymbol{A}$, for any i, $\mu(\boldsymbol{A}) \geqq \tilde{\mu}_i$. However, from [Lemma 4.2], $f(\tilde{\mu}_i) > 0$ but this again contradicts (4.41). Accordingly, from the assumptions for the induction approach and (4.38),

$$\text{for any } \mu \geqq \mu(\boldsymbol{A}), f_{ij}(\mu) = \sum_{k \neq i} a_{ki} M^i_{kj}(\mu) \geqq 0 \quad (i \neq j). \tag{4.42}$$

Furthermore, if we consider the fact that for any $i \in \boldsymbol{N}$, $\tilde{\mu}_i$ is the maximum real root of $f_{ii}(\mu) = 0$ and the fact that $\mu(\boldsymbol{A}) \geqq \tilde{\mu}_i$ holds for any $i \in \boldsymbol{N}$, from [Lemma 4.2] again,

$$\begin{gathered} \text{for any } \mu \geqq \mu(\boldsymbol{A}), f_{ii}(\mu) \geqq 0 \quad (i = 1, \cdots, n) \\ \text{and} \\ \text{for any } \mu > \mu(\boldsymbol{A}), f_{ii}(\mu) > 0 \quad (i = 1, \cdots, n). \end{gathered} \tag{4.43}$$

Consequently, from (4.42) and (4.43), the conclusion is evident.

(iv)

Assume $((\rho \boldsymbol{I}_n - \boldsymbol{A})^{-1} \geq [\boldsymbol{0}] \Rightarrow \rho > \mu(\boldsymbol{A}))$ $(\rho \boldsymbol{I}_n - \boldsymbol{A})^{-1} \geq [\boldsymbol{0}]$ and $\rho \leqq \mu(\boldsymbol{A})$. From (i), it corresponds to $\mu(\boldsymbol{A})$ and an eigenvector $\bar{\boldsymbol{x}} \geq \boldsymbol{0}$ is defined so $(\rho \boldsymbol{I}_n - \boldsymbol{A})^{-1} \bar{\boldsymbol{x}} \geqq \boldsymbol{0}$. If

$(\rho \boldsymbol{I}_n - \boldsymbol{A})^{-1}\overline{\boldsymbol{x}} = \boldsymbol{0}$, $\overline{\boldsymbol{x}} \geq \boldsymbol{0}$, thus, this contradicts the nonsingularity of $(\rho \boldsymbol{I}_n - \boldsymbol{A})^{-1}$. Therefore,

$$(\rho \boldsymbol{I}_n - \boldsymbol{A})^{-1}\overline{\boldsymbol{x}} \geq \boldsymbol{0} \tag{4.44}$$

and

$$\begin{aligned} \boldsymbol{0} \leq \overline{\boldsymbol{x}} &= (\rho \boldsymbol{I}_n - \boldsymbol{A})^{-1}(\rho \boldsymbol{I}_n - \boldsymbol{A})\overline{\boldsymbol{x}} \\ &= (\rho \boldsymbol{I}_n - \boldsymbol{A})^{-1}[(\rho - \mu(\boldsymbol{A})\boldsymbol{A})\boldsymbol{I}_n + (\mu(\boldsymbol{A})\boldsymbol{I}_n - \boldsymbol{A})]\overline{\boldsymbol{x}} \\ &= (\rho - \mu(\boldsymbol{A}))(\rho \boldsymbol{I}_n - \boldsymbol{A})^{-1}\overline{\boldsymbol{x}}. \end{aligned} \tag{4.45}$$

Corresponding (4.45),

$$\begin{aligned} \boldsymbol{0} \leq \overline{\boldsymbol{x}} &= (\rho - \mu(\boldsymbol{A}))(\rho \boldsymbol{I}_n - \boldsymbol{A})^{-1}\overline{\boldsymbol{x}} \\ &\leqq \boldsymbol{0} \end{aligned} \tag{4.46}$$

are obtained, however, these are self-contradictory.

If $(\rho \boldsymbol{I}_n - \boldsymbol{A})^{-1} \geq [\boldsymbol{0}] \Leftarrow \rho > \mu(\boldsymbol{A}) \quad \rho > \mu(\boldsymbol{A})$[9]

$$\varphi(\rho) \equiv det(\rho \boldsymbol{I}_n - \boldsymbol{A}) \neq 0. \tag{4.47}$$

Moreover, for any $\rho \geqq \mu(\boldsymbol{A})$,

$$\varphi'(\rho) = \sum_{i=1}^{n} det\left[\rho \boldsymbol{I}_n - \boldsymbol{A}\begin{pmatrix} i \\ i \end{pmatrix}\right] \geqq 0. \tag{4.48}$$

However, the equality in (48), even if it holds, is limited to the case when $\rho = \mu(\boldsymbol{A})$. Thus, $\varphi(\rho)$ is monotonically non-decreasing for $\rho \geqq \mu(\boldsymbol{A})$ and $\varphi(\rho) \neq 0$ is derived in respect of $\varphi(\mu(\boldsymbol{A})) = 0$ and any $\rho > \mu(\boldsymbol{A})$. Therefore, for all $\rho > \mu(\boldsymbol{A})$, $\varphi(\rho) > 0$ holds. Combining the fact above and (iii), under $\rho > \mu(\boldsymbol{A})$, $(\rho \boldsymbol{I}_n - \boldsymbol{A})^{-1} \geq [\boldsymbol{0}]$ is guaranteed.

(v) If for some $\mu > \mu(\boldsymbol{A})$, there exists $\boldsymbol{x} \geq \boldsymbol{0}$ which satisfies $\mu \boldsymbol{x} \leqq \boldsymbol{A}\boldsymbol{x}$, the following self-contradiction is obtained.

---

[9]From (i), if μ is a real eigenvalue, $\mu = Re(\mu)$. Therefore $\mu(\boldsymbol{A}) \geqq Re(\mu) = \mu$ holds and $\mu(\boldsymbol{A})$ becomes the maximum eigenvalue of $\boldsymbol{A}$. Therefore, when $\rho > \mu(\boldsymbol{A})$, $\varphi(\rho) \neq 0$.

$$\begin{aligned}\mathbf{0} &\geqq (\mu \boldsymbol{I}_n - \boldsymbol{A})^{-1}[(\mu \boldsymbol{I}_n - \boldsymbol{A})\boldsymbol{x}] \\ &= \boldsymbol{x} \\ &\geq \mathbf{0}.\end{aligned}$$

Therefore for any $\mu \in \tilde{L}_A$, $\mu \leqq \mu(\boldsymbol{A})$. However, from (i), for any $\overline{\boldsymbol{x}} \geq \mathbf{0}$, $\mu(\boldsymbol{A})\overline{\boldsymbol{x}} = \boldsymbol{A}\overline{\boldsymbol{x}}$. Therefore $\mu(\boldsymbol{A}) \in \tilde{L}_A$ has also been shown.

(vi)

$({}^{\exists}\boldsymbol{x} \geq \mathbf{0} | \boldsymbol{A}\boldsymbol{x} \geqq \mathbf{0} \Rightarrow \mu(\boldsymbol{A}) \geqq 0)$ Assume that $\boldsymbol{A}\boldsymbol{x} \geqq \mathbf{0}$ has a solution $\boldsymbol{x} \geq \mathbf{0}$ but $\mu(\boldsymbol{A}) < 0$. Here if we let it be (iv)'s $\rho = 0$,

$$(0 \cdot \boldsymbol{I}_n - \boldsymbol{A})^{-1} = -\boldsymbol{A}^{-1} \geq [\mathbf{0}]. \tag{4.49}$$

Thus, the following self-contradictions

$$\begin{aligned}\mathbf{0} &\leqq -\boldsymbol{A}^{-1}\boldsymbol{A}\boldsymbol{x} \\ &= -\boldsymbol{x} \\ &\leq \mathbf{0}\end{aligned}$$

are derived.

$({}^{\exists}\boldsymbol{x} \geq \mathbf{0} | \boldsymbol{A}\boldsymbol{x} \geqq \mathbf{0}) \Leftarrow \mu(\boldsymbol{A}) \geqq 0)$ If $\mu(\boldsymbol{A}) \geqq 0$, from (i), there exists some $\overline{\boldsymbol{x}} \geq \mathbf{0}$ and $\boldsymbol{A}\overline{\boldsymbol{x}} = \mu(\boldsymbol{A})\overline{\boldsymbol{x}} \geqq \boldsymbol{0}$ holds. QED

(Notes on [Theorem 4.5]): When we let $f(\boldsymbol{x}) = \boldsymbol{A}\boldsymbol{x}$, it is clear that it can be expressed as $\boldsymbol{L}_f = \boldsymbol{L}_A = \{\mu \in \boldsymbol{R} | {}^{\exists}\boldsymbol{x} \in \boldsymbol{S}_n | \boldsymbol{A}\boldsymbol{x} \geqq \mu\boldsymbol{x}\}$ in [Theorem 4.1] and that this $\boldsymbol{L}_A$ matches $\tilde{\boldsymbol{L}}_A$ in this theorem's (v). However, the gross substitution matrix $\boldsymbol{A}$ does not necessarily satisfy (NN) hence $\mu(\boldsymbol{A}) \geqq 0$ is not guaranteed. Thus, care should be taken to note that the statement in (v) is simply a part of the conclusion in [Theorem 4.1; (i)]. Additionally, semipositive eigenvector $\overline{\boldsymbol{x}}$ of $\boldsymbol{A}$ in respect of the theorem's $\mu(\boldsymbol{A})$, as shown in the proof process of it, corresponds to $\boldsymbol{y}^f$ in [Theorem 4.1], [Theorem 4.2]'s $\boldsymbol{x}^f$ and [Theorem 4.3]'s $\boldsymbol{y}^f$. Thus, henceforth, $\mu(\boldsymbol{A})$ and $\overline{\boldsymbol{x}}$ shall be represented as gross substitution matrix $\boldsymbol{A}$'s **Frobenius root** and **Frobenius vector** and expressed as μ_A and $\boldsymbol{x}^A$, respectively.

When a gross substitution matrix $\boldsymbol{A}$ is indecomposable, [Theorem 4.5] is strengthened by the following theorem.

[Theorem 4.6] If the n-order gross substitution matrix $\boldsymbol{A}$ is indecomposable, the following hold.

(i) $\boldsymbol{x}^A > \mathbf{0}$.
(ii) $rank(\mu_A \boldsymbol{I}_n - \boldsymbol{A}) = n - 1$.
(iii) $adj(\mu_A \boldsymbol{I}_n - \boldsymbol{A}) > \mathbf{0}$.
(iv) $\mu_A > \mu_{A\binom{i}{i}} \quad (i = 1, \cdots, n)$.

(v) μ_A is an eigen equation of A and the single root of $\varphi(\mu_A) = det(\mu_A I_n - A) = 0$.
(vi) $\rho > \mu_A$ and $(\rho I_n - A)^{-1} > [0]$ are mutually equivalent.
(vii) When $Ax > 0$ has a solution $x \geq 0$, and only then, $\mu_A > 0$.

(Proof)

(i) Define as $B = [dI_n + A]$ and choose an appropriate $d \geqq [0]$ satisfying $B \geqq [0]$ and consider fixing it. If B is indecomposable, there exists a nonempty proper subset J of N and $A_{JJ^c} = [0]$. However, this contradicts the assumed indecomposability of A hence B is indecomposable. Accordingly, it can be applied to [Theorem 4.2]'s B. Consequently, if [Theorem 4.2]'s μ_f and $x^f > 0$ are expressed as μ_B and x^B, respectively, and the proof of [Theorem 4.5; (i)] is considered, $\mu_A = \mu_B - d$ and $x^A = x^B > 0$.
(ii) From [Theorem 4.2; (iii)], $(\mu_A I_n - A)x = 0$ has a fundamental solution formed from one vector x^A. Thus, $rank(\mu_A I_n - A) = n - 1$.[10]
(iii) From [Theorem 4.5; (ii)], for any $i \in N$, $\mu_A > \mu_{A\binom{i}{i}}$. Moreover, from [Theorem 4.5; (i)], $\mu_{A\binom{i}{i}}$ is the maximum real root of A's $(n - 1)$-order principal submatrix $A\binom{i}{i}$ and $\varphi(\mu) = det(\mu I_n - A)$'s the maximum order number μ^n's coefficient $=1$. Thus, from [Lemma 4.2], for any $i \in N$,

$$\begin{aligned}[adj(\mu_A I_n - A)]_{ii} &= (-1)^{2i} det\left[\mu_A I_{n-1} - A\binom{i}{i}\right] \\ &= det\left[\mu_A I_{n-1} - A\binom{i}{i}\right] \\ &\geqq 0.\end{aligned} \tag{4.50}$$

Though, for any $i, j \in N$, $[adj(\mu_A I_n - A)]_i$, $[adj(\mu_A I_n - A)]^j$ and $[adj(\mu_A I_n - A)]_{ij}$ represent $adj(\mu_A I_n - A)$'s ith row, jth column and the (i,j)th element, respectively. In addition,

$$(\mu_A I_n - A)[adj(\mu_A I_n - A)] = 0$$

and

$$rank(\mu_A I_n - A) = n - 1.$$

Therefore, there exists $\alpha \in R$ which becomes $p, q \in N$ and $p, q \in N$ which satisfies $[adj(\mu_A I_n - A)]_{pq} \neq 0$. Therefore,

[10] See ⟨Definition 18; (iii)⟩ and [Theorem 19; (iv-ii)] (Appendix).

$$0 \leqq [adj(\mu_A \boldsymbol{I}_n - \boldsymbol{A})]_{qq} \\ = \alpha x_q^A \tag{4.51}$$

and

$$0 \neq [adj(\mu_A \boldsymbol{I}_n - \boldsymbol{A})]_{pq} \\ = \alpha x_p^A. \tag{4.52}$$

From (i), $\boldsymbol{x}^A > \boldsymbol{0}$, hence, from (4.51) and (4.52), $\alpha > 0$, thus,

$$[adj(\mu_A \boldsymbol{I}_n - \boldsymbol{A})]^q = \alpha \boldsymbol{x}^A > \boldsymbol{0} \tag{4.53}$$

holds. In addition, if (i) is used again in $\boldsymbol{A}^t$, $\boldsymbol{A}$ has an eigen-row vector $\boldsymbol{y} > \boldsymbol{0}$ that belongs to μ_A. Hence, if this fact and (4.53) are combined, as per the argument above, there exists $\alpha_i \in \boldsymbol{R}$ for any $i \in \boldsymbol{N}$, and

$$[adj(\mu_A \boldsymbol{I}_n - \boldsymbol{A})]_i = \alpha_i \boldsymbol{y}. \tag{4.54}$$

Therefore for all $i \in \boldsymbol{N}$,

$$0 < [adj(\mu_A \boldsymbol{I}_n - \boldsymbol{A})]_{iq} = \alpha_i y_q$$

holds and from $\boldsymbol{y} > \boldsymbol{0}$, becomes each $\alpha_i > 0$. Therefore, $adj(\mu_A \boldsymbol{I}_n - \boldsymbol{A}) > \boldsymbol{0}$.
(iv) This can be obtained from (iii) and [Theorem 4.5; (v)].
(v) Using (iii),

$$\varphi'(\mu_A) = \sum_{i=1}^n det\left[\mu_A \boldsymbol{I}_n - \boldsymbol{A}\binom{i}{i}\right] > 0. \tag{4.55}$$

Here, if we take the multiplicity of μ_A to be $\geqq 2$, $\varphi'(\mu)$ has at least one, $(\mu - \mu_A)$ as a factor, thus, $\varphi'(\mu_A) = 0$ and contradicts (4.55). Therefore, μ_A is a single root.
(vi)
$(\rho > \mu_A) \Rightarrow (\rho \boldsymbol{I}_n - \boldsymbol{A})^{-1} > [\boldsymbol{0}]$ If $\rho > \mu_A$, From [Theorem 4.5; (iv)], $(\rho \boldsymbol{I}_n - \boldsymbol{A})^{-1} \geq [\boldsymbol{0}]$. Additionally, the fact that when $\rho > \mu_A$, $\varphi(\rho) = det\,(\rho \boldsymbol{I}_n - \boldsymbol{A}) > 0$ has been shown in the proof for [Theorem 4.5; (iv)]. Hence,

$$[\boldsymbol{0}] \leq (\rho \boldsymbol{I}_n - \boldsymbol{A})^{-1} = \frac{adj(\rho \boldsymbol{I}_n - \boldsymbol{A})}{det(\rho \boldsymbol{I}_n - \boldsymbol{A})}.$$

Then, if we see $adj(\rho \boldsymbol{I}_n - \boldsymbol{A}) > [\boldsymbol{0}]$, $(\rho \boldsymbol{I}_n - \boldsymbol{A}) > [\boldsymbol{0}]$ is shown. Contradict the conclusion to assume that there exists some $\rho > \mu_A$ and it is the $q \equiv [adj(\rho \boldsymbol{I}_n - \boldsymbol{A})]^q \ngtr 0$ of $adj(\rho \boldsymbol{I}_n - \boldsymbol{A})$,

$$[adj(\rho \boldsymbol{I}_n - \boldsymbol{A})]^q \geq 0. \tag{4.56}$$

Thus, if we define $\boldsymbol{I}_q = \{i \in \boldsymbol{N} | [adj(\rho \boldsymbol{I}_n - \boldsymbol{A})]_{iq} = 0\}$, $\boldsymbol{I}_q \neq \phi$ and $\boldsymbol{I}_q \subset \boldsymbol{N}$. However,

$$(\rho \boldsymbol{I}_n - \boldsymbol{A})[adj(\rho \boldsymbol{I}_n - \boldsymbol{A})]^q \geqq \boldsymbol{0}. \tag{4.57}$$

Therefore for any $i \in \boldsymbol{I}_q$,

$$\begin{aligned} 0 &= (\rho - a_{ii})[adj(\rho \boldsymbol{I}_n - \boldsymbol{A})]_{iq} \\ &\geqq \sum\nolimits_{j \notin \boldsymbol{I}_q} a_{ij}[adj(\rho \boldsymbol{I}_n - \boldsymbol{A})]_{iq} \\ &\geqq 0 \end{aligned}$$

holds and for any $i \in \boldsymbol{I}_q$ and $j \notin \boldsymbol{I}_q$, $a_{ij} = 0$ is derived. Though, this contradicts the indecomposability of $\boldsymbol{A}$.
$(\rho > \mu_{\boldsymbol{A}}) \Leftarrow (\rho \boldsymbol{I}_n - \boldsymbol{A})^{-1} > [\boldsymbol{0}]$ If $(\rho \boldsymbol{I}_n - \boldsymbol{A})^{-1} > [\boldsymbol{0}]$, $[\rho \boldsymbol{I}_n - \boldsymbol{A}]^{-1} \geq [\boldsymbol{0}]$. Therefore, from [Theorem 4.5; (iv)], it is evident that $\rho > \mu_{\boldsymbol{A}}$.
(vii)
$({}^\exists \boldsymbol{x} \geq \boldsymbol{0} | \boldsymbol{Ax} > \boldsymbol{0}) \Rightarrow (\mu_{\boldsymbol{A}} > 0)$ If there exists $\boldsymbol{x} \geq \boldsymbol{0}$ which satisfies $\boldsymbol{Ax} > \boldsymbol{0}$,

$$\boldsymbol{Ax} > \boldsymbol{0}. \tag{4.58}$$

If (4.58) is multiplied on the left by the eigenvector $\boldsymbol{y} > \boldsymbol{0}$ belonging to $\mu_{\boldsymbol{A}}$,

$$\begin{aligned} \mu_{\boldsymbol{A}}(\boldsymbol{yx}) &= \boldsymbol{yAx} \\ &> 0. \end{aligned}$$

Therefore, from $\boldsymbol{x} \geq \boldsymbol{0}$ and $\boldsymbol{y} > \boldsymbol{0}$, $\mu_{\boldsymbol{A}} > 0$ is derived.
$({}^\exists \boldsymbol{x} \geq \boldsymbol{0} | \boldsymbol{Ax} > \boldsymbol{0}) \Leftarrow (\mu_{\boldsymbol{A}} > 0)$ From (i), $\boldsymbol{x}^A > \boldsymbol{0}$. Therefore, if $\mu_{\boldsymbol{A}} > 0$, $0 < \mu_{\boldsymbol{A}} \boldsymbol{x}^A = \boldsymbol{Ax}^A$ holds. QED

[Theorem 4.5] and [Theorem 4.6] discussed gross substitution matrices. Hence, if we define $\boldsymbol{B} = -\boldsymbol{A}$, $\boldsymbol{B}$ becomes a n-order square matrix whose all nondiagonal elements are nonnegative. Accordingly, Frobenius' theorem in respect of the gross substitution matrix $\boldsymbol{A}$ derives the following theorem.

[Theorem 4.7][11] Let matrix $\boldsymbol{B} = [b_{ij}]$ $(i, j = 1, \cdots, n)$ be a n-order square matrix that satisfies $b_{ij} \leqq 0 (i \neq j)$. Here, the following (i)–(v) are mutually equivalent.

(i) $\boldsymbol{Bx} > 0$ has a solution $\boldsymbol{x} > \boldsymbol{0}$.
(ii) $\mu_{(-\boldsymbol{B})} < 0$, however, $\mu_{(-\boldsymbol{B})}$ is a Frobenius root of $-\boldsymbol{B}$.

[11] See Hawkins and Simon (1949).

(iii) $det\begin{bmatrix} b_{11} & \cdots & b_{1k} \\ \vdots & & \vdots \\ b_{n1} & \cdots & b_{kk} \end{bmatrix} > 0 \quad (k = 1, \cdots, n).$[12]

(iv) For any $\boldsymbol{c} \geqq \boldsymbol{0}$, there exists some $\boldsymbol{x} \geqq \boldsymbol{0}$ and $\boldsymbol{Bx} = \boldsymbol{c}$.
(v) $\boldsymbol{B}^{-1} \geq [\boldsymbol{0}]$.

Additionally, when $\boldsymbol{B}$ is indecomposable, (iv) and (v) are further strengthened.

(iv)′ For any $\boldsymbol{c} \geq \boldsymbol{0}$, there exists some $\boldsymbol{x} > \boldsymbol{0}$ and $\boldsymbol{Bx} = \boldsymbol{c}$.
(v)′ $\boldsymbol{B}^{-1} > [\boldsymbol{0}]$.

(Proof) Show in the order (i) ⇒ (ii) ⇒ (iii) ⇒ (iv) ⇒ (v) ⇒ (i).

(i) ⇒ (ii) Let $\boldsymbol{A} = -\boldsymbol{B}$, then $a_{ij} \geqq 0 (i \neq j)$. Thus, $\boldsymbol{A}$ is a gross substitution matrix. Here if an eigen-row vector of $\boldsymbol{A}$ that corresponds to $\mu_{\boldsymbol{A}}$ is taken to be $\widehat{\boldsymbol{y}} \in \boldsymbol{R}^n, \widehat{\boldsymbol{y}} \geq \boldsymbol{0}$[13] and $\mu_{\boldsymbol{A}} \widehat{\boldsymbol{y}} = \widehat{\boldsymbol{y}} \boldsymbol{A}$. Therefore

$$\begin{aligned} \mu_{\boldsymbol{A}} \widehat{\boldsymbol{y}} \boldsymbol{x} &= \widehat{\boldsymbol{y}} \boldsymbol{A} \boldsymbol{x} \\ &= -\widehat{\boldsymbol{y}} \boldsymbol{B} \boldsymbol{x} \\ &< 0 \end{aligned} \tag{4.59}$$

holds. However, by $\widehat{\boldsymbol{y}} \geq \boldsymbol{0}$ and $\boldsymbol{x} > \boldsymbol{0}$, $\widehat{\boldsymbol{y}} \boldsymbol{x} > 0$. Combining this with (4.59) gives $\mu_{\boldsymbol{A}} = \mu_{(-\boldsymbol{B})} < 0$.
(ii) ⇒ (iii) If we define

$$\boldsymbol{A} = -\boldsymbol{B} \quad \text{and} \quad \boldsymbol{B}_k = \begin{bmatrix} b_{11} & \cdots & b_{1k} \\ \vdots & & \vdots \\ b_{k1} & \cdots & b_{kk} \end{bmatrix} = -\boldsymbol{A}_k \quad (k = 1, \cdots, n),$$

utilizing (ii) and [Theorem 4.5; (ii)],

$$\mu_{\boldsymbol{A}_k} \leqq \mu_{\boldsymbol{A}} \leqq \mu_{(-\boldsymbol{B})} < 0.$$

As a result,

$$\begin{aligned} det(\boldsymbol{B}_k) &= det(-\boldsymbol{A}_k) \\ &= det(0 \cdot \boldsymbol{I}_n - \boldsymbol{A}_k) \\ &> 0. \end{aligned}$$

(iii) ⇒ (iv) This can be proven using the induction approach on the matrix $\boldsymbol{B}$'s order number n.

[12] It is the matrix equation of k-order leading principal minor matrix made by choosing k rows and k columns from the top left corner of $\boldsymbol{B}$.

[13] By applying [Theorem 4.5; (i)] to $\boldsymbol{A}'$, the existence of $\widehat{\boldsymbol{y}} \geq \boldsymbol{0}$ is apparent.

If $n = 1$, from (iii), $b_{11} > 0$. Therefore for any $c_1 \geqq 0$, $b_{11}x_1 = c_1 \geqq 0$ has a solution $x_{11} = c_1/b_{11} \geqq 0$.
Next, assume that (iv) holds up to ($n = 1$)-order and show that it also holds for n-order square matrix $\boldsymbol{B}$. Therefore, with $b_{11} > 0$ as the axis, if sweep out approach is used, $\boldsymbol{Bx} = \boldsymbol{c} \geqq \boldsymbol{0}$ can be changed into the form in (60).

$$\begin{bmatrix} 1 & \tilde{b}_{12} & \cdots & \tilde{b}_{1n} \\ 0 & \tilde{b}_{22} & \cdots & \tilde{b}_{2n} \\ \vdots & \vdots & & \vdots \\ 0 & \tilde{b}_{n2} & \cdots & \tilde{b}_{nn} \end{bmatrix} \boldsymbol{x} = \begin{bmatrix} \tilde{c}_1 \\ \tilde{c}_2 \\ \vdots \\ \tilde{c}_n \end{bmatrix}. \tag{4.60}$$

Where

$$\tilde{b}_{1j} = \frac{b_{1j}}{b_{11}} \quad (j = 2, \cdots, n),$$

$$\tilde{c}_1 = \frac{c_1}{b_{11}},$$

$$\tilde{b}_{ij} = b_{ij} - \tilde{b}_{1j} b_{i1} \quad (i = 2, \cdots, n; j = 1, \cdots, n),$$

$$\tilde{c}_i = c_i - \tilde{c}_1 b_{i1} \quad (i = 2, \cdots, n).$$

Thus,

$$det \begin{bmatrix} b_{11} & \cdots & b_{1k} \\ \vdots & & \vdots \\ b_{k1} & \cdots & b_{kk} \end{bmatrix} = b_{11} det \begin{bmatrix} \tilde{b}_{22} & \cdots & \tilde{b}_{2k} \\ \vdots & & \vdots \\ \tilde{b}_{k2} & \cdots & \tilde{b}_{kk} \end{bmatrix} > 0$$

$$(k = 2, \cdots, n).$$

Hence,

$$det \begin{bmatrix} \tilde{b}_{22} & \cdots & \tilde{b}_{2k} \\ \vdots & & \vdots \\ \tilde{b}_{k2} & \cdots & \tilde{b}_{kk} \end{bmatrix} > 0 \quad (k = 2, \cdots, n)$$

is obtained. In addition, from $b_{ij} \leqq 0$ $(i \neq j; i, j = 1, \cdots, n)$, $c_i \geqq 0$ $(i = 1, \cdots, n)$ and $b_{11} > 0$, $\tilde{b}_{ij} \leqq 0$ $(i \neq j; i, j = 2, \cdots, n)$ and $\tilde{c}_i \geqq 0$ $(i = 1, \cdots, n)$ are evident. Consequently, from the assumption for the induction approach, there exists $\boldsymbol{y} = (y_2, \cdots, y_n)' \geqq \boldsymbol{0}$ and

$$\tilde{\boldsymbol{B}}\boldsymbol{y} = \tilde{\boldsymbol{c}},$$

where

$$\widetilde{\boldsymbol{B}} = \begin{bmatrix} \widetilde{b}_{22} & \cdots & \widetilde{b}_{2n} \\ \vdots & & \vdots \\ \widetilde{b}_{n2} & \cdots & \widetilde{b}_{nn} \end{bmatrix}, \quad \widetilde{\boldsymbol{c}} = \begin{bmatrix} \widetilde{c}_2 \\ \vdots \\ \widetilde{c}_n \end{bmatrix}$$

hold. Here if we define $y_1 = \frac{1}{b_{11}}\left(c_1 - \sum_{j=2}^{n} b_{1j}y_j\right)$ and $\widehat{\boldsymbol{y}} = (y_1, y_2, \cdots, \boldsymbol{y}_n)'$,

$$y_1 \geqq 0.$$

Thus,

$$\widehat{\boldsymbol{y}} \geqq \boldsymbol{0}$$

and

$$\begin{bmatrix} 1 & \widetilde{b}_{12} \cdots \widetilde{b}_{1n} \\ \boldsymbol{0} & \widetilde{\boldsymbol{B}} \end{bmatrix} \widehat{\boldsymbol{y}} = \begin{bmatrix} y_1 + \sum_{j=2}^{n} \widetilde{b}_{1j}y_j \\ \widetilde{\boldsymbol{B}}\boldsymbol{y} \end{bmatrix} = \begin{pmatrix} \widetilde{c}_1 \\ \widetilde{\boldsymbol{c}} \end{pmatrix}$$

are obtained. Accordingly, for any $\boldsymbol{c} \geq \boldsymbol{0}$, there exists some $\widehat{\boldsymbol{y}} \geq \boldsymbol{0}$ and $\boldsymbol{B}\widehat{\boldsymbol{y}} = \boldsymbol{c}$ holds.
(iv) ⇒ (v) From condition (iv), $\boldsymbol{Bx} = \boldsymbol{e}^i$ has a solution $\boldsymbol{x}^i \geqq \boldsymbol{0}$ for the $i(i \in \boldsymbol{N})$th unit vector $\boldsymbol{e}^i$. Here, to show that for all $i \in \boldsymbol{N}$, $\boldsymbol{x}^i \geq \boldsymbol{0}$, if the existence of $i_0 \in \boldsymbol{N}$, which satisfies $\boldsymbol{x}^{i_0} = \boldsymbol{0}$ is assumed,

$$\begin{aligned} \boldsymbol{0} &= \boldsymbol{Bx}^{i_0} \\ &= \boldsymbol{e}^{i_0} \\ &\geq \boldsymbol{0}. \end{aligned}$$

This is self-contradictory. Therefore for all $i \in \boldsymbol{N}$, $\boldsymbol{x}^i \geq \boldsymbol{0}$.

On the other hand,

$$\begin{aligned} \boldsymbol{B}\left[\boldsymbol{x}^1, \cdots, \boldsymbol{x}^n\right] &= \left[\boldsymbol{e}^1, \ldots, \boldsymbol{e}^n\right] \\ &= \boldsymbol{I}_n. \end{aligned}$$

If both sides of the equation above are multiplied on the left by $\boldsymbol{B}^{-1}$, $[\boldsymbol{x}^1, \cdots, \boldsymbol{x}^n] = \boldsymbol{B}^{-1}$. Thus, from each $\boldsymbol{x}^i \geq \boldsymbol{0}$, $\boldsymbol{B}^{-1} \geq [\boldsymbol{0}]$ is shown.
(v) ⇒ (i) If $\boldsymbol{B}^{-1} \geq [\boldsymbol{0}]$, there exists at least one positive element in each row of $\boldsymbol{B}^{-1}$.

Therefore, for any $\boldsymbol{z} > \boldsymbol{0}$,

$$\boldsymbol{0} < \boldsymbol{B}^{-1}\boldsymbol{z}$$

holds. If we define $\boldsymbol{B}^{-1}\boldsymbol{z} = \boldsymbol{x}$ $\boldsymbol{x} > \boldsymbol{0}$. In addition, multiplying the above equation on the left by $\boldsymbol{B}$ gives $\boldsymbol{0} < \boldsymbol{z} = \boldsymbol{Bx}$. Thus, (i) has been shown.

Next, the case when $\boldsymbol{B}$ is indecomposable is considered. The predicate of (i) ⇒ (ii) ⇒ (iii) holds whether $\boldsymbol{B}$ is an indecomposable matrix or not. Additionally, it is also clear that (iv)′ ⇒ (v)′ ⇒ (i) hence only (iii) ⇒ (iv)′ must be proven.

(iii) ⇒ (iv)′ (iii) ⇒ (iv) has already been shown so for any $\boldsymbol{c} \geq \boldsymbol{0}$, there exists some $\boldsymbol{x} \geqq \boldsymbol{0}$ and $\boldsymbol{Bx} = \boldsymbol{c}$ holds. If $\boldsymbol{x} = \boldsymbol{0}$, the next self-contradiction

$$\boldsymbol{0} \leq \boldsymbol{c} = \boldsymbol{Bx} = \boldsymbol{0}$$

is obtained. Therefore, let us assume $\boldsymbol{x} \geq \boldsymbol{0}$ and, define $\boldsymbol{I}_x = \{i \in N | x_i = 0\}$. $\boldsymbol{I}_x = \phi$, therefore, if $\boldsymbol{x} > \boldsymbol{0}$, the proof is complete so let $\boldsymbol{I}_x \neq \phi$. Here, for any $i \in \boldsymbol{I}_x$,

$$\begin{aligned} 0 &\geqq \sum\nolimits_{j=1}^{n} b_{ij} x_j \\ &= \sum\nolimits_{j \notin \boldsymbol{I}_x} b_{ij} x_j \\ &= c_i \geqq 0. \end{aligned}$$

Consequently, for any $i \in \boldsymbol{I}_x$ and $j \notin \boldsymbol{I}_x$, $b_{ij} = 0$, hence contradicts the indecomposability of $\boldsymbol{B}$.

(Notes on [Theorem 4.7])

Note on (i): (i) is equivalent to (i)′ [for any $\boldsymbol{c} > \boldsymbol{0}$, there exists some $\boldsymbol{x} > \boldsymbol{0}$ and $\boldsymbol{Bx} = \boldsymbol{c}$ holds] hence (i) is equivalent to (i)′.

Note on (iii): Generally, (iii) is known as the **Hawkins-Simon condition**. Additionally, in the proof of (ii) ⇒ (iii), we defined a k-order leading principal minor matrix taking k rows and k columns from the top left corner of $\boldsymbol{B}$ as $\boldsymbol{B}_k$, but it is fine to replace $\boldsymbol{B}_k$ with any $k(k \in N)$-order leading principal minor matrix of $\boldsymbol{B}$. This is because it is possible to take any k-order leading principal minor matrix of $\boldsymbol{B}$ by the identical interchanging of rows and columns as $\boldsymbol{B}_k$ furthermore, this maintains $\mu_{(-\boldsymbol{B})}$ unchanged.

4.5 Indecomposability and Decomposability

The fact that the nonlinear transformation's indecomposability plays a crucial role in the process of proving Frobenius' theorem can be easily understood from Sect. 4.3 too. However, there are two types of definitions of indecomposability of nonlinear transformation f one proposed by Morishima (1964) and the other by Nikaido (1968) and in this book, the definition by Morishima has been used. Therefore, the following clarifies these definitions as well as shows their equivalence.

⟨**Definition 4.1**⟩[14] Let $\boldsymbol{J}$ be any nonempty proper subset of $\boldsymbol{N} = \{1, \cdots, n\}$. If when $\boldsymbol{x} \in \boldsymbol{R}^n$ $\boldsymbol{y} \in \boldsymbol{R}^n$ satisfy $\boldsymbol{x_J} = \boldsymbol{y_J}$ and $\boldsymbol{x}_{\boldsymbol{J}^C} > \boldsymbol{y}_{\boldsymbol{J}^C}$, there exists $k \in \boldsymbol{J}$ that is $f_k(\boldsymbol{x}) \neq f_k(\boldsymbol{y})$, the nonlinear transformation f is said to be **indecomposable**.

[14] See Morishima (1964).

⟨**Definition 4.2**⟩[15] When $k \notin N(\boldsymbol{x},\boldsymbol{y})$ exists that is $f_k(\boldsymbol{x}) \neq f_k(\boldsymbol{y})$ for a pair $(\boldsymbol{x},\boldsymbol{y})$ of any vectors such that $\boldsymbol{x} \geqq \boldsymbol{y}$ and $\boldsymbol{N}(\boldsymbol{x},\boldsymbol{y}) = \{j \in \boldsymbol{N}|x_j > y_j\}$ become nonempty subsets of $\boldsymbol{N}$, the nonlinear transformation f is referred to as **indecomposablet**.

Before we move on to the equivalence of definitions, we will add a little about the differences and similarities of the two definitions as they appear.

Firstly, ⟨Definition 4.1⟩ takes as premise that $\boldsymbol{S_J} = \{(\boldsymbol{x},\boldsymbol{y})|\boldsymbol{x_J} = \boldsymbol{y_J} \text{ and } \boldsymbol{x}_{J^C} > \boldsymbol{y}_{J^C}\} \neq \phi$ can be defined for any subset $\boldsymbol{J} \neq \phi$ of $\boldsymbol{N}$ and means that for all $(\boldsymbol{x},\boldsymbol{y}) \in \boldsymbol{S_J} \subset \boldsymbol{N}$, there exists $k \in \boldsymbol{J}$ that satisfies $f_k(\boldsymbol{x}) \neq f_k(\boldsymbol{y})$. In this way, depending on the set $\boldsymbol{J}$, in respect of ⟨Definition 4.1⟩ that defines the vector pair $(\boldsymbol{x},\boldsymbol{y})$, ⟨Definition 4.2⟩ depends on $(\boldsymbol{x},\boldsymbol{y})$ and set $\boldsymbol{N}(\boldsymbol{x},\boldsymbol{y}) = \{j \in \boldsymbol{N}|x_j > y_j\} \neq \phi$ is defined to be a proper subset of $\boldsymbol{N}$.

Although there are differences in format both definitions are meaningful and $\boldsymbol{x} \geq \boldsymbol{y}$ is obvious.

[Theorem 4.8] ⟨Definition 4.1⟩ and ⟨Definition 4.2⟩ are equivalent.

(Proof) If both definitions are represented using the notation to simplify, they become as follows.

⟨**Definition 4.1**⟩

$$^{\forall}\boldsymbol{J}(\neq\phi) \subset \boldsymbol{N},\, ^{\forall}(\boldsymbol{x},\boldsymbol{y}) \in \boldsymbol{S_J},\, ^{\exists}K \in \boldsymbol{J}|f_k(\boldsymbol{x}) \neq f_k(\boldsymbol{y})$$

$$\text{However}, \boldsymbol{S_J} = \{(\boldsymbol{x},\boldsymbol{y})|\boldsymbol{x_J} = \boldsymbol{y_J} \ \textit{and}\ \boldsymbol{x}_{J^c} > \boldsymbol{y}_{J^c}\}$$

⟨**Definition 4.2**⟩

$$^{\forall}(\boldsymbol{x},\boldsymbol{y})\left|\begin{array}{c}\boldsymbol{N}(\boldsymbol{x},\boldsymbol{y})(\neq\phi)\subset\boldsymbol{N}\\ \text{and}\\ \boldsymbol{x}\geqq\boldsymbol{y}\end{array}\right.,\ ^{\exists}k \in \boldsymbol{N}(\boldsymbol{x},\boldsymbol{y})|f_k(\boldsymbol{x}) \neq f_k(\boldsymbol{y})$$

$$\text{However}, \boldsymbol{N}(\boldsymbol{x},\boldsymbol{y}) = \{j \in \boldsymbol{N}|x_j > y_j\}$$

⟨Definition 4.1⟩ ⇒ ⟨Definition 4.2⟩ If under ⟨Definition 4.1⟩, ⟨Definition 4.2⟩ is denied,

$$^{\exists}(\bar{\boldsymbol{x}},\bar{\boldsymbol{y}})\left|\begin{array}{c}\boldsymbol{N}(\bar{\boldsymbol{x}},\bar{\boldsymbol{y}})(\neq\phi)\subset\boldsymbol{N}\\ \text{and}\\ \bar{\boldsymbol{x}}\geqq\bar{\boldsymbol{y}}\end{array}\right|\, ^{\forall}i \in \boldsymbol{N}(\boldsymbol{x},\boldsymbol{y}), f_i(\bar{\boldsymbol{x}}) \neq f_i(\bar{\boldsymbol{y}})$$

Using $\boldsymbol{N}(\bar{\boldsymbol{x}},\bar{\boldsymbol{y}}) \neq \phi$ and $\boldsymbol{N}(\bar{\boldsymbol{x}},\bar{\boldsymbol{y}}) \subset \boldsymbol{N}$, $\boldsymbol{N}^C(\bar{\boldsymbol{x}},\bar{\boldsymbol{y}}) = \{j \in \boldsymbol{N}|\bar{x}_j = \bar{y}_j\} \subset \boldsymbol{N}$ and $\boldsymbol{N}^C(\bar{\boldsymbol{x}},\bar{\boldsymbol{y}}) \neq \phi$. Therefore, if we view ⟨Definition 4.1⟩'s $\boldsymbol{J}$ as $\boldsymbol{N}^C(\bar{\boldsymbol{x}},\bar{\boldsymbol{y}})$ and $(\boldsymbol{x},\boldsymbol{y})$ as $(\bar{\boldsymbol{x}},\bar{\boldsymbol{y}})$, it is certainly the case that $\bar{\boldsymbol{x}}_{N^C} = \bar{\boldsymbol{y}}_{N^C}$ and $\bar{\boldsymbol{x}}_N > \bar{\boldsymbol{y}}_N$. Thus, from ⟨Definition 4.1⟩, there exists $k \in \boldsymbol{N}^C(\bar{\boldsymbol{x}},\bar{\boldsymbol{y}})$ which is $f_k(\bar{\boldsymbol{x}}) \neq f_k(\bar{\boldsymbol{y}})$

[15] See Nikaido (1968).

However, from the assumption of proof by contradiction, $f_i(\overline{x}) = f_i(\overline{y})$ holds for all $i \in N^C(\overline{x}, \overline{y})$ hence a contradiction arises.

⟨Definition 4.1⟩ ⇐ ⟨Definition 4.2⟩ It can be shown in the same way as per the previous proof. QED

The two definitions' equivalence has been shown above. Hence, let us use this to define the decomposability of nonlinear transformations.

⟨**Definition 4.3**⟩ When a nonlinear transformation f is indecomposable, in other words, when there exists a nonempty proper subset $\boldsymbol{J}$ of $\boldsymbol{N} = \{1, \cdots, n\}$ and even when the vectors $\boldsymbol{x}$ and $\boldsymbol{y}$ satisfy $\boldsymbol{x}_J = \boldsymbol{y}_J$ and $\boldsymbol{x}_{J^C} > \boldsymbol{y}_{J^C}$, $f_i(\boldsymbol{x}) = f_i(\boldsymbol{y})$ for any $i \in \boldsymbol{J}$, the nonlinear transformation f is said to be **decomposable.**[16]

The indecomposability of n-order square matrix $\boldsymbol{A}$ can be sought in the following way by applying the definition of indecomposability of nonlinear transformation f to $f(\boldsymbol{x}) = \boldsymbol{A}\boldsymbol{x}$.

From ⟨Definition 4.1⟩, in respect of any nonempty proper subset $\boldsymbol{J}$ of $\boldsymbol{N}$, the vector pair $(\boldsymbol{x}, \boldsymbol{y})$ satisfies $\boldsymbol{x}_J = \boldsymbol{y}_J$ and $\boldsymbol{x}_{J^C} > \boldsymbol{y}_{J^C}$. For this $(\boldsymbol{x}, \boldsymbol{y})$, $k \in \boldsymbol{J}$ exists and $f_k(\boldsymbol{x}) \neq f_k(\boldsymbol{y})$. However, $f_k(\boldsymbol{x})$, $f_k(\boldsymbol{y})$ can be expressed as

$$f_k(\boldsymbol{x}) = \sum_{j=1}^{n} a_{kj}x_j = \sum_{j \in \boldsymbol{J}^C} a_{kj}x_j + \sum_{j \in \boldsymbol{J}} a_{kj}x_j$$

$$f_k(\boldsymbol{y}) = \sum_{j=1}^{n} a_{kj}y_j = \sum_{j \in \boldsymbol{J}^C} a_{kj}y_j + \sum_{j \in \boldsymbol{J}} a_{kj}y_j$$

respectively. However, if $\boldsymbol{x}_J = \boldsymbol{y}_J$ is considered, $f_k(\boldsymbol{x}) \neq f_k(\boldsymbol{y})$ is equivalent to

$$\sum_{j \in \boldsymbol{J}^C} a_{kj}x_j \neq \sum_{j \in \boldsymbol{J}^C} a_{kj}y_j. \tag{4.61}$$

Additionally, it is $\boldsymbol{x}_{J^C} > \boldsymbol{y}_{J^C}$, but if for all $j \in \boldsymbol{J}^C$, $a_{kj} = 0$, then (4.61) does not hold. Accordingly, (4.61) predicates the fact that there exists $k \in \boldsymbol{J}$ and $j \in \boldsymbol{J}^C$ that satisfy $a_{kj} \neq 0$.

Thus, the following definition is shown.

⟨**Definition 4.4**⟩ Let $\boldsymbol{A}$ be a n-order square matrix. When there exist $i_0 \in \boldsymbol{J}$ and $j_0 \in \boldsymbol{J}^c$ that satisfy $a_{i_0 j_0} \neq 0$ for any nonempty proper subset $\boldsymbol{J}$ of $\boldsymbol{N} = \{1, \cdots, n\}$, $\boldsymbol{A}$ is called **indecomposable**.

⟨**Definition 4.5**⟩ When n-order square matrix $\boldsymbol{A}$ is not indecomposable, in other words, when there exists a nonempty subset $\boldsymbol{J}$ of $\boldsymbol{N}$ and for any $i \in \boldsymbol{J}$ and $j \in \boldsymbol{J}^C$, $a_{ij} = 0$, $\boldsymbol{A}$ is called **indecomposable**.

Characteristics of a decomposable matrix can be summarized below.

[16] Here, decomposability was defined based on ⟨Definition 4.1⟩.

[Theorem 4.9] When $\boldsymbol{A}$ is a n-order square matrix, the following holds.

(i) For $\boldsymbol{A}$ to be decomposable, the necessary condition is that for some real number μ, there exist $\boldsymbol{x} \geq \boldsymbol{0}$ and $\boldsymbol{x} \not> \boldsymbol{0}$ that satisfy $\mu\boldsymbol{x} \geqq \boldsymbol{A}\boldsymbol{x}$.
(ii) For $\boldsymbol{A}$ to be decomposable, the sufficient condition is that $a_{ij} \geqq 0$ $(i \neq j; i, j = 1,\ldots,n)$ holds and also for $\boldsymbol{x} \geq \boldsymbol{0}$ and $\boldsymbol{x} \not> \boldsymbol{0}$ to exist that satisfy $\mu\boldsymbol{x} \geqq \boldsymbol{A}\boldsymbol{x}$ for some real number μ.
(iii) For $\boldsymbol{A}$ to be decomposable, the sufficient condition is that there exists $i_0 \in \boldsymbol{N} = \{1,\ldots,n\}$ and that for all $j \in \boldsymbol{N}$, which are different from i_0, $a_{i_0 j} = 0$ is satisfied.
(iv) For $\boldsymbol{A}$ to be decomposable, the necessary and sufficient condition is that $\boldsymbol{A}^t$ is decomposable.
(v) For $\boldsymbol{A}$ to be decomposable, the necessary and sufficient condition is that for some substitution matrix $\boldsymbol{P}$,

$$\boldsymbol{P}\boldsymbol{A}\boldsymbol{P}^t = \begin{bmatrix} \boldsymbol{A}_{11} & \boldsymbol{A}_{12} \\ \boldsymbol{0} & \boldsymbol{A}_{22} \end{bmatrix}$$

holds. However, $\boldsymbol{A}_{ii}$ $(i = 1, 2)$ is a square matrix.

(Proof)

(i) $\boldsymbol{A}$ is decomposable so there exists a proper subset $\boldsymbol{J} \neq \varphi$ of $\boldsymbol{N}$ and for any $i \in \boldsymbol{J}$ and $j \in \boldsymbol{J}^C$, $a_{ij} = 0$. Here if correspondingly to this $\boldsymbol{J}$, $\boldsymbol{x}^t = \left(\boldsymbol{x}^t_{J^C}, \boldsymbol{x}^t_J\right)^t$ is defined as $\boldsymbol{x}_{J^C} > \boldsymbol{0}$ and $\boldsymbol{x}_J = \boldsymbol{0}$, $\boldsymbol{x} \not> \boldsymbol{0}$ and $\boldsymbol{x} \geq \boldsymbol{0}$. Moreover, for this $\boldsymbol{x}$ and any real number μ,

$$\begin{aligned}(\mu - \boldsymbol{A})\boldsymbol{x} &= \begin{bmatrix} \mu\boldsymbol{I}_{J^C} - \boldsymbol{A}_{J^C J^C} & -\boldsymbol{A}_{J^C J} \\ -\boldsymbol{A}_{J J^C} & \mu\boldsymbol{I}_J - \boldsymbol{A}_{JJ} \end{bmatrix} \begin{bmatrix} \boldsymbol{x}_{J^C} \\ \boldsymbol{x}_J \end{bmatrix} \\ &= \begin{bmatrix} (\mu\boldsymbol{I}_{J^C} - \boldsymbol{A}_{J^C J^C})\boldsymbol{x}_{J^C} \\ \boldsymbol{0} \end{bmatrix} \begin{bmatrix} \boldsymbol{x}_{J^C} \\ \boldsymbol{x}_J \end{bmatrix}\end{aligned} \tag{4.62}$$

holds. Therefore if for any $i \in \boldsymbol{J}^C$, we select $\tilde{\mu} = \max_{i \in J^C} \left(\sum_{j \in J^C} a_{ij} x_j / x_i\right)$,

$$(\tilde{\mu}\boldsymbol{I}_{J^C} - \boldsymbol{A}_{J^C J^C})\boldsymbol{x}_{J^C} \geqq \boldsymbol{0}. \tag{4.63}$$

Therefore from (4.62) and (4.63), $(\tilde{\mu} - \boldsymbol{A})\boldsymbol{x} \geqq \boldsymbol{0}$, in other words, $\tilde{\mu}\boldsymbol{x} \geqq \boldsymbol{A}\boldsymbol{x}$ is obtained.
(ii) If we define $\boldsymbol{Q} = \{i \in N | x_i = 0\}$, from $\boldsymbol{x} \not> \boldsymbol{0}$ and $\boldsymbol{x} \geq \boldsymbol{0}$, $\boldsymbol{Q} \neq \phi$ and $\boldsymbol{Q} \subset \boldsymbol{N}$. Therefore for all $i \in \boldsymbol{Q}$,

$$
\begin{aligned}
0 &= (\mu - a_{ii})x_i \\
&\geqq \sum_{\substack{j\neq i \\ j\in Q}} a_{ij}x_j + \sum_{\substack{j\neq i \\ j\in Q^C}} a_{ij}x_j \\
&= \sum_{\substack{j\neq i \\ j\in Q^c}} a_{ij}x_j \\
&\geqq 0.
\end{aligned}
$$

Therefore $\sum_{\substack{j\neq i \\ j\in Q^C}} a_{ij}x_j = 0$ is derived but from the definition of $\boldsymbol{Q}$, $x_i > 0$. Therefore for any $i \in \boldsymbol{Q}$ and $j \in \boldsymbol{Q}^C$, $a_{ij} = 0$.

(iii) If the conclusion is contradicted, $\boldsymbol{A}$ is an indecomposable matrix. Therefore when the set with i_0 removed from $\boldsymbol{N}$ is set to be $\boldsymbol{N}(i_0) = \{1, \cdots, i_0 - 1, i_0 + 1, \cdots, n\}$, $\boldsymbol{N}(i_0) \neq \phi$. Moreover, for all $j \in \boldsymbol{N}$ other than $i_0 \in (\boldsymbol{N}(i_0))^C \subset \boldsymbol{N}$ and $i_0, j \notin (\boldsymbol{N}(i_0))^C$. Therefore, $a_{i_0 j} \neq 0$. However, this contradicts the assumptions hence $\boldsymbol{A}$ is decomposable.

(iv)

(Necessity) If the (j, i)th element of $\boldsymbol{A}^t$ is expressed as a^t_{ji},

$$a^t_{ji} = a_{ij} \; (i, j = 1, \cdots, n). \tag{4.64}$$

Considering the assumed decomposability of $\boldsymbol{A}$, there exists a proper subset $\boldsymbol{J} \neq \phi$ of $\boldsymbol{N}$ and for all $i \in \boldsymbol{J}$ and $j \notin \boldsymbol{J}$, $a_{ij} = 0$. Then. from $\boldsymbol{J} \subset \boldsymbol{N}$, $\boldsymbol{J}^C \neq \phi$ and from $\boldsymbol{J} \neq \phi$, $\boldsymbol{J}^C \subset \boldsymbol{N}$ are derived. Furthermore, from (4.64), for any $j \in \boldsymbol{J}^C$ and $i \notin \boldsymbol{J}^C$, $a^t_{ji} = 0 = a_{ij}$ is obtained. Thence $\boldsymbol{A}^t$ is also decomposable.

(Sufficiency) It can be proven in the same way as for necessity.

(v)

(Necessity) $\boldsymbol{A}$ is a decomposable matrix hence there exists a proper subset of $\boldsymbol{N}$ which is not empty $\boldsymbol{J}$ exists and $\boldsymbol{A}_{\boldsymbol{J}\boldsymbol{J}^C} = [\boldsymbol{0}]$ holds. Hence if we define $\boldsymbol{J} = \{k_1, \cdots, k_{\#J}\}$ and $\boldsymbol{J}^C = \{k_{\#J+1}, \cdots, k_n\}$ and let $k_{\#J+j}$ be $j (j = 1, \cdots, n - \#\boldsymbol{J})$, and k_i, $n - \#\boldsymbol{J} + i$ $(i = 1, \cdots, \#\boldsymbol{J})$, $\boldsymbol{A}$ can be transformed by using permutation matrix $\boldsymbol{P}$[17] corresponding to the numbering of $k_i \in \boldsymbol{J}$.

$$\boldsymbol{PAP}^t = \begin{bmatrix} \boldsymbol{A}_{11} & \boldsymbol{A}_{12} \\ \boldsymbol{0} & \boldsymbol{A}_{22} \end{bmatrix}$$

(Sufficiency) If we direct our attention to the set of numbers in a row (or a column) of $\boldsymbol{A}_{22}$, the conclusion is evident. QED

[17] By multiplying the substitution matrix to the left (right) of $\boldsymbol{A}$, the positions of the rows (columns) of $\boldsymbol{A}$ can be switched.

[Corollary 4.9] Let $\boldsymbol{A}$ be a n-order square matrix and $a_{ij_0} \geqq 0 (i \neq j)$. In that case, the necessary and sufficient condition for $\boldsymbol{A}$ to be decomposable is that for some real number μ, there exists $\boldsymbol{x} \geq \boldsymbol{0}$ and $\boldsymbol{x} \not> \boldsymbol{0}$ which satisfy $\mu\boldsymbol{x} \geqq \boldsymbol{A}\boldsymbol{x}$.

(Proof) It is evident from [Theorem 4.9; (i), (ii)]. QED

If $\boldsymbol{A}$ is (in)decomposable, since $\boldsymbol{A}$ is indecomposable (decomposable) and [Theorem 4.9], the characteristics relating to the indecomposable matrices can also be easily derived.

[Theorem 4.10] Let $\boldsymbol{A}$ be a n-order square matrix. Then the following hold.

(i) The sufficient condition for $\boldsymbol{A}$ to be indecomposable is that there are no $\boldsymbol{x} \geq \boldsymbol{0}$ and $\boldsymbol{x} \not> \boldsymbol{0}$ that satisfy $\mu\boldsymbol{x} \geqq \boldsymbol{A}\boldsymbol{x}$ for all real numbers μ.

(ii) The necessary condition for $\boldsymbol{A}$ to be indecomposable is that $a_{ij} < 0$ $(i \neq j; i, j = 1, \cdots, n)$ holds or that there are no $\boldsymbol{x} \geq \boldsymbol{0}$ and $\boldsymbol{x} \not> \boldsymbol{0}$ which satisfy $\mu\boldsymbol{x} \geqq \boldsymbol{A}\boldsymbol{x}$ for all real numbers μ.

(iii) The necessary condition for $\boldsymbol{A}$ to be indecomposable is that there exists $j_0 \in \boldsymbol{N}$ which is different to i and $a_{ij_0} \neq 0$ is satisfied for all $i \in \boldsymbol{N} = (1, \cdots, n)$.

(iv) The necessary and sufficient condition for $\boldsymbol{A}$ to be indecomposable is that in respect of all substitution matrices $\boldsymbol{P}$,

$$\boldsymbol{P}\boldsymbol{A}\boldsymbol{P}^t = \begin{bmatrix} \boldsymbol{A}_{11} & \boldsymbol{A}_{12} \\ \boldsymbol{0} & \boldsymbol{A}_{22} \end{bmatrix}$$

does not hold.

References

Farina, L., & Rinaldi, S. (2000). *Positive linear systems: Theory and applications*. John Wiley and Sons.

Frobenius, G. (1908). Über Martrizen aus posotiven Elementen I. *Sitzungsberichte der Königlichpreussischen Akademie der Wissenschaften*, 471–476.

Frobenius, G. (1909). Über Martrizen aus posotiven Elementen II. *Sitzungsberichte der Königlichpreussischen Akademie der Wissenschaften*, 514–518.

Hawkins, D., & Simon, H. A. (1949). Note: Some conditions of macroeconomic stability. *Econometrica, 17*, 245–248.

Kemp, M. C., & Kimura, Y. (1978). *Introduction to mathematical economics*. Springer-Verlag. ISBN-13: 978-0387903040.

Metzler, L. (1945). Stability of multiple markets: The Hicks conditions. *Econometrica, 13*, 277–292.

Morishima, M. (1964). *Equilibrium, stability, and growth—A multi-sectoral analysis*. Oxford University Press.

Narendra, K. S., & Shorten, R. (2009). A characterization of the Hurwitz stability of Metzler matrices. *Proceedings of the American Control Conference, 3*, 1833–1837.

Nikaido, H. (1968). *Convex structures and economic theory* (Mathematics in science and engineering: A series of monographs and textbooks) (Vol. 51). Academic Press. ISBN-13: 978-0125194501.
Takayama, A. (2010). Mathematical economics (2nd revised ed.). Cambridge University Press. ISBN-13: 978-0521314985.
Uekawa, Y., & Kimura, Y. (1987). *Riron Keizaigaku [Theoretical economics]*. Toyo Keizai Inc.. (in Japanese). ISBN-13: 978-4492812228.

Chapter 5
Dominant Diagonal Matrix and the Quadratic Form

5.1 Introduction

In the previous section, Frobenius' theorem relating to Metzler matrix ([Theorem 4.5]) was used to derive the conditions to obtain nonnegative solutions of the Leontief system ([Theorem 4.7]) but from a slightly different viewpoint, this condition, from Mckenzie (1960, p. 7), is closely related to the dominant diagonal matrix's characteristics, whose usefulness was acknowledged in economic analysis. Thus, in this section, the definition and basic characteristics of the dominant diagonal matrix—the so-called Hadamard's theorem and its expansion—will be described, and using these, some theorems relating to stability in linear dynamic systems will be demonstrated. In addition, the quadratic form, which plays a crucial role will also be explained.

5.2 Dominant Diagonal Matrix

In this section, unless otherwise stated, n-order square matrix $\boldsymbol{A} = [a_{ij}]$ with complex numbers as elements shall be taken as in scope for analysis. Notations to be used are defined below.

(Notation)

$\boldsymbol{N} = \{1, \cdots, n\}$: a set of numbers

$\boldsymbol{J}(k) = \boldsymbol{J} - \{k\}$: A set with k removed from $\boldsymbol{J}$

$\boldsymbol{l} = (1, \cdots, 1)^t$: n-dimensional column vector composed of 1 as all elements

$$\boldsymbol{B}(\boldsymbol{A}) = \begin{cases} |a_{ii}|(i = 1, \ldots, n) \\ -|a_{ij}|(i \neq j; i, j = 1, \ldots, n) \end{cases}$$

© Springer Nature Singapore Pte Ltd. 2024

K. Nakayama, *Eigenvalue Problem and Nonlinear Programming Problem*, New Frontiers in Regional Science: Asian Perspectives 70,
https://doi.org/10.1007/978-981-99-2943-6_5

Since Mckenzie (1960), concepts that are equivalent to dominant diagonal matrices presented by Hadamard (1903) have been shown by Henderson and Quandt (1980) or Tarr (1977) and at present, there are four equivalent concepts in existence regarding dominant diagonal matrix. One reason for this is because Mckenzie (1960) defined a quasidominant diagonal matrix as the same concept as the dominant diagonal matrix and applied it to economic analysis. Hence in this section, a unified concept incorporating the advantages of each definition is proposed then the equivalence of the various concepts is shown in simple ways.

Below we list the concepts and facts required for future arguments.

⟨**Definition 5.1**⟩ The matrix $\boldsymbol{A}$ has a **dominant diagonal** (hereafter abbreviated as **d.d.**) when $\boldsymbol{x} = (x_1, \cdots, x_n)^t > 0$ that satisfies $[\boldsymbol{B}(\boldsymbol{A})]\boldsymbol{x} > \boldsymbol{0}$ exists.[1] Especially, if all diagonal elements of $\boldsymbol{A}$ are positive (negative), $\boldsymbol{A}$ is said to have a **positive dominant diagonal (p.d.d.)** (**negative dominant diagonal (n.d.d.)**). Note that $\boldsymbol{A}$ with d.d. may be called a **diagonally dominant diagonal matrix**.

⟨**Definition 5.2**⟩ The matrix $\boldsymbol{A}$ is said to have **semidominant diagonal (s.d.d.)**[2] or to be a **semidominant matrix** when $\boldsymbol{x} > \boldsymbol{0}$[3] that satisfies $[\boldsymbol{B}(\boldsymbol{A})]\boldsymbol{x} \geq \boldsymbol{0}$ exists.

(Notes on <Definition 5.1>): From this definition and how to define $\boldsymbol{B}(\boldsymbol{A})$, if [Theorem 4.7] is applied to the matrix $\boldsymbol{B}(\boldsymbol{A})$, (i) in [Theorem 4.7] means that $\boldsymbol{A}$ has d.d.

[Theorem 5.1] Hadamard's Theorem[4] If $\boldsymbol{A}$ has a d.d., $\boldsymbol{A}$ is nonsingular.

(Proof) By the assumption,

$$[\boldsymbol{B}(\boldsymbol{A})]\boldsymbol{x} \equiv \boldsymbol{b} > \boldsymbol{0} \tag{5.1}$$

has a solution $\boldsymbol{x} > \boldsymbol{0}$. Assuming $\boldsymbol{A}$ is singular, there exists $\boldsymbol{q} = (q_1, \cdots, q_n)^t \neq \boldsymbol{0}$ that satisfies $q_{ii}a_{ii} = -\sum_{\substack{j=1 \\ j \neq i}}^{n} q_{ij}a_{ji}$. The absolute values on both sides of this equation are expressed as

$$|q_i a_{ii}| = |q_i||a_{ii}| = |\sum_{\substack{j=1 \\ j \neq i}}^{n} q_j a_{ji}| \leqq \sum_{\substack{j=1 \\ j \neq i}}^{n} |q_i||a_{ii}|, \quad (i = 1, \ldots, n). \tag{5.2}$$

[1] Strictly speaking, this definition assumes a matrix of second order or higher. In the case $\boldsymbol{A}$ is a first order matrix, that is $\boldsymbol{A} = [a_{11}]$, if there exists some $x_1 > 0$ and $|a_{11}|x_1 > 0$ is satisfied, then $\boldsymbol{A}$ is said to have d.d.

[2] Furthermore, when there is a need to demonstrate this $\boldsymbol{x}$, $\boldsymbol{A}$ can be said to have an s.d.d. that has $\boldsymbol{x}$ as a coefficient vector.

[3] Similar to <Definition 5.1>, <Definiton 5.2> is meaningful only for a second or higher order matrix, and when A is a first order matrix, d.d. and s.d.d. are same.

[4] See Hadamard (1903).

Therefore, if $z_i = |q_i|\ i = (1, \cdots, n)$, then $\boldsymbol{z} = (z_1, \cdots, z_n)' \geq \boldsymbol{0}$ by $\boldsymbol{q} \neq \boldsymbol{0}$. Consequently, using $\boldsymbol{b} > \boldsymbol{0}$, $\boldsymbol{z} \geq \boldsymbol{0}$ and (5.2),

$$\boldsymbol{b}'\boldsymbol{z} > \boldsymbol{0} \text{ and } [\boldsymbol{B}(\boldsymbol{A})]'\boldsymbol{z} \leqq 0 \tag{5.3}$$

hold. Though, according to Farkas's theorem ([Theorem 1.7]), the establishment of (5.3) means that (5.1) has no nonnegative solution, so that a contradiction arises. Thus, $\boldsymbol{A}$ is nonsingular. QED

[Theorem 5.2] Taussky's Theorem[5] If an indecomposable $\boldsymbol{A}$ has an s.d.d., $\boldsymbol{A}$ is nonsingular.

(Proof) Firstly, to show that all the diagonal elements of $\boldsymbol{A}$ are nonzero, if we assume that there exists $k \in \boldsymbol{N}$ that is $a_{kk} = 0$, for $\boldsymbol{x} = (x_1, \cdots, x_n)' > \boldsymbol{0}$ that is $[\boldsymbol{B}(\boldsymbol{A})]$ $\boldsymbol{x} \geqq \boldsymbol{0}$,

$$0 = |a_{kk}|x_k \geqq \sum_{\substack{j=1 \\ j \neq i}}^{n} |a_{kj}|x_j \geqq 0. \tag{5.4}$$

Therefore, from (5.4), it is $a_{kj} = 0$ for any $j \in \boldsymbol{N}$ which is different from k. However, this means $\boldsymbol{A}_{\boldsymbol{J}\boldsymbol{J}^C} = [\boldsymbol{0}]$ if $\boldsymbol{J} = \{k\} \subset \boldsymbol{N}$ and contradicts the indecomposability of $\boldsymbol{A}$. Therefore for any $i \in \boldsymbol{N}$, $a_{ii} \neq 0$.

Thus, if we define $\boldsymbol{J} = \left\{ i \in \boldsymbol{N} \,\middle|\, |a_{ii}|x_i > \sum_{\substack{j=1 \\ j \neq i}}^{n} |a_{ij}|x_j \right\}$, from the definition of s.d.d., $\boldsymbol{J} \neq \phi$. If $\boldsymbol{J} = \boldsymbol{N}$, $\boldsymbol{A}$ has a d.d. Hence, the conclusion is evident from [Theorem 5.1]. Therefore, if $\boldsymbol{J} \subset \boldsymbol{N}$, $\boldsymbol{J}^C = \boldsymbol{N} - \boldsymbol{J} = \left\{ i \in \boldsymbol{N} \,\middle|\, |a_{ii}|x_i = \sum_{\substack{j=1 \\ j \neq i}}^{n} |a_{ij}|x_j \right\} \neq \phi$. $\boldsymbol{A}$ is indecomposable, hence there exist some numbers $i_0 \in \boldsymbol{J}^C$ and $j_0 \in \boldsymbol{J}$, and $a_{i_0 j_0} \neq 0$. Accordingly, if the j_0th element of $\boldsymbol{x}$ is rewritten as $\tilde{x}_{j_0} \in \left(\frac{1}{|a_{j_0 j_0}|} \sum_{\substack{j=1 \\ j \neq j_0}}^{n} |a_{j_0 j}|x_j,\ x_{j_0} \right) (\leqq \#\boldsymbol{J}^C)$, to correspond with the number $j_0 \in \boldsymbol{J}$, at least the i_0th equality can be transformed to be an inequality equation while maintaining the existing inequality belonging to $\boldsymbol{J}$. If this procedure is repeated a limited number of times ($\leqq \#\boldsymbol{J}^C$), $\boldsymbol{A}$ satisfies the condition for [Theorem 5.1]. Thus, $\boldsymbol{A}$ is certainly nonsingular. QED

Next, let us think about applying the results we have up to this point to decomposable matrices. For this, the following lemma is useful.

[Lemma 5.1] When a matrix $\boldsymbol{A}$ is decomposable, there exists a partition $\equiv\{\boldsymbol{Q}_1, \cdots, \boldsymbol{Q}_m\}$ of $\boldsymbol{N}(=\{1, \cdots, n\})$ that satisfies all of the conditions below, ①–⑤.

---

[5] See Taussky (1949).

① $\boldsymbol{A}_{N-\cup_{t=1}^{i}\boldsymbol{Q}_t\boldsymbol{Q}_i} = [\boldsymbol{0}] \quad (i = 1, \cdots, m-1)$
② $\bigcup_{i=1}^{m}\boldsymbol{Q}_i = \boldsymbol{N}$
③ $\boldsymbol{Q}_i \cap \boldsymbol{Q}_j = \phi \quad (i \neq j)$
④ $\boldsymbol{Q}_i \neq \phi \quad (i = 1, \cdots, m)$
⑤ $\boldsymbol{A}_{\boldsymbol{Q}_i\boldsymbol{Q}_i}(i = 1, \cdots, m)$ is indecomposable

(Proof) Prove by induction on the number of order of a matrix.

Firstly, if $n = 2$, from the decomposability, assumed, one of the nondiagonal elements of $\boldsymbol{A} = \begin{bmatrix} a_{11} & a_{12} \\ a_{21} & a_{22} \end{bmatrix}$ is 0. $a_{21} = 0$ or, in response to $a_{12} = 0$, let $\boldsymbol{Q}_1 = \{1\}$ or $\boldsymbol{Q}_2 = \{2\}$ and let $\boldsymbol{Q}_2 \equiv \boldsymbol{N} - \boldsymbol{Q}_1$. Here, it is evident that $\{\boldsymbol{Q}_1, \boldsymbol{Q}_2\}$ satisfies all of the conclusions other than ⑤. Moreover, $\boldsymbol{A}_{\boldsymbol{Q}_i\boldsymbol{Q}_i}(i = 1, 2)$ is a linear matrix so from the definition of indecomposability, ⑤ is also satisfied.

Next, let $n \geqq 3$. From the decomposability of A there exists $\boldsymbol{K} \subset \boldsymbol{N}$ that becomes $\boldsymbol{A}_{\boldsymbol{LK}} = [\boldsymbol{0}]$. However, $\boldsymbol{L} \equiv \boldsymbol{N} - \boldsymbol{K}$. If $\boldsymbol{A}_{\boldsymbol{KK}}$, $\boldsymbol{A}_{\boldsymbol{LL}}$ are both indecomposable, is the class of $\boldsymbol{N}$ sought by $\{\boldsymbol{K}, \boldsymbol{L}\}$. Therefore, below, the proof will be given for the cases when

(I) $\boldsymbol{A}_{\boldsymbol{KK}}$ and $\boldsymbol{A}_{\boldsymbol{LL}}$ are both decomposable.
(II) $\boldsymbol{A}_{\boldsymbol{KK}}$ or $\boldsymbol{A}_{\boldsymbol{LL}}$ is decomposable.

(I) If the induction hypothesis is applied to each $\boldsymbol{A}_{\boldsymbol{KK}}$, $\boldsymbol{A}_{\boldsymbol{LL}}$, there exist classifications of $\boldsymbol{K}$ and $\boldsymbol{L}$, $\{\boldsymbol{K}_1, \cdots, \boldsymbol{K}_p\}$ and $\{\boldsymbol{L}_1, \cdots, \boldsymbol{L}_q\}$, and the following expressions hold.

①′ $\boldsymbol{A}_{\boldsymbol{K}-\cup_{t=1}^{i}\boldsymbol{K}_t\boldsymbol{K}_i} - [\boldsymbol{0}] \quad (i - 1, \cdots, p-1)$
②′ $\bigcup_{i=1}^{p}\boldsymbol{K}_i = K$
③′ $\boldsymbol{K}_i \cap \boldsymbol{K}_j = \phi \quad (i \neq j)$
④′ $\boldsymbol{K}_i \neq \phi \quad (i = 1, \cdots, p)$
⑤′ $\boldsymbol{A}_{\boldsymbol{K}_i\boldsymbol{K}_i}(i = 1, \cdots, p)$ is indecomposable
and
①″ $\boldsymbol{A}_{\boldsymbol{L}-\cup_{t=1}^{i}\boldsymbol{L}_t\boldsymbol{L}_i} = [\boldsymbol{0}] \quad (i = 1, \cdots, q-1)$
②″ $\bigcup_{i=1}^{q}\boldsymbol{L}_i = \boldsymbol{L}$
③″ $\boldsymbol{L}_i \cap \boldsymbol{L}_j = \phi \quad (i \neq j)$
④″ $\boldsymbol{L}_i \neq \phi \quad (i = 1, \cdots, q)$
⑤″ $\boldsymbol{A}_{\boldsymbol{L}_i\boldsymbol{L}_i}(i = 1, \cdots, q)$ is indecomposable

Defining

$$m \equiv p + q \text{ and } \boldsymbol{Q}_i \equiv \begin{cases} \boldsymbol{K}_i & (i = 1, \cdots, p) \\ \boldsymbol{L}_{i-p} & (i = p+1, \cdots, m) \end{cases},$$

it is easy to see that m certainly has a limit and $\{\boldsymbol{Q}_1, \cdots, \boldsymbol{Q}_m\}$ satisfy ②–④. Below, the establishment of ① is shown.

Firstly, if $1 \leqq i \leqq p-1$,[6]

$$\begin{aligned} A_{N-\cup_{t=1}^{i} Q_t Q_i} &= A_{N-\cup_{t=1}^{i} K_t K_i} \\ &= A_{(N-K)\cup\left(K-\cup_{t=1}^{i} K_t\right) K_i} \\ &= \begin{bmatrix} A_{N-K\ K_i} \\ A_{K-\cup_{t=1}^{i} K_t\ K_i} \end{bmatrix} \end{aligned}$$

Here, if we note the assumption from the approach of conduction in respect of A_{KK}, ①′, and the fact that $A_{N-K\ K_i}$ is a subset of $A_{N-KK} = [\mathbf{0}]$, $A_{N-\cup_{t=1}^{i} Q_t\ Q_i} = [\mathbf{0}] (i = 1, \cdots, p-1)$.

Moreover, from the definitions of $K_p = Q_p \subset K = \cup_{t=1}^{p} Q_i$ and K, $A_{N-\cup_{t=1}^{p} Q_t\ Q_p} = A_{N-K\ K_p}$ is a submatrix of $A_{N-KK} = [\mathbf{0}]$. Therefore

$$A_{N-\cup_{t=1}^{i} Q_t\ Q_p} = [\mathbf{0}], \quad (i = 1, \cdots, p). \tag{5.5}$$

Next, when $p + 1 \leqq i \leqq m - 1$, let everything be as above and

$$\begin{aligned} A_{N-\cup_{t=1}^{i} Q_t\ Q_i} &= A_{N-\left(\cup_{t=1}^{p} Q_t \cup \cup_{s=1}^{i} Q_s\right)\ Q_i} \\ &= A_{N-\left(K \cup_{s=1}^{i-p} L_s\right)\ L_{i-p}} \\ &= A_{N-K-\cup_{s=1}^{i-p} L_s\ L_{i-p}} \\ &= [\mathbf{0}], \quad (i = p+1, \cdots, m-1). \end{aligned} \tag{5.6}$$

Combining (5.5) and (5.6) above derives ①, in other words, $A_{N-\cup_{t=1}^{i} Q_t\ Q_i} = [\mathbf{0}]\ (i = 1, \cdots, m-1)$.

(II) If only $A_{KK}(A_{LL})$ is decomposable, we just need to set $p = 1$ and $Q_1 = K(q = 1$ and $Q_m = L)$ in the proof of (I). This concludes the proof. QED

(Notes on [Lemma 5.1]): It is fine to replace condition ① $A_{N-\cup_{t=1}^{i} Q_t\ Q_i} = [\mathbf{0}]$ $(i = 1, \cdots, m-1)$ with $A_{Q_i\ N-\cup_{t=1}^{i} Q_t} = [\mathbf{0}]$.

Let us call the leading principal minor matrix $\mathbf{A}_{Q_i Q_i} (i = 1, \cdots, m)$ corresponding to the types of N shown in [Lemma 5.1] going forward as **indecomposable principal submatrix characterizing the decomposability (IPSD)** in the sense that it is a leading principal minor matrix that is decomposable and characterizes the decomposability of the matrix. Moreover, using this terminology, the IPSD of a decomposable matrix is itself the matrix and also that the opposite is true. Thinking in this way, the following definition should be very natural.

---

[6] Generally, $U \supset T \supset S \Rightarrow (U - T) \cup (T - S) = U - S$. Therefore, $N \supset K \supset \cup_{t=1}^{i} K_t$.

⟨**Definition 5.3**⟩ When all of the IPSDs of the matrix $\boldsymbol{A}$ has s.d.d., $\boldsymbol{A}$ is said to have a **synthesized quasidominant diagonal (s.q.d.d.).**[7]

To compare with this new concept, s.q.d.d., let us bring up an existing concept.

⟨**Definition 5.4**⟩[8] The matrix $\boldsymbol{A}$, when there exists $\boldsymbol{x} > \boldsymbol{0}$ which satisfies the following condition or one of (a) or (b), is called a **modified quasidominant diagonal (m.q.d.d.).**

(a) $\boldsymbol{A}$ is indecomposable and $[\boldsymbol{B}(\boldsymbol{A})]\boldsymbol{x} \geq \boldsymbol{0}$
(b) For all $\boldsymbol{J} \subset \boldsymbol{N}$ that are $\boldsymbol{A}_{\boldsymbol{J}\ N\ -\ \boldsymbol{J}} = [\boldsymbol{0}]$, $[\boldsymbol{B}(\boldsymbol{A}_{\boldsymbol{JN}})]\boldsymbol{x} > \boldsymbol{0}$ and $[\boldsymbol{B}(\boldsymbol{A})]\boldsymbol{x} \geqq \boldsymbol{0}$

⟨**Definition 5.5**⟩[9] When some $\boldsymbol{x} > \boldsymbol{0}$ exists and $[\boldsymbol{B}(\boldsymbol{A}_{\boldsymbol{JJ}})]\boldsymbol{x}_{\boldsymbol{J}} \geq \boldsymbol{0}$ is satisfied for any $\boldsymbol{J} \subseteq \boldsymbol{N}$, the matrix $\boldsymbol{A}$ is said to have a **quasidominant diagonal (q.d.d.).**

⟨**Definition 5.6**⟩ **Tarr's Theorem**[10] The matrix $\boldsymbol{A}$, when there exists $\boldsymbol{x} > \boldsymbol{0}$ that satisfies the following condition or (a) or (b), is said to have a **quasidominant diagonal by Tarr (t.q.d.d.).**[11,12]

(a) $\boldsymbol{A}$ is indecomposable and $[\boldsymbol{B}(\boldsymbol{A})]\boldsymbol{x} \geqq \boldsymbol{0}$
(b) There exists some permutation matrix $\boldsymbol{P}$ such that

$$\boldsymbol{P}^t\boldsymbol{A}\boldsymbol{P} = \begin{bmatrix} \boldsymbol{A}_{11} & \boldsymbol{A}_{12} & \cdots & \boldsymbol{A}_{1m} \\ \boldsymbol{0} & \boldsymbol{A}_{21} & \cdots & \boldsymbol{A}_{2m} \\ \boldsymbol{0} & \boldsymbol{0} & & \vdots \\ \vdots & \vdots & & \\ \boldsymbol{0} & \boldsymbol{0} & \boldsymbol{0} & \boldsymbol{A}_{mm} \end{bmatrix}$$

Each $\boldsymbol{A}_{ii}(i = 1, \cdots, m)$ is $\boldsymbol{A}$'s indecomposable leading principal minor matrix and $[\boldsymbol{B}(\boldsymbol{A}_{ii})]\boldsymbol{x}_{J_i} \geq \boldsymbol{0}$, however J_i is a number set forming $\boldsymbol{A}_{ii}$.

---

[7] It is clear that $(\boldsymbol{x}_{Q_1}, \ldots, \boldsymbol{x}_{Q_m})$ on s.q.d.d. does not become $\boldsymbol{x} > \boldsymbol{0}$ in the other three definitions shown below(⟨Definition 5.4⟩–⟨Definition 5.6⟩). However, $\boldsymbol{x}_{Q_i}$ is similar to the definition of s.q.d.d.

[8] See Henderson and Quandt (1980).

[9] See Mckenzie (1960, p. 48) and Uekawa (1971, p. 198).

[10] See Tarr (1977, p. 172).

[11] Tarr (1977) named⟨Definition 5.5⟩as revised quasi-dominant diagonal and⟨Definition 5.6⟩as quasi-dominant diagonal. However, as it was Mckenzie (1960) that had presented the concept quasi-dominant diagonal and to clarify that⟨Definition 5.6⟩is a concept of Tarr, in this book, we've renamed⟨Definition 5.5⟩ and⟨Definition 5.6⟩.

[12] Woodbury (1954, p.357, Corollary 3.6) proposes substantially the same concept as t.q.d.d. regarding the Leontief system. For details on this, see Nakayama (1993b).

[Lemma 5.2.1] A matrix with $\boldsymbol{A}$ s.q.d.d. has a d.d.

(Proof) If $\boldsymbol{A}$ is decomposable, $\boldsymbol{A}$ has an s.d.d. Hence from Taussky's theorem ([Theorem 5.2]), it is clear that $\boldsymbol{A}$ has a d.d.

If $\boldsymbol{A}$ is decomposable, there exists a class in $\boldsymbol{N}$ in [Lemma 5.1], $\{\boldsymbol{Q}_1, \cdots, \boldsymbol{Q}_m\}$, and each of $\boldsymbol{A}_{\boldsymbol{Q}_i\boldsymbol{Q}_i}(i=1, \cdots, m)$ is indecomposable. Each $\boldsymbol{A}_{\boldsymbol{Q}_i\boldsymbol{Q}_i}$ has an s.d.d. so these have d.d. In other words, for all numbers k constructing any $\boldsymbol{Q}_i$, there exists $\boldsymbol{x}_{\boldsymbol{Q}_i} > \boldsymbol{0}$, which satisfies $|a_{kk}|x_k > \sum_{i\in \boldsymbol{Q}_i(k)}^{n} |a_{ki}|x_i$.

The equation above can be rewritten as $x_k| - |a_{kk}|| > \sum_{i\in \boldsymbol{Q}_i(k)}^{n} |a_{ki}|x_i$ but this means $-\left[\boldsymbol{B}\left(\boldsymbol{A}_{\boldsymbol{Q}_i\boldsymbol{Q}_i}\right)\right]$ having n.d.d. Accordingly, from [Theorem 5.12], which will be explained later, the real part of any eigenvalue of $-\left[\boldsymbol{B}\left(\boldsymbol{A}_{\boldsymbol{Q}_i\boldsymbol{Q}_i}\right)\right]$ becomes negative. Additionally, from the definition of $\left[\boldsymbol{B}\left(\boldsymbol{A}_{\boldsymbol{Q}_i\boldsymbol{Q}_i}\right)\right]$, the nondiagonal elements of $-\left[\boldsymbol{B}\left(\boldsymbol{A}_{\boldsymbol{Q}_i\boldsymbol{Q}_i}\right)\right]$ are all nonnegative. Hence, for any $\boldsymbol{Q}_i$, $\lambda^*_{\boldsymbol{Q}_i} < 0$. However, $\lambda^*_{\boldsymbol{Q}_i}$ is the Frobenius root of $-\left[\boldsymbol{B}\left(\boldsymbol{A}_{\boldsymbol{Q}_i\boldsymbol{Q}_i}\right)\right]$. Furthermore, if the Frobenius root of $-[\boldsymbol{B}(\boldsymbol{A})]$ is expressed as λ^*, from the matrix equation characteristics, class, and the form of decomposition with $\{\boldsymbol{Q}_1, \cdots, \boldsymbol{Q}_m\}$,

$$det(\lambda^* \boldsymbol{I}_n - \boldsymbol{A})^* = \Pi_{i=1}^{m} det\left(\lambda^*_{\boldsymbol{Q}_i} \boldsymbol{I}_{\#\boldsymbol{Q}_i} - \boldsymbol{A}_{\boldsymbol{Q}_i\boldsymbol{Q}_i}\right).$$

Therefore, $\lambda^* = \max_{i=1,\cdots,m} \lambda^*_{\boldsymbol{Q}_i} < 0$. Hence from Frobenius theorem [Theorem 4.5; (iv)],

$$\boldsymbol{0} \leq (0 \cdot \boldsymbol{I}_n) - [-\boldsymbol{B}(\boldsymbol{A})]^{-1} = [\boldsymbol{B}(\boldsymbol{A})]^{-1}.$$

Accordingly, if for any $\boldsymbol{y} > \boldsymbol{0}$, $\boldsymbol{x} \equiv [\boldsymbol{B}(\boldsymbol{A})]^{-1}\boldsymbol{y} > \boldsymbol{0}$ is defined, $[\boldsymbol{B}(\boldsymbol{A})]\boldsymbol{x} > \boldsymbol{0}$. This shows that $\boldsymbol{A}$ has a d.d. QED

Hadamar's theorem that provides nonsingularity of a matrix with d.d. ([Theorem 5.1]) is widely known, but according to that, the nonsingularity of a matrix with an s.q.d.d. is also self-evident. Thus, let us capture this implication as [Lemma 5.2.2].

[Lemma 5.2.2] A matrix with s.q.d.d. is nonsingular.

Now, let us mention the direct differences between s.q.d.d. and other concepts ahead of the main theorems related to the equivalence of the various concepts aforementioned.

Looking at ⟨Definition 5.4⟩ and ⟨Definition 5.6⟩, condition (b) for t.q.d.d. And m.q.d.d. takes the decomposability of $\boldsymbol{A}$ as a premise. Therefore, even if we leave aside the concept of d.d., it was only Mckenzie (1960) who provided definitions without categorizing them into various decomposability and indecomposability. However, he had to pay the price of having to deal with all leading principal minor matrices in his analyses. Although Henderson and Quandt (1980) or Tarr (1977) succeeded in specifying the leading principal minor matrices to target, their definitions had to be separated depending on whether or not a matrix was decomposable.

In respect of the above, the condition for s.q.d.d. can be denoted in a unified format regardless of a matrix's decomposability or indecomposability. In addition, a submatrix, which is the key for there being a d.d. is shown as IPSD. Thus, s.q.d.d. can be considered as a refined version of the amended concept of Mckenzie (1960).

Additionally, s.q.d.d. is slightly more advantageous in practice, compared to m.q. d.d., q.d.d. and t.q.d.d. This is because the presence of s.q.d.d. only necessitates confirming that each IPSD has an s.d.d., whereas in the case of the latter three, checks must be made that there exists some n-dimensional vector $\boldsymbol{x} > \boldsymbol{0}$ and the leading principal minor matrix under consideration has an s.d.d. with a subvector of a corresponding $\boldsymbol{x}$ as a coefficient.

[Theorem 5.3] The following (i)–(v) are mutually equivalent.

(i) The matrix $\boldsymbol{A}$ has a d.d. ⟨Definition 5.1⟩
(ii) The matrix $\boldsymbol{A}$ has an m.q.d.d. ⟨Definition 5.4⟩
(iii) The matrix $\boldsymbol{A}$ has a q.d.d. ⟨Definition 5.5⟩
(iv) The matrix $\boldsymbol{A}$ has a t.q.d.d. ⟨Definition 5.6⟩
(v) The matrix $\boldsymbol{A}$ has a s.q.d.d. ⟨Definition 5.3⟩

(Proof) The fact that (i) means (iii) ~ (v) and that (ii) is derived from (iii) is evident. Therefore, if it is shown that (iv) ⇒ (i), (v) ⇒ (i), and (ii) ⇒ (v), the proof is complete. However, the first two are direct consequences of [Lemma 5.2] hence only (ii) ⇒ (v) needs to be proven.

(ii) ⇒ (v) The definition of m.q.d.d. when $\boldsymbol{A}$ is indecomposable is no other than $\boldsymbol{A}$'s only IPSD, $\boldsymbol{A}$ having an s.d.d., i.e., $\boldsymbol{A}$ has an s.q.d.d. Here below, just the case when $\boldsymbol{A}$ is decomposable is proved using proof by contradiction.

$\boldsymbol{A}$ is decomposable so from the assumption of proof by contradiction, there exists $\boldsymbol{Q}_k$ that exist in a class of $\boldsymbol{N}$, $\{\boldsymbol{Q}_1, \cdots, \boldsymbol{Q}_m\}$, and for any $\boldsymbol{y}_{\boldsymbol{Q}_k}$, $[\boldsymbol{B}(\boldsymbol{A}_{\boldsymbol{Q}_k\boldsymbol{Q}_k})]\boldsymbol{y}_{\boldsymbol{Q}_k} \not\geq \boldsymbol{0}$. Therefore from the definition of m.q.d.d., for the partial vector $\boldsymbol{x}_{\boldsymbol{Q}_k} > \boldsymbol{0}$ of $\boldsymbol{x}$ for which such existence is guaranteed,

$$[\boldsymbol{B}(\boldsymbol{A}_{\boldsymbol{Q}_k\boldsymbol{Q}_k})]\boldsymbol{x}_{\boldsymbol{Q}_k} \not\geq \boldsymbol{0} \tag{5.7}$$

or

$$[\boldsymbol{B}(\boldsymbol{A}_{\boldsymbol{Q}_k\boldsymbol{Q}_k})]\boldsymbol{x}_{\boldsymbol{Q}_k} = \boldsymbol{0} \tag{5.8}$$

hold. If (5.7) holds, for some number i_0 belonging to $\boldsymbol{Q}_k$,

$$|a_{i_0 i_0}|x_{i_0} < \sum_{j \in Q_k(i_0)} |a_{i_0 j}|x_j. \tag{5.9}$$

However, from the definition of m.q.d.d., $[B(A)]x \geqq 0$. Therefore the i_0th inequality is

$$|a_{i_0 i_0}|x_{i_0} \geqq \sum_{j \in N(i_0)} |a_{i_0 j}|x_j = \sum_{j \in Q_k(i_0)} |a_{i_0 j}|x_j + \sum_{j \in N - Q_k} |a_{i_0 j}|x_j. \tag{5.10}$$

(5.9) and (5.10) immediately leads to the following self-contradiction, $|a_{i_0 i_0}|x_{i_0} < \sum_{j \in Q_k(i_0)} |a_{i_0 j}|x_j \leqq |a_{i_0 i_0}|x_{i_0}$. Further, when (5.8) holds, for any number i of Q_k, $|a_{ii}|x_i = \sum_{j \in Q_k(i)} |a_{ij}|x_j \leqq \sum_{j \in N(i)} |a_{ij}|x_j \leqq |a_{ii}|x_i$. However, the inequality on the furthest right is due to the definition of m.q.d.d., $[B(A)]x \geqq 0$. Therefore

$$|a_{ii}|x_i = \sum_{j \in Q_k(i)} |a_{ij}|x_j = \sum_{j \in N(i)} |a_{ij}|x_j = |a_{ii}|x_i. \tag{5.11}$$

(5.11) means $\sum_{j \in N - Q_k} |a_{ij}|x_j = 0$, therefore for any number j of $(N - Q_k)$ and any number i of Q_k, $a_{ij} = 0$. In other words, $A_{Q_k\ N - Q_k} = [0]$. If this is combined with (5.8) and the definition of m.q.d.d., the following self-contradiction is derived yet again.

$$0 \leq [B(A_{Q_k N})]x = [B(A_{Q_k Q_k})]x_{Q_k} + [B(A_{Q_k\ N - Q_k})]x_{N - Q_k} = 0. \qquad \text{QED}$$

(Notes on [Theorem 5.3]): It is easy to understand that Tarr's theorem (1977, pp. 172–174) can be obtained as a system of this theorem but what needs to be noted is the fact that the actual proof of this theorem is a separate proof which greatly simplifies a part of the proof of Tarr. In reality, Tarr showed the equivalence of the four concepts cyclically [(iv) t.q.d.d. ⇒ Hawkins-Simon condition ⇒ (i) d.d. ⇒ (iii) q.d.d. ⇒ (ii) m.q.d.d. ⇒ (iv) t.q.d.d.] but (ii) ⇒ (v) in the proof of this theorem is one which replaces Tarr's (ii) ⇒ (iv).

Combining [Theorem 5.1] and [Theorem 5.2] with the theorem above, it can be shown that all of <Definition 5.1> to <Definition 5.6> are sufficient conditions to guarantee the regularity of the **weakly dominant diagonal matrix**(complex square matrix whose absolute values of each diagonal element are no lower than the absolute total of the nondiagonal elements of the row or the column), in other words, that they are necessary conditions for the weakly dominant diagonal matrix to become singular.

Furthermore, Kojima (1914) and Woodbury (1954) presented the necessary and sufficient condition for a weakly dominant diagonal matrix to be singular and it is obvious that studying the necessary and sufficient conditions are more desirable compared to discussions regarding just the necessary conditions for a weakly dominant diagonal matrix to be singular, which is what has been done since Hadamard (1903), such as by Mckenzie (1960), Henderson and Quandt (1980) and Tarr (1977). However, the description in Kojima's conclusion is slightly vague and Woodbury's theorem can be expanded a little more.[13] Thus, henceforth, these results will be refined and generalized to present the necessary and sufficient conditions for the weakly dominant diagonal matrix to be singular (or to be nonsingular).

[Lemma 5.3] When a matrix $\boldsymbol{A}$ satisfies the following conditions

(A1) $\boldsymbol{A}$ is not decomposable,
(A2)′ $[\boldsymbol{B}(\boldsymbol{A})]\boldsymbol{l} = \boldsymbol{0}$

[L 1-1] All $(n-1)$-order leading principal minor matrices of $\boldsymbol{A}$ are nonsingular
and
[L 1-2] $rank(\boldsymbol{A}) \geqq n-1$
hold.

(Proof)

[L 1-1] From (A2)′, $0 \cdot \boldsymbol{l} = -[\boldsymbol{B}(\boldsymbol{A})]\boldsymbol{l} = \boldsymbol{0}$ is a matter of course. Therefore, 0 is a real eigenvalue of $-[\boldsymbol{B}(\boldsymbol{A})]$ and $\boldsymbol{l}$ is an eigenvector of $-[\boldsymbol{B}(\boldsymbol{A})]$ corresponding to this 0. Moreover, from the definition of $[\boldsymbol{B}(\boldsymbol{A})]$ and (A1), $-[\boldsymbol{B}(\boldsymbol{A})]$ is also nondecomposable and from Frobenius theorem, none of the eigenvectors of $-[\boldsymbol{B}(\boldsymbol{A})]$ corresponding to a different eigenvalue of $\lambda^*_{-\boldsymbol{B}(\boldsymbol{A})}$ are semipositive. However, since $\boldsymbol{l} > \boldsymbol{0}$, $\boldsymbol{l} \geq \boldsymbol{0}$ is evident. Therefore, if we take the contraposition of the above, $0 - \lambda^*_{-[\boldsymbol{B}(\boldsymbol{A})]}$. Accordingly, again, from Frobenius theorem [Theorem 4.6; (iv)], $0 > \lambda^*_{-\boldsymbol{B}\left(\boldsymbol{A}\binom{i}{i}\right)}$ $(i = 1, \cdots, n)$ and $\boldsymbol{0} < \left\{0 \cdot \boldsymbol{I}_{n-1} - \left(-\boldsymbol{B}\left(\boldsymbol{A}\binom{i}{i}\right)\right)\right\}^{-1} = \boldsymbol{B}\left(\boldsymbol{A}\binom{i}{i}\right)^{-1}$.

If for any $\boldsymbol{y} > \boldsymbol{0}$, $\boldsymbol{x} = \boldsymbol{B}\left(\boldsymbol{A}\binom{i}{i}\right)^{-1}\boldsymbol{y} > \boldsymbol{0}$ is defined, $\boldsymbol{x} > \boldsymbol{0}$ certainly exists and $\boldsymbol{B}\left(\boldsymbol{A}\binom{i}{i}\right)\boldsymbol{x}(=\boldsymbol{y}) > \boldsymbol{0}$.

This is none other than that $\boldsymbol{B}\left(\boldsymbol{A}\binom{i}{i}\right)$ is a dominant diagonal matrix. Hence, from Hadamard's theorem ([Theorem 5.1]), $\boldsymbol{A}\binom{i}{i}$ is nonsingular for any $i \in \boldsymbol{N}$.

[L1-2] [L1-1] means $rank\boldsymbol{A}\binom{i}{i} = n-1$ $(i = 1, \cdots, n)$.

Also, if $rank(\boldsymbol{A}) \geqq$ *rank (a submatrix of* $\boldsymbol{A}$) is considered, $rank(\boldsymbol{A}) \geqq n - 1$. QED

[Lemma 5.4] When matrix $\boldsymbol{A}$ satisfies the following conditions:

(A1) A is indecomposable
(A2) $[\boldsymbol{B}(\boldsymbol{A})]\boldsymbol{l} \geqq \boldsymbol{0}$
(A3) $\boldsymbol{A}$ is specific,
the following hold.

[13] For details about this, see Nakayama (1993a, 1993b).

[L2-1] $[\boldsymbol{B}(\boldsymbol{A})]\boldsymbol{l} = \boldsymbol{0}$ (A2)′
[L2-2] For all $\boldsymbol{x} \neq \boldsymbol{0}$ which is $\boldsymbol{Ax} = \boldsymbol{0}$, each element of $\boldsymbol{x}$ is nonzero.
[L2-3] $rank(\boldsymbol{A}) = n - 1$
and
[L2-4] simultaneous equations

$$\varphi_{kk} + \theta_k + \pi = \varphi_{kj} + \theta_j (mod.2\pi) \quad \left({}^{\forall}k \in \boldsymbol{N},\ {}^{\forall}j \in \boldsymbol{Z}_k\right)$$

has a solution $\boldsymbol{\Theta}_N = (\theta_1, \cdots, \theta_n)^t \in \boldsymbol{R}^n$.

However, φ_{rs}, θ_s are arguments $\boldsymbol{Z}_k = \{j \in \boldsymbol{N}(k) | a_{ij} \neq 0\}$ of a_{rs}, x_s, respectively.

(Proof)

[L2-1] (A2) means $[\boldsymbol{B}(\boldsymbol{A})]\boldsymbol{l} \geq \boldsymbol{0}$ or $[\boldsymbol{B}(\boldsymbol{A})]\boldsymbol{l} = \boldsymbol{0}$. In the former case, from Taussky's theorem ([Theorem 5.2]), $\boldsymbol{A}$ is nonsingular. Consequently, under (A3), $[\boldsymbol{B}(\boldsymbol{A})]\boldsymbol{l} = \boldsymbol{0}$.
[L2-2] When $n = 1$, from (A3), $x_1 \neq 0$. Thus, the case for $n \geqq 2$ is proven. Let us suppose that some number k belongs to $\boldsymbol{N}$ and that, in respect of that k, $x_k = 0$. From this assumption for proof by contradiction and (A3),

$$\boldsymbol{Ax} = \sum\nolimits_{j=1}^{n} \boldsymbol{a}^j x_j = \sum\nolimits_{j \in \boldsymbol{N}(k)} \boldsymbol{a}^j x_j = \boldsymbol{0}. \tag{5.12}$$

Now, from [L1-1], we know that any of $\boldsymbol{A}$'s $(n - 1)$-order leading principal minor matrix is nonsingular, hence $\boldsymbol{A}\binom{k}{k}$ is also nonsingular. In other words, each column vector of $\boldsymbol{A}\binom{k}{k}$ is linearly independent. Accordingly, the remaining column vector of $\boldsymbol{A}$, excluding $\boldsymbol{a}^k$, also becomes linearly independent. From this, (5.12) and $x_k = 0$, all columns of $\boldsymbol{A}$ including $\boldsymbol{a}^k$ are linearly independent, in other words, $\boldsymbol{A}$ is nonsingular and contradicts (A3).
[L2-3] From [L1-2], $rank(\boldsymbol{A}) \geqq n - 1$. On the other hand, from (A3), $rank(\boldsymbol{A}) \leqq n - 1$. Hence, $rank(\boldsymbol{A}) = n - 1$.
[L2-4] From the definition of $\boldsymbol{Z}_k$, and from (A1), $\boldsymbol{N} \supseteq \boldsymbol{Z}_k \neq \phi$. Additionally, from (A3), $\boldsymbol{N}(\boldsymbol{A}) - \{\boldsymbol{0}\} = \{\boldsymbol{x} \neq \boldsymbol{0} | \boldsymbol{Ax} = \boldsymbol{0}\} \neq \phi$. However, $\boldsymbol{N}(\boldsymbol{A}) \equiv \{\boldsymbol{x} | \boldsymbol{Ax} = \boldsymbol{0}\}$. Here, if for $\boldsymbol{x} \neq \boldsymbol{0}$ that is $\boldsymbol{Ax} = \boldsymbol{0}$, $\boldsymbol{M}_x = \left\{ r \in N \middle| |x_\gamma| = \max_{j \in \boldsymbol{N}} |x_j| \right\}$ is defined, $\boldsymbol{N} \supseteq \boldsymbol{M}_x \neq \phi$ is also evident.

From above and (A3) again, for any number $k \in \boldsymbol{M}_x$,

$$\begin{aligned} |a_{kk}||x_k| &= \left| -\sum\nolimits_{j \in \boldsymbol{N}(k)} a_{kj} x_j \right| \\ &= \left| \sum\nolimits_{j \in \boldsymbol{N}(k)} a_{kj} x_j \right| \end{aligned}$$

$$
\begin{aligned}
&= \left| \sum\nolimits_{j \in Z_k} a_{kj} x_j a_{kj} x_j \right| \\
&\leqq \sum\nolimits_{j \in Z_k} |a_{kj}||x_j| \\
&\leqq \sum\nolimits_{j \in Z_k} |a_{kj}||x_k| \\
&\leqq |a_{kk}||x_k|.
\end{aligned}
$$

Hence, for all $k \in \boldsymbol{M}_x$, $a_{kk} \neq 0$ and

$$
|a_{kk}||x_k| = \left| \sum\nolimits_{j \in Z_k} a_{kj} x_j \right| = \sum\nolimits_{j \in Z_k} |a_{kj}||x_j| = \sum\nolimits_{j \in Z_k} |a_{kj}||x_k| = |a_{kk}||x_k|. \tag{5.13}
$$

Let us suppose now that there exists a $k \in \boldsymbol{M}_x$ that becomes $\boldsymbol{Z}_k \nsubseteq \boldsymbol{M}_x$. Here, there exists some number j_0, which does not belong to $\boldsymbol{M}_x$ and $|a_{kj0}||x_{j0}| < |a_{kj0}||x_k|$. Adding (5.13) to this derives the following self-contradiction.

$$
\begin{aligned}
|a_{kk}||x_k| &= \sum\nolimits_{j \in Z_k} |a_{kj}||x_j| \\
&= \sum\nolimits_{j \in Z_k(j_0)} |a_{kj}||x_j| + |a_{kj_0}||x_{j_0}| \\
&< \sum\nolimits_{j \in Z_k(j_0)} |a_{kj}||x_j| + |a_{kj_0}||x_k| \\
&= \sum\nolimits_{j \in Z_k} |a_{kj}||x_k| \\
&= |a_{kk}||x_k|.
\end{aligned}
$$

Therefore, for any number $k \in \boldsymbol{M}_x$, $\boldsymbol{Z}_k \subseteq \boldsymbol{M}_x$. Accordingly, $\boldsymbol{Z}_k^C \supseteq \boldsymbol{M}_x^C$.
Next, if we assume $\boldsymbol{M}_x \subset \boldsymbol{N}$, $\boldsymbol{M}_x^C \neq \phi$. If we remind ourselves that for any $k \in \boldsymbol{M}_x$, $\boldsymbol{Z}_k^C \supseteq \boldsymbol{M}_x$, $\boldsymbol{A}_{\boldsymbol{M}_x \boldsymbol{M}_x^C} = [\boldsymbol{0}]$, in other words, $\boldsymbol{A}$ becomes decomposable and contradicts (A1). Hence $\boldsymbol{M}_x = \boldsymbol{N}$ is shown and for any $k \in \boldsymbol{N}$, (5.13) holds. Moreover, a part of the equality $\left| \sum\nolimits_{j \in Z_k} a_{kj} x_j \right| = \sum\nolimits_{j \in Z_k} |a_{kj}||x_j|$ $(k = 1, \cdots, n)$ is equivalent to the following.
For any $k \in \boldsymbol{N}$, there exists $\tau_k \in \boldsymbol{R}$ and for any $j \in \boldsymbol{Z}_k$,

$$
\tau_k = \varphi_{kj} + \theta_j \tag{5.14}
$$

holds. However, τ_k is the angle of deviation of $a_{kj} x_j$.
Now, from $\boldsymbol{M}_x = \boldsymbol{N}$ and (A3) again, for any $k \in \boldsymbol{N}$,

$$
- a_{kk} x_k = \sum\nolimits_{j \in Z_k} a_{kk} x_j. \tag{5.15}
$$

If we express (5.15) in a polar form,

$$-|a_{kk}||x_k||e^{i(\varphi_{kk}+\theta_k)} = \sum_{j\in Z_k} |a_{kj}||x_j|e^{i\tau_k}. \tag{5.15$'$}$$

Utilizing $-1 = e^{i\pi}$, the equation above is:

$$|a_{kk}||x_k|e^{i(\varphi_{kk}+\theta_k+\pi)} = \sum_{j\in Z_k} |a_{kj}||x_j|e^{i\tau_k}. \tag{5.15$''$}$$

Additionally, from (A2)$'$ and $\boldsymbol{M}_x = \boldsymbol{N}$,

$$|a_{kk}||x_k| = \sum_{j\in Z_k} |a_{kj}||x_k| \tag{5.16}$$

is obtained. Therefore, (5.15)$''$ can be rearranged based on (5.16) and (5.14).

$$e^{i(\varphi_{kk}+\theta_k+\pi)} = e^{i\tau_k} = e^{i(\varphi_{kj}+\theta_j)}$$

This has demonstrated that $\boldsymbol{x}$'s angle of deviation vector $\boldsymbol{\Theta}_N = (\theta_1, \cdots, \theta_n) \in R^n$ is the solution for the following simultaneous equations, $\varphi_{kk} + \theta_k + \pi = \varphi_{kj} + \theta_j \; (mod.2\pi)$ ($^\forall k \in \boldsymbol{N}$, $^\forall j \in \boldsymbol{Z}_k$). QED

The argument of this section can be summarized using the following theorem.

[Theorem 5.4] When matrix $\boldsymbol{A}$ satisfies

(A1) $\boldsymbol{A}$ is not decomposable

and

(A2) $[\boldsymbol{B}(\boldsymbol{A})]\boldsymbol{l} \geqq \boldsymbol{0}$,

the necessary and sufficient condition for $\boldsymbol{A}$ to be a singular matrix is

$$[\boldsymbol{B}(\boldsymbol{A})]\boldsymbol{l} = \boldsymbol{0}$$

and for the simultaneous equation

$$\text{(i)} \quad \varphi_{kk} + \theta_k + \pi = \varphi_{kj} + \theta_j (mod.2\pi) \quad \left(^\forall k \in \boldsymbol{N},\ ^\forall j \in \boldsymbol{Z}_k\right)$$

to have solution $\boldsymbol{\Theta}_N = (\theta_1, \cdots, \theta_n)^t \in \boldsymbol{R}^n$.

However, the notation shall be as per those in [L2-4].

(Proof)

(Necessity) The conclusion sought is evident from [Lemma 5.4].

(Sufficiency) If we use (I)'s solution $\boldsymbol{\Theta}_N = (\theta_1, \cdots, \theta_n)^t \in R^n$ to define $x_k = e^{i\theta_k}$ $(k = 1, \cdots, n)$, it is certain that $\boldsymbol{x} = \left(e^{i\theta_1}, \cdots, e^{i\theta_n}\right)^t \neq \boldsymbol{0}$ and from the direct calculation, for all $k \in \boldsymbol{N}$,

$$a_{kk}x_k + \sum_{j\in N(k)} a_{kj} = a_{kk}x_k + \sum_{j\in Z_k} a_{kj}x_j$$

$$= |a_{kk}x_k|e^{i(\varphi_{kk}+\theta_k)} + \sum_{j\in Z_k} |a_{kj}x_j|e^{i(\varphi_{kj}+\theta_j)}$$

$$= |a_{kk}x_k|e^{i(\varphi_{kk}+\theta_k)} + \sum_{j\in Z_k} |a_{kj}x_j|e^{i(\varphi_{kk}+\theta_k+\pi)}$$

$$= e^{i(\varphi_{kk}+\theta k)}\left\{|a_{kk}||x_k| - \left(\sum_{j\in Z_k} |a_{kj}||x_j|\right)\right\}$$

$$= e^{i(\varphi_{kk}+\theta k)}\left(|a_{kk}| - \sum_{j\in Z_k} |a_{kj}|\right)$$

$$= e^{i(\varphi_{kk}+\theta k)} \cdot 0$$

$$= 0.$$

Therefore, the specificity of $\boldsymbol{A}$ has been shown. QED

Moreover, as a system of [Theorem 5.4], Woodbury's theorem (1954, pp. 350–352, Theorem 2) is derived.

[Corollary 5.4.1] When n-order complex square matrix $\boldsymbol{C} = [c_{ij}]$ satisfies:

(W1) $\boldsymbol{C}$ is indecomposable
(W2) $s_j = \sum_{i=1}^{n} |c_{ij}| \leqq 1 \quad (i = 1, \cdots, n)$
the necessary and sufficient conditions for $(\boldsymbol{I}_n - \boldsymbol{C})$ to become a singular matrix is that

$$s_j = 1$$

and there exists a real vector $\boldsymbol{\Theta}_N = (\theta_1, \cdots, \theta_n)^t$ that

$$\text{(II)} \quad c_{kj} = |c_{kj}|e^{i(\theta_j - \theta_k)} \quad ({}^{\forall}k \in \boldsymbol{N},\ {}^{\forall}j \in \boldsymbol{N}(k))$$

holds.

(Proof) For all $k \in \boldsymbol{N}$,

$$|1 - c_{kk}| \geqq 1 - |c_{kk}| \geqq \sum_{j=1}^{n} |c_{jk}| - |c_{kk}| = \sum_{j\in N(k)}^{n} |c_{jk}|. \tag{5.17}$$

Therefore, $[\boldsymbol{B}(\boldsymbol{I}_n - \boldsymbol{C})^t]\boldsymbol{l} \geqq \boldsymbol{0}$.[14] In addition, from the indecomposability of $\boldsymbol{C}$ and the singularity of $(\boldsymbol{I}_n - \boldsymbol{C})$, $(\boldsymbol{I}_n - \boldsymbol{C})^t$ is also an indecomposable and singular matrix. Therefore, from the necessary condition of [Theorem 5.4],

[14] If $\boldsymbol{A}$ in [Theorem 5.4] is regarded as $(\boldsymbol{I}_n - \boldsymbol{C})^t$, $\boldsymbol{B}(\boldsymbol{I}_n - \boldsymbol{C})^t = \begin{cases} |1 - c_{ii}| \\ -|-c_{ji}| = |c_{ji}|(i \neq j) \end{cases}$.

$$|1-c_{kk}| = \sum_{j\in N(k)} |c_{jk}| \quad (k=1,\cdots,n). \tag{5.18}$$

By (5.17) and (5.18),

$$|1-c_{kk}| = 1-|c_{kk}| = \sum_{j\in N(k)}^{n} |c_{jk}| \quad (k=1,\cdots,n).$$

Therefore, $c_{kk} \in \boldsymbol{R}$[15] $(k = 1,\cdots,n)$ and $s_k = \sum_{j=1}^{n} |c_{jk}| = 1\ (k=1,\cdots,n)$. If $\varphi_{jk}\ (j,\ k = 1,\cdots,n)$ is taken to be the argument of c_{jk}, from above, $\varphi_{kk} = 0$ $(k=1,\cdots,n)$. Additionally, If the fact that the argument of $-c_{jk}$ is given by $(\varphi_{jk}+\pi)$ is considered, as per [Theorem 5.4], there exists a solution to the following simultaneous equation $\theta_k = \varphi_{jk} = \theta_j(mod.2\pi)(^\forall k \in \boldsymbol{N},\ ^\forall j \in \boldsymbol{Z}_k)$, $\boldsymbol{\Theta}_N = (\theta_1,\cdots,\theta_n)^t \in R^n$. Therefore, $c_{jk} = |c_{jk}| e^{i(\theta_k-\theta_j)} \left(^\forall k \in \boldsymbol{N},\ ^\forall j \in \boldsymbol{Z}_k\right)$. Here, if we define

$$|c_{jk}| = \begin{bmatrix} |c_{jk}| & (c_{jk} \neq 0) \\ 0 & (c_{jk} = 0) \end{bmatrix} \quad (j \neq k),$$

$c_{jk} = |c_{jk}| e^{i(\theta_k-\theta_j)}\ (j \neq k)$. This shows the necessary condition. The sufficient condition is also apparent so the proof will be omitted. QED

Furthermore, the following corollary is obtained when the consideration's target is replaced by decomposable matrices.

[Corollary 5.4.2] When a matrix $\boldsymbol{A}$ satisfies

(A1)′ $\boldsymbol{A}$ is decomposable
and
(A2) $[\boldsymbol{B}(\boldsymbol{A})]\boldsymbol{l} \geqq \boldsymbol{0}$,
the necessary and sufficient condition for $\boldsymbol{A}$ to be a singular matrix is that in respect of some $\boldsymbol{Q}_p$ in a class of $\boldsymbol{N}$, $\{\boldsymbol{Q}_1,\cdots,\boldsymbol{Q}_m\}$ in **[Lemma 5.1]**, the following equation,

$$\left[\boldsymbol{B}\left(\boldsymbol{A}_{\boldsymbol{Q}_p\boldsymbol{Q}_p}\right)\right]\boldsymbol{l}_{\boldsymbol{Q}_p} = \boldsymbol{0}$$

holds and the simultaneous equation,

$$\text{(III)} \quad \varphi_{kk} + \theta_k + \pi = \varphi_{kj} + \theta_j(mod.2\pi) \quad \left(^\forall k \in \boldsymbol{Q}_p,\ ^\forall j \in \boldsymbol{Z}_{\boldsymbol{Q}_p(k)}\right)$$

has the solution $\boldsymbol{\Theta}_{\boldsymbol{Q}_p} = \left(\theta_1^p,\cdots,\theta_{\#\boldsymbol{Q}_p}^p\right)^t \in \boldsymbol{R}^{\#\boldsymbol{Q}_p}$, where $\boldsymbol{Q}_p = \left\{q_1^p,\cdots,q_{\#\boldsymbol{Q}_p}^p\right\}$ and $\boldsymbol{Z}_{\boldsymbol{Q}_p(k)} = \left\{j \in \boldsymbol{Q}_{p(k)} \middle| a_{kj} \neq 0\right\}$, other notations shall be taken to be the same as in [L2-4].

[15] $|1-c_{kk}| = 1-|c_{kk}|$ and the law of cosines mean that c_{kk} is a real number.

(Proof) If classification $=\{\boldsymbol{Q}_1,\cdots,\boldsymbol{Q}_m\}$ is used in the matrix equation of $\boldsymbol{A}$, $det(\boldsymbol{A})=\prod_{i=1}^{m} det(\boldsymbol{A}_{Q_iQ_i})$. Therefore, $\boldsymbol{A}$ being singular is equivalent to $\boldsymbol{A}_{Q_pQ_p}$ being singular in respect of some $\boldsymbol{Q}_p$ in $\{\boldsymbol{Q}_1,\cdots,\boldsymbol{Q}_m\}$. Additionally, it is evident that this $\boldsymbol{A}_{Q_pQ_p}$ satisfies the conditions of the theorem hence the conclusion can be obtained immediately. QED

5.3 Quadratic Form

In this section, we describe the characteristics of the quadratic form or Hermite form, which will be needed for the next section's arguments. To do so, we define notation and assumptions as below.

⟨Definition 5.7⟩

⟨D5.7-1⟩ A matrix $[\bar{a}_{ij}]$ in which each of the element a_{ij} of $m \times n$ complex matrix $\boldsymbol{A} = [a_{ij}]$ is replaced by its conjugate complex number is called the **conjugate matrix** of $\boldsymbol{A}$ and is represented as $\bar{\boldsymbol{A}}$.

⟨D5.7-2⟩ $\left(\overline{\boldsymbol{A}^t}\right) = (\bar{\boldsymbol{A}})^t$ always holds so this $n \times m$ matrix is called a **transposed conjugate matrix** of $\boldsymbol{A}$ and is expressed as $\boldsymbol{A}^*$.

$$\boldsymbol{A}^* = \left(\overline{\boldsymbol{A}^t}\right) = (\bar{\boldsymbol{A}})^t.$$

(Notes on ⟨Definition 5.7⟩)

① $\bar{\boldsymbol{A}}$ and $\boldsymbol{A}^*$ have several characteristics.

$$(\bar{\bar{\boldsymbol{A}}}) = \boldsymbol{A}, \overline{(\alpha \boldsymbol{A})} = \bar{\alpha}\bar{\boldsymbol{A}}, \overline{(\boldsymbol{A}+\boldsymbol{B})} = \bar{\boldsymbol{A}} + \bar{\boldsymbol{B}}, \overline{\boldsymbol{AB}} = \bar{\boldsymbol{A}}\bar{\boldsymbol{B}},$$

$$(\boldsymbol{A}^*)^* = \boldsymbol{A}, (\alpha \boldsymbol{A})^* = \bar{\alpha}\boldsymbol{A}^*, (\boldsymbol{A}+\boldsymbol{B})^* = \boldsymbol{A}^* + \boldsymbol{B}^*,$$

$$(\boldsymbol{AB})^* = \boldsymbol{B}^*\boldsymbol{A}^*.$$

However, α is a complex number.

② If each element of $\boldsymbol{A}$ is a real number, in other words, if $\boldsymbol{A}$ is a real matrix, $\boldsymbol{A}^* = \boldsymbol{A}^t$.

⟨Definition 5.8⟩ Let $\boldsymbol{A}$ be a n-order complex square matrix.

⟨D5.8-1⟩ When $\boldsymbol{AA}^* = \boldsymbol{A}^*\boldsymbol{A}$ is satisfied, $\boldsymbol{A}$ is called a **normal matrix**.
⟨D5.8-2⟩ When $\boldsymbol{AA}^* = \boldsymbol{A}^*\boldsymbol{A} = \boldsymbol{I}_n$ is satisfied, $\boldsymbol{A}$ is called a **unitary matrix**.
⟨D5.8-3⟩ When $\boldsymbol{A}^* = \boldsymbol{A}$ is satisfied, $\boldsymbol{A}$ is called a **Hermitian matrix**.
⟨D5.8-4⟩ When $\boldsymbol{A}^* = -\boldsymbol{A}$ is satisfied, $\boldsymbol{A}$ is called a **skew Hermitian matrix**.

(Notes on ⟨Definition 5.8⟩)

② A unitary matrix, a Hermitian matrix, and a skew Hermitian matrix are all normal matrices.

① When $\boldsymbol{A}$ is a real square matrix, its unitary matrix, Hermitian matrix, and skew Hermitian matrix are referred to as an **orthogonal matrix**, **symmetric matrix**, and **skew-symmetric matrix**, respectively. Of course, these fall under the category of the normal matrix.

⟨Definition 5.9⟩ Let $\boldsymbol{A}$ be a n-order complex square matrix and $\boldsymbol{x} = (x_1, \cdots, x_n)^t$ a n-dimensional vector and let

$$q(\boldsymbol{x}, \boldsymbol{A}) = \boldsymbol{x}^t \boldsymbol{A} \boldsymbol{x}.$$

Here, $q(\boldsymbol{x}, \boldsymbol{A})$ is called a **quadratic form** of the variable $x_1, \cdots, x_n$ and $\boldsymbol{A}$ is called its **coefficient matrix**.

(Notes on ⟨Definition 5.9⟩)

① A second-order coefficient matrix can be thought of as a symmetric matrix. The reasons for this shall be demonstrated below. $\boldsymbol{A} = \left(\frac{\boldsymbol{A}+\boldsymbol{A}^t}{2}\right) + \left(\frac{\boldsymbol{A}-\boldsymbol{A}^t}{2}\right)$ hence for any $\boldsymbol{x}$, $q(\boldsymbol{x}, \boldsymbol{A}) = q\left(\boldsymbol{x}, \frac{\boldsymbol{A}+\boldsymbol{A}^t}{2}\right) + q\left(\boldsymbol{x}, \frac{\boldsymbol{A}-\boldsymbol{A}^t}{2}\right)$. Thus, $\left(\frac{\boldsymbol{A}+\boldsymbol{A}^t}{2}\right)^t = \frac{\boldsymbol{A}+\boldsymbol{A}^t}{2}$ and $\left(\frac{\boldsymbol{A}-\boldsymbol{A}^t}{2}\right)^t = -\left(\frac{\boldsymbol{A}-\boldsymbol{A}^t}{2}\right)$. Therefore, for any $\boldsymbol{x}$, $q\left(\boldsymbol{x}, \frac{\boldsymbol{A}-\boldsymbol{A}^t}{2}\right) = -q\left(\boldsymbol{x}, \frac{\boldsymbol{A}-\boldsymbol{A}^t}{2}\right)$, therefore $q\left(\boldsymbol{x}, \frac{\boldsymbol{A}-\boldsymbol{A}^t}{2}\right) = 0$. Accordingly, for any $\boldsymbol{x}$, $q(\boldsymbol{x}, \boldsymbol{A}) = q\left(\boldsymbol{x}, \frac{\boldsymbol{A}+\boldsymbol{A}^t}{2}\right)$. Consequently, if $\frac{\boldsymbol{A}+\boldsymbol{A}^t}{2}$ is viewed as $\boldsymbol{A}$, it is certainly the case that $\boldsymbol{A} = \boldsymbol{A}^t$.

② Particularly, depending on whether a square matrix $\boldsymbol{A}$ is a complex or a real matrix, $q(\boldsymbol{x}, \boldsymbol{A})$ is called a **complex quadratic form** or a **real quadratic form**.

⟨Definition 5.10⟩ For n-order Hermitian matrix $\boldsymbol{A} = [a_{ij}]$ and n-order complex vector $\boldsymbol{x} = (x_1, \cdots, x_n)^t$,

$$h(\boldsymbol{x}, \boldsymbol{A}) = \boldsymbol{x}^* \boldsymbol{A} \boldsymbol{x} = \sum_{i=1}^{n} \sum_{j=1}^{n} a_{ij} \bar{x}_i x_j, \quad (a_{ij} = \bar{a}_{ji})$$

is called a **quadratic form of the Hermitian of a variable** of $x_1, \cdots, x_n$ or simply the **Hermitian form** and $\boldsymbol{A}$ is called its coefficient matrix.

(Notes on ⟨Definition 5.10⟩)

① $\overline{\boldsymbol{x}^* \boldsymbol{A} \boldsymbol{x}} = (\boldsymbol{x}^* \boldsymbol{A} \boldsymbol{x})^* = \boldsymbol{x}^* \boldsymbol{A}^* \boldsymbol{x} = \boldsymbol{x}^* \boldsymbol{A} \boldsymbol{x}$ holds. Thus, for any $\boldsymbol{x}$, $h(\boldsymbol{x}, \boldsymbol{A}) = \boldsymbol{x}^* \boldsymbol{A} \boldsymbol{x}$ is a real number.

② In addition, when $\boldsymbol{A}$ is an n-order complex square matrix that is not necessarily a Hermitian and $\boldsymbol{x}$ is an n-order complex vector, this real part of this quadratic form can be expressed as a Hermitian's quadratic form. Generally, $(\boldsymbol{A}^*)^* = \boldsymbol{A}$ holds, thus,

$$A = \frac{A + A^*}{2} + \frac{A - A^*}{2} \tag{5.19}$$

$$\left(\frac{A + A^*}{2}\right)^* = \frac{A + A^*}{2} \tag{5.20}$$

and

$$\left(\frac{A - A^*}{2}\right)^* = -\frac{A - A^*}{2}. \tag{5.21}$$

Therefore, from the direct calculation,

$$q(\boldsymbol{x}, \boldsymbol{A}) = q\left(\boldsymbol{x}, \frac{\boldsymbol{A} + \boldsymbol{A}^*}{2}\right) + q\left(\boldsymbol{x}, \frac{\boldsymbol{A} - \boldsymbol{A}^*}{2}\right) \tag{5.22}$$

and

$$\begin{aligned} \overline{q(\boldsymbol{x}, \boldsymbol{A})} &= q(\boldsymbol{x}, \boldsymbol{A}^*) \\ &= q\left(\boldsymbol{x}, \frac{\boldsymbol{A}^* + \boldsymbol{A}}{2}\right) + q\left(\boldsymbol{x}, \frac{\boldsymbol{A}^* - \boldsymbol{A}}{2}\right) \\ &= q\left(\boldsymbol{x}, \frac{\boldsymbol{A} + \boldsymbol{A}^*}{2}\right) - q\left(\boldsymbol{x}, \frac{\boldsymbol{A} - \boldsymbol{A}^*}{2}\right). \end{aligned} \tag{5.23}$$

Thus,

$$\frac{1}{2}\left\{q(\boldsymbol{x}, \boldsymbol{A}) + \overline{q(\boldsymbol{x}, \boldsymbol{A})}\right\} = Re\{q(\boldsymbol{x}, \boldsymbol{A})\} = h\left(\boldsymbol{x}, \frac{\boldsymbol{A} + \boldsymbol{A}^*}{2}\right). \tag{5.24}$$

③ The Hermitian form contains the real quadratic form limiting the domain of $\boldsymbol{x}$ to be $\boldsymbol{R}^n$.

⟨**Definition 5.11**⟩ Through⟨D5.11-1⟩–⟨D5.11-3⟩, $\boldsymbol{A}$ is taken to be an $n-$ order Hermitian matrix and $\boldsymbol{x}$ is an n-order complex column vector and $h(\boldsymbol{x}, \boldsymbol{A})$ denotes the Hermitian form.

⟨D5.11-1⟩ When for any $\boldsymbol{x} \neq \boldsymbol{0}$, $h(\boldsymbol{x}, \boldsymbol{A}) > (<) 0$ holds, $h(\boldsymbol{x}, \boldsymbol{A})$ is called a **positive definite** (a **negative definite**), or that it is positive (negative).

⟨D5.11-2⟩ When for any $\boldsymbol{x} \neq \boldsymbol{0}$, $h(\boldsymbol{x}, \boldsymbol{A}) \geqq (\leqq) 0$ holds, $h(\boldsymbol{x}, \boldsymbol{A})$ is called a **nonnegative definite** (**nonpositive definite**).

⟨D5.11-3⟩ If due to $\boldsymbol{x}$, $h(\boldsymbol{x}, \boldsymbol{A})$ can take both positive and negative values, $h(\boldsymbol{x}, \boldsymbol{A})$ is said to be **indefinite**.

⟨D5.11-4⟩ Particularly, if $\boldsymbol{A}$ is an n-order square matrix, which is not necessarily symmetric, for any n-order real vector $\boldsymbol{x} \neq \boldsymbol{0}$, if $q(\boldsymbol{x}, \boldsymbol{A}) > (<)0$, $q(\boldsymbol{x}, \boldsymbol{A})$ is called **quasi-positive definite (quasi-negative definite)**.

(Notes on ⟨Definition 5.11⟩)

(Notes on ⟨D5.11-1⟩): $h(\boldsymbol{x}, \boldsymbol{A})$ being a positive definite can simply be referred to as $\boldsymbol{A}$ being a positive definite or a positive definite matrix.

(Notes on ⟨D5.11-2⟩): For any $\boldsymbol{x} \neq \boldsymbol{0}$, $h(\boldsymbol{x}, \boldsymbol{A}) \geqq (\leqq)0$ may be referred to as **positive semi-definite** or **negative semi-definite**.

(Notes on ⟨D5.11-3⟩): A real square matrix, which is not necessarily symmetric $\boldsymbol{A}$ being a quasi-positive definite (quasi-negative definite) is mutually equivalent to a real symmetric matrix $(\boldsymbol{A} + \boldsymbol{A}^t)$ being a positive definite (negative definite).

The following holds for eigenvalues of complex matrices.

[Theorem 5.5] Let λ be an eigenvalue of n-order complex square matrix $\boldsymbol{A}$ and $\mu_i(i = 1, \ldots, n)$ an eigenvalue of $\dot{\boldsymbol{A}} = \frac{1}{2}(\boldsymbol{A} + \boldsymbol{A}^*)$. Here, if we define $\underset{i \in \boldsymbol{N}}{min}\, \mu_i = \mu_{min}$ and $\underset{i \in \boldsymbol{N}}{max}\, \mu_i = \mu_{max}$,

$$\mu_{min} \leqq Re(\lambda) \leqq \mu_{max}.$$

(Proof) If $\boldsymbol{x}$ is assumed to be an eigenvector for λ of $\boldsymbol{A}$, from the direct calculation,

$$\begin{aligned} \boldsymbol{x}^*\dot{\boldsymbol{A}}\boldsymbol{x} &= Re(\boldsymbol{x}^*\boldsymbol{A}\boldsymbol{x}) \\ &= Re(\lambda\boldsymbol{x}^*\boldsymbol{x}). \end{aligned}$$

$\dot{\boldsymbol{A}}$ is a Hermitian matrix hence from [Theorem 20] (Appendix), there exists a unitary matrix $\boldsymbol{P}$ that becomes $\boldsymbol{P}^*\dot{\boldsymbol{A}}\boldsymbol{P} = \widehat{\boldsymbol{M}}$. However, $\widehat{\boldsymbol{M}} = diag(\mu_1, \cdots, \mu_n)$. Thus, if we let $\boldsymbol{x} = \boldsymbol{P}\boldsymbol{y}$, by definition, $\boldsymbol{x} \neq \boldsymbol{0}$ and $\boldsymbol{P}$ are nonsingular hence $\boldsymbol{y} \neq \boldsymbol{0}$. Additionally,

$$Re(\lambda\boldsymbol{x}^*\boldsymbol{x}) = Re(\lambda\boldsymbol{y}^*\boldsymbol{y}) = Re\left(\lambda\sum_{j=1}^{n}|y_j|^2\right) = Re(\lambda)\sum_{j=1}^{n}|y_j|^2$$

and

$$Re(\lambda\boldsymbol{x}^*\boldsymbol{x}) = \boldsymbol{x}^*\dot{\boldsymbol{A}}\boldsymbol{x} = \boldsymbol{y}^*\widehat{\boldsymbol{M}}\boldsymbol{y} = \sum_{j=1}^{n}\mu_j|y_j|^2.$$

Thus,

$$\mu_{min}\sum_{j=1}^{n}|y_j|^2 \leqq Re(\lambda)\sum_{j=1}^{n}|y_j|^2 \leqq \mu_{max}\sum_{j=1}^{n}|y_j|^2$$

is established. $\boldsymbol{y} \neq \boldsymbol{0}$ means $\sum_{j=1}^{n}|y_j|^2 > 0$, hence it is obvious that $\mu_{min} \leqq Re(\lambda) \leqq \mu_{max}$.

QED

Next, let us describe the basic characteristics of the Hermitian form.

[Theorem 5.6] When $\boldsymbol{A}$ is a n-order Hermitian matrix, the following (i)–(v) are mutually equivalent.

(i) $\boldsymbol{A}$ is positive definite (negative definite).
(ii) Any $\boldsymbol{A}$'s leading principal minor matrix is a positive definite (negative definite).
(iii) All eigenvalues of any leading principal minor matrix of $\boldsymbol{A}$ are positive (negative).
(iv) For any $k \in \boldsymbol{N}$, $det(\boldsymbol{A}(k)) > 0$ $((-1)^k \, det\,(\boldsymbol{A}(k)) > 0)$.
(v) For any $k \in \boldsymbol{N}$, $det\left(\boldsymbol{A}_{\boldsymbol{K}_k\boldsymbol{K}_k}\right) > 0 \left((-1)^k det\left(\boldsymbol{A}_{\boldsymbol{K}_k\boldsymbol{K}_k}\right) > 0\right)$.

Where $\boldsymbol{A}(k)$ is $\boldsymbol{A}$'s k-order leading principal minor matrix, $\boldsymbol{A}_{\boldsymbol{K}_k\boldsymbol{K}_k}$ is k-order principal minor matrix taken from the top left corner of $\boldsymbol{A}$,[16] $\boldsymbol{K}_k = \{1, \ldots, k\}$.

(Proof) Show for the situation when $\boldsymbol{A}$ is a positive definite in the order, (i) $\Rightarrow$ (ii) $\Rightarrow$ (iii) $\Rightarrow$ (iv) $\Rightarrow$ (v) $\Rightarrow$ (i).

(i) $\Rightarrow$ (ii) Let $\boldsymbol{J}$ be a nonempty subset of $\boldsymbol{N}$ and let the complementary set of $\boldsymbol{J}$ about $\boldsymbol{N}$ for $\boldsymbol{J}$ be $\boldsymbol{J}^C$ and any nonzero vector of order $\#\boldsymbol{J}$ to be $\boldsymbol{x}_J$. If we set $\boldsymbol{x}^t = \left(\boldsymbol{x}_J^t \; \boldsymbol{x}_{J^C}^t\right) = \left(\boldsymbol{x}_J^t \; \boldsymbol{0}\right)$, from (i), $0 < h(\boldsymbol{x}, \boldsymbol{A}) = h(\boldsymbol{x}_J, \boldsymbol{A}_{JJ})$.

(ii) $\Rightarrow$ (iii) For any $k \in \boldsymbol{N}$, if we let $\boldsymbol{A}(k)$ be $\boldsymbol{A}$'s k-order principal minor matrix and $\boldsymbol{x}_k \neq \boldsymbol{0}$ a k-dimensional vector, from (ii),

$$\boldsymbol{x}_k^* \boldsymbol{A}(k) \boldsymbol{x}_k > 0. \tag{5.25}$$

$\boldsymbol{A}(k)$ is a Hermite matrix hence from [Theorem 20] (Appendix), there exists a k-order unitary matrix $\boldsymbol{P}_k$ that becomes

$$\boldsymbol{P}_k^* \boldsymbol{A}(k) \boldsymbol{P}_k = \boldsymbol{\Lambda}_k = diag(\lambda_1, \cdots, \lambda_k) \tag{5.26}$$

However, $\lambda_j (j = 1, \cdots, k)$ is an eigenvalue of $\boldsymbol{A}(k)$. For any $j \in \{1, \cdots, k\}$, if $\boldsymbol{P}_k$'s jth row is expressed as $\boldsymbol{p}_k^i$, $\boldsymbol{p}_k^i \neq \boldsymbol{0}$. Therefore, for any $j \in \{1, \cdots, k\}$,

$$0 < \left(\boldsymbol{p}_k^i\right)^* \boldsymbol{A}(k) \boldsymbol{p}_k^i = \lambda_j. \tag{5.27}$$

(iii) $\Rightarrow$ (iv) Eigenvalues $\lambda_1, \ldots, \lambda_k$ of *any* k − order leading principal minor matrix $\boldsymbol{A}(k)$ of $\boldsymbol{A}$ are solutions of $det(\lambda \boldsymbol{I}_k - \boldsymbol{A}(k)) = 0$ hence they can be factorized as $\det(\lambda \boldsymbol{I}_k - \boldsymbol{A}(k)) = \Pi_{j=1}^k (\lambda - \lambda_j)$. Substituting in $\lambda = 0$ into both sides of this gives, for any $k \in \boldsymbol{N}$,

[16] See the list of symbols.

$$det(\boldsymbol{A}(k)) = \Pi_{j=1}^{k}\lambda_j > 0.$$

(iv) ⇒ (v) This is immediately evident.
(v) ⇒ (i) To simplify notation, if the $(n-1)$-order leading principal minor matrix taken from the top left of $\boldsymbol{A}, \boldsymbol{A}_{K_{n-1}K_{n-1}}$, is expressed as $\boldsymbol{A}_{n-1}$ and we let $\boldsymbol{a}^n = (a_{1n}, \cdots, a_{n-1,\,n})^t$, from (v), $det(\boldsymbol{A}_{n-1}) > 0$. Hence if we define:

$$\boldsymbol{E} = \begin{bmatrix} \boldsymbol{I}_{n-1} & \boldsymbol{A}_{n-1}^{-1}\boldsymbol{a}^n \\ \boldsymbol{0} & -1 \end{bmatrix}, \qquad (5.28)$$

$$\boldsymbol{EE} = \boldsymbol{I}_n = \boldsymbol{E}^*\boldsymbol{E}^*$$

and

$$\boldsymbol{E}^*\boldsymbol{AE} = \begin{bmatrix} \boldsymbol{A}_{n-1} & \boldsymbol{0} \\ \boldsymbol{0} & a_{nn} - h\left(\boldsymbol{a}^n,\ \boldsymbol{A}_{n-1}^{-1}\right) \end{bmatrix}. \qquad (5.29)$$

Therefore, if $\boldsymbol{E}^*\boldsymbol{AE} = \boldsymbol{D}$, for any $\boldsymbol{x}$,

$$\begin{aligned} h(\boldsymbol{x},\ \boldsymbol{A}) &= h(\boldsymbol{x},\ \boldsymbol{E}^*\boldsymbol{DE}) \\ &= h(\boldsymbol{Ex},\ \boldsymbol{D}) \\ &= h(\boldsymbol{y},\ \boldsymbol{A}_{n-1}) + \left[a_{nn} - h\left(\boldsymbol{a}^n,\ \boldsymbol{A}_{n-1}^{-1}\right)\right]x_n^2 \end{aligned} \qquad (5.30)$$

is established. However, $\boldsymbol{y}$ is a vector made up of the first $(n-1)$ elements of $\boldsymbol{Ex}$. Furthermore, if the matrix equations on both sides of (5.29) are used,

$$\begin{aligned} 0 < det(\boldsymbol{A}) &= det(\boldsymbol{E}^*\boldsymbol{AE}) \\ &= det(\boldsymbol{A}_{n-1})\left[a_{nn} - h\left(\boldsymbol{a}^n,\ \boldsymbol{A}_{n-1}^{-1}\right)\right]. \end{aligned} \qquad (5.31)$$

From (v), $det(\boldsymbol{A}_{n-1}) > 0$, (5.31) means $\left[a_{nn} - h\left(\boldsymbol{a}^n,\ \boldsymbol{A}_{n-1}^{-1}\right)\right] > 0$. Hence, to show the conclusion, if the induction's approach is used for the order of $\boldsymbol{A}$, n, $\boldsymbol{A}_{n-1}$ is a positive definite and from the definition, $\boldsymbol{E}$ is regular. Thus, for any $\boldsymbol{x} \neq \boldsymbol{0}$, $\boldsymbol{y} \neq \boldsymbol{0}$ or $x_n \neq 0$. Therefore (5.30) gives $h(\boldsymbol{x},\ \boldsymbol{A}) > 0$.

When $\boldsymbol{A}$ is a negative definite, the above results should be applied to a positive definite matrix $-\boldsymbol{A}$. QED

[Theorem 5.7] When $\boldsymbol{A}$ is a n-order Hermitian matrix, the following (i)–(iii) are mutually equivalent.

(i) $\boldsymbol{A}$ is positive definite (negative definite).
(ii) There exists a n-order nonsingular matrix $\boldsymbol{B}$ which is $\boldsymbol{A} = \boldsymbol{B}^*\boldsymbol{B}(\boldsymbol{A} = -\boldsymbol{B}^*\boldsymbol{B})$.
(iii) $\boldsymbol{A}^{-1}$ is a positive definite (negative definite).

(Proof) For only the case when $\boldsymbol{A}$ is a positive definite, shown in the order (i) $\Rightarrow$ (ii) $\Rightarrow$ (iii) $\Rightarrow$ (i).

(i) $\Rightarrow$ (ii) From [Theorem 20] (Appendix),

$$\boldsymbol{PAP}^* = \boldsymbol{\Lambda} = diag(\lambda_1, \cdots, \lambda_n) \tag{5.32}$$

There exists a n-order unitary matrix $\boldsymbol{P}$ that satisfies $\boldsymbol{PAP}^* = \boldsymbol{\Lambda} = diag\ (\lambda_1, \cdots, \lambda_n)$. However, λ_j is $\boldsymbol{A}$'s $j(j = 1, \cdots, n)$th eigenvalue. From [Theorem 5.6; (i) $\Rightarrow$ (iii)], $\lambda_j > 0$. Therefore, if we define $\boldsymbol{\Lambda}^{\frac{1}{2}} = diag\left(\sqrt{\lambda_1}, \cdots, \sqrt{\lambda_n}\right)$ and $\boldsymbol{B} = \boldsymbol{\Lambda}^{\frac{1}{2}}\boldsymbol{P}$, $\left(\boldsymbol{\Lambda}^{\frac{1}{2}}\right)^* = \boldsymbol{\Lambda}^{\frac{1}{2}}$ and $\boldsymbol{B}$ are nonsisngular. Thus, from (5.32),

$$\boldsymbol{A} = \boldsymbol{P}^*\left(\boldsymbol{\Lambda}^{\frac{1}{2}}\right)^*\boldsymbol{\Lambda}^{\frac{1}{2}}\boldsymbol{P} = \boldsymbol{B}^*\boldsymbol{B}.$$

In addition, if we note that all the eigenvalues of $\boldsymbol{A}$ are positive, from the definition of $\boldsymbol{B}$,

$$det(\boldsymbol{B}) = det\left(\boldsymbol{\Lambda}^{\frac{1}{2}}\boldsymbol{P}\right) = det\left(\boldsymbol{\Lambda}^{\frac{1}{2}}\right)det(\boldsymbol{P}) \neq 0.$$

Therefore, $\boldsymbol{B}$ is nonsingular.
(ii) $\Rightarrow$ (iii) From the direct calculation,

$$\begin{aligned}\boldsymbol{A}^{-1} &= (\boldsymbol{B}^*\boldsymbol{B})^{-1}\\ &= \boldsymbol{B}^{-1}(\boldsymbol{B}^*)^{-1}.\end{aligned} \tag{5.33}$$

If there exists $\tilde{\boldsymbol{x}} \neq \boldsymbol{0}$ that is $\left(\boldsymbol{B}^{-1}\right)^*\tilde{\boldsymbol{x}} = \boldsymbol{0}$, it goes against the assumed nonsingularity of $\boldsymbol{B}$. Hence, for any $\boldsymbol{x} \neq \boldsymbol{0}$, $(\boldsymbol{B}^{-1})^*\boldsymbol{x} = \boldsymbol{0}$. Therefore in for any $\boldsymbol{x} \neq \boldsymbol{0}$,

$$\begin{aligned}\boldsymbol{x}^*\boldsymbol{A}^{-1}\boldsymbol{x} &= \left[\left(\boldsymbol{B}^{-1}\right)^*\boldsymbol{x}\right]^*\left[\left(\boldsymbol{B}^{-1}\right)^*\boldsymbol{x}\right]\\ &> 0.\end{aligned}$$

(iii) $\Rightarrow$ (i) If $\boldsymbol{y} = \boldsymbol{Ax}$ is set for all $\boldsymbol{x} \neq \boldsymbol{0}$, from the nonsingularity of $\boldsymbol{A}$, $\boldsymbol{y} \neq \boldsymbol{0}$. Therefore

$$\begin{aligned} 0 &< y^* A^{-1} y \\ &= x^* A^* A^{-1} A x \\ &= x^* A x \end{aligned}$$

holds. QED

[Theorem 5.8] When A is a n-order Hermitian matrix, the following (i)–(iv) are mutually equivalent.

(i) A is a nonnegative definite (nonpositive definite).
(ii) Any leading principal minor matrices of A are nonnegative definite (nonpositive definite).
(iii) All eigenvalues of any leading principal minor matrices of A are nonnegative (nonpositive).
(iv) For any $k \in N$, $det(A(k)) \geqq 0((-1)^k\, det\,(A(k)) \geqq 0)$.

(Proof) Only the case when A is a nonnegative definite is shown. (i) ⇒ (ii) ⇒ (iii) ⇒ (iv) can be shown in a similar way to [Definition 5.6; (i) ⇒ (ii) ⇒ (iii) ⇒ (iv)] hence only (iv) ⇒ (i) must be proven.

(iv) ⇒ (i) Firstly, show that under (iv), for any $\alpha > 0$, all k-order leading principal minor matrices taken from the top left corner of $(\alpha I_n + A) = det((\alpha I_n + A)_{AkAk})$ $I_n + A) = det((\alpha I_n + A)_{AkAk})$ are positive.

If $k - 1$, the conclusion is evident. Hence, here below, induction approach on $k(\geqq 2)$ will be used for the proof. If for any $\alpha > 0$, $g(\alpha) = det\left((\alpha I_n + A)_{K_k K_k}\right)$ is set,

$$g(0) = det(A_{K_k K_k}) \geqq 0$$

and

$$g'(\alpha) = \sum_{j=1}^{k} det\left(\alpha I_{k-1} + A_{K_k K_k}\binom{j}{j}\right).$$

However, $A_{K_k K_k}\binom{j}{j}$ is a $(k-1)$-order square matrix, which is the matrix $A_{K_k K_k}$ with the jth column and jth row removed. From the assumption, all leading principal minor matrices taken from the top left corner of $A_{K_k K_k}\binom{j}{j}$ are nonnegative. Accordingly, the assumption of the induction approach means $det\left(\alpha I_{k-1} + A_{K_k K_k}\binom{j}{j}\right) > 0$ hence for all $\alpha > 0$, $g'(\alpha) > 0$. Therefore if $\alpha \geqq 0$, $g(\alpha) \geqq 0$. Next. Let us assume there exists an $\alpha_1 > 0$ such that $g(\alpha_1) = 0$. Then, due to the continuity of $g'(\alpha)$ and $g'(\alpha_1) > 0$, for $\alpha_2 > 0$ which is $\alpha_2 < \alpha_1$, $g(\alpha_2) < 0$. However, this goes against $g(\alpha) \geqq 0$ shown earlier. Thence, for any $\alpha > 0$, $g(\alpha) > 0$

holds. From above, $(\alpha \boldsymbol{I}_n + \boldsymbol{A})$ satisfies the condition [Theorem 5.6; (v)] for any $\alpha > 0$ hence it is a positive definite.

If the conclusion sought is assumed to not hold and assume that there exists $\boldsymbol{x}$ which is $\boldsymbol{x}^*\boldsymbol{Ax} < 0$, from $\boldsymbol{x} \neq \boldsymbol{0}$, $\boldsymbol{x}^*\boldsymbol{x} > 0$. Therefore for any $\alpha \in \left(0, \dfrac{|\boldsymbol{x}^*\boldsymbol{Ax}|}{\boldsymbol{x}^*\boldsymbol{x}}\right)$, the following self-contradiction is derived.

$$0 = \frac{|\boldsymbol{x}^*\boldsymbol{Ax}|}{\boldsymbol{x}^*\boldsymbol{x}}(\boldsymbol{x}^*\boldsymbol{x}) - |\boldsymbol{x}^*\boldsymbol{Ax}| > \alpha(\boldsymbol{x}^*\boldsymbol{x}) + \boldsymbol{x}^*\boldsymbol{Ax} = h(\boldsymbol{x},\ \alpha\boldsymbol{I}_n + \boldsymbol{A}) > 0.$$

Accordingly, $\boldsymbol{A}$ is a negative definite. $\boldsymbol{A}$ being a nonnegative definite and nonpositive definite are equivalent so the remaining conclusion is also clear. QED

(Notes on [Theorem 5.8]): We saw in [Theorem 5.6], that $\boldsymbol{A}$ being a positive definite and $det(\boldsymbol{A}_{K_kK_k}) > 0$ holding for any $k \in \boldsymbol{N}$ are equivalent. However, $det(\boldsymbol{A}_{K_kK_k}) \geqq 0$ $(k = 1, \cdots, n)$ is a necessary condition for $\boldsymbol{A}$ to be a negative definite but it is not necessarily a sufficient condition. This is clear from the following example.

Example 5.1 If $\boldsymbol{A} = \begin{bmatrix} 1 & 0 & -2 \\ 0 & 0 & 0 \\ -2 & 0 & -1 \end{bmatrix}$, $det(\boldsymbol{A}_{K_kK_k}) = det(1) = 1$, $det(\boldsymbol{A}_{K_kK_k}) = det\begin{bmatrix} 1 & 0 \\ 0 & 0 \end{bmatrix} = 0$ and $det(\boldsymbol{A}_{K_kK_k}) - det(\boldsymbol{A}) - 0$. However, for $\boldsymbol{x} = (x_1, x_2, x_3)'$, we can calculate $\boldsymbol{x}'\boldsymbol{Ax} = (x_1 - 2x_3)^2 - 5x_3^2$, so if $\boldsymbol{x} = (2, 1, 1)'$ is selected for example, $\boldsymbol{x}'\boldsymbol{Ax} = -5 < 0$.

[Theorem 5.9] If $\boldsymbol{A}$ is a n-order Hermitian matrix, the following holds.

(i) The necessary and sufficient condition for $\boldsymbol{A}$ to be a nonnegative definite (nonpositive definite) is that there exists an n-order square matrix $\boldsymbol{B}$ such that $\boldsymbol{A} = \boldsymbol{B}^*\boldsymbol{B}(\boldsymbol{A} = -\boldsymbol{B}^*\boldsymbol{B})$.
(ii) The necessary and sufficient condition for $\boldsymbol{x}^*\boldsymbol{Ax} = 0$ when $\boldsymbol{A}$ is a nonnegative definite (nonpositive definite) is $\boldsymbol{x} \in N_{\boldsymbol{A}} = \{\boldsymbol{x} | \boldsymbol{Ax} = \boldsymbol{0}\}$.

(Proof) Show only the case when $\boldsymbol{A}$ is negative definite.

(i) (Necessity) Similarly to [Theorem 5.7; (i) $\Rightarrow$ (ii)], if we define $\boldsymbol{B} = \boldsymbol{\Lambda}^{\frac{1}{2}}\boldsymbol{P}$, $\boldsymbol{A} = \boldsymbol{B}^*\boldsymbol{B}$. However, from [Theorem 5.8; (i) $\Rightarrow$ (iii)], the eigenvalue of $\boldsymbol{A}$ is nonnegative hence there is a possibility that the diagonal elements of $\boldsymbol{\Lambda}^{\frac{1}{2}}$ may contain nonzero elements. Therefore, in this case, the row=$\boldsymbol{0}$ of $\boldsymbol{B}$. Hence the nonsingularity of $\boldsymbol{B}$ is not necessarily guaranteed.
(Sufficiency) As $\boldsymbol{A} = \boldsymbol{B}^*\boldsymbol{B}$, for any $\boldsymbol{x}$,

$$\boldsymbol{x}^*\boldsymbol{Ax} = \boldsymbol{x}^*\boldsymbol{B}^*\boldsymbol{Bx} = (\boldsymbol{Bx})^*\boldsymbol{Bx} \geqq 0.$$

(ii) (Sufficiency) is self-evident hence, only necessity will be shown.

(Necessity) $\boldsymbol{A}$ is a nonnegative definite hence from (i), there exists a square matrix $\boldsymbol{B}$ such that $\boldsymbol{A} = \boldsymbol{B}^*\boldsymbol{B}$. Hence, if $\boldsymbol{x}^*\boldsymbol{A}\boldsymbol{x} = 0$,

$$0 = \boldsymbol{x}^*\boldsymbol{A}\boldsymbol{x} = \boldsymbol{x}^*\boldsymbol{B}^*\boldsymbol{B}\boldsymbol{x} = (\boldsymbol{B}\boldsymbol{x})^*(\boldsymbol{B}\boldsymbol{x}). \tag{5.34}$$

Here, if $\boldsymbol{Bx} \neq \boldsymbol{0}$, the term on the furthest right of (34) is positive. Therefore, $\boldsymbol{Bx} = \boldsymbol{0}$. If both sides of this are multiplied on the left by $\boldsymbol{B}^*$ and $\boldsymbol{A} = \boldsymbol{B}^*\boldsymbol{B}$ is used, $\boldsymbol{Ax} = \boldsymbol{0}$ is obtained immediately. QED

Next, we describe the necessary and sufficient condition for something to be a positive (negative) definite.

In addition, unless specifically mentioned otherwise, real matrices and real vectors are considered.

[Theorem 5.10] Let $\boldsymbol{A}$ be a n-order symmetric matrix, $m \times n$ matrix $\boldsymbol{C} = [\boldsymbol{C}_1 \ \boldsymbol{C}_2]$ and assume $rank(\boldsymbol{C}_1) = m < n$.

Here, for all $\boldsymbol{x} \neq \boldsymbol{0}$ that satisfy

$$\boldsymbol{Cx} = [\boldsymbol{C}_1 \ \boldsymbol{C}_2]\begin{bmatrix}\boldsymbol{x}_1\\ \boldsymbol{x}_2\end{bmatrix} = \boldsymbol{0},$$

(i) The necessary and sufficient condition for

$$\boldsymbol{x}^t\boldsymbol{A}\boldsymbol{x} = \begin{bmatrix}\boldsymbol{x}_1^t & \boldsymbol{x}_2^t\end{bmatrix}\begin{bmatrix}\boldsymbol{A}_{11} & \boldsymbol{A}_{12}\\ \boldsymbol{A}_{21} & \boldsymbol{A}_{22}\end{bmatrix}\begin{bmatrix}\boldsymbol{x}_1\\ \boldsymbol{x}_2\end{bmatrix} > 0$$

to hold is the following.

$$(-1)^m det(\boldsymbol{D}_k)\boldsymbol{C}_1^t = (-1)^m det\begin{bmatrix}\boldsymbol{0} & \boldsymbol{C}_1 & \boldsymbol{C}_2^{(k)}\\ \boldsymbol{C}_1^t & \boldsymbol{A}_{22} & \boldsymbol{A}_{12}^{(k)}\\ \left(\boldsymbol{C}_2^{(k)}\right)^t & \boldsymbol{A}_{21}^{(k)} & \boldsymbol{A}_{22}^{(k)}\end{bmatrix} > 0$$

$$(k = 1, \cdots, n - m)$$

(ii) The necessary condition for $\boldsymbol{x}^t\boldsymbol{A}\boldsymbol{x} \geqq 0$ is $(-1)^m \, det\,(\boldsymbol{D}_k) \geqq 0 \; (k = 1, \cdots, \boldsymbol{n} - \boldsymbol{m})$.
The notation is as defined below.

$\boldsymbol{C}_2^{(k)}$: Matrix constructed from $\boldsymbol{C}_2$'s first row to the kth row

$\boldsymbol{A}_{12}^{(k)}$: Matrix constructed from $\boldsymbol{A}_{12}$'s first row to the kth row

$\boldsymbol{A}_{12}^{(k)} = A_{21}^{(k)}$

$\boldsymbol{A}_{22}^{(k)}$: principal minor matrix of $\boldsymbol{A}_{22}$ that takes column k and row k from the top left corner of $\boldsymbol{A}_{22}$

(Proof)

(i) From $det(\boldsymbol{C}_1) = m, \boldsymbol{x}_1 = -\boldsymbol{C}_1^{-1}\boldsymbol{C}_2\boldsymbol{x}_2$. Therefore, $\boldsymbol{x}$ that is $\boldsymbol{Cx} = \boldsymbol{0}$ is expressed as

$$\boldsymbol{x}=\begin{bmatrix}\boldsymbol{x}_1\\ \boldsymbol{x}_2\end{bmatrix}=\begin{bmatrix}-\boldsymbol{C}_1^{-1}\boldsymbol{C}_2\boldsymbol{x}_2\\ \boldsymbol{x}_2\end{bmatrix}=\begin{bmatrix}-\boldsymbol{C}_1^{-1}\boldsymbol{C}_2\\ \boldsymbol{I}_{n-m}\end{bmatrix}\boldsymbol{x}_2. \tag{5.35}$$

Substituting (5.35) into gives, through direct calculation,

$$\begin{aligned}\boldsymbol{x}^t\boldsymbol{A}\boldsymbol{x}&=\boldsymbol{x}_2^t\begin{bmatrix}-\boldsymbol{C}_2^t(\boldsymbol{C}_1^{-1})^t & \boldsymbol{I}_{n-m}\end{bmatrix}\begin{bmatrix}\boldsymbol{A}_{11} & \boldsymbol{A}_{12}\\ \boldsymbol{A}_{21} & \boldsymbol{A}_{22}\end{bmatrix}\begin{bmatrix}-\boldsymbol{C}_1^{-1}\boldsymbol{C}_2\\ \boldsymbol{I}_{n-m}\end{bmatrix}\boldsymbol{x}_2\\ &=\boldsymbol{x}_2^t\left[\boldsymbol{C}_2^t(\boldsymbol{C}_1^{-1})^t\boldsymbol{A}_{11}\boldsymbol{C}_1^{-1}\boldsymbol{C}_2-\boldsymbol{C}_2^t(\boldsymbol{C}_1^{-1})^t\boldsymbol{A}_{12}-\boldsymbol{A}_{21}\boldsymbol{C}_1^{-1}\boldsymbol{C}_2+\boldsymbol{A}_{22}\right]\boldsymbol{x}_2.\end{aligned} \tag{5.36}$$

Here, if we set it as the matrix $=\boldsymbol{B}$ in the [ ] on the right-hand side of (5.36), for all $\boldsymbol{x}\neq\boldsymbol{0}$ satisfying $\boldsymbol{C}\boldsymbol{x}=\boldsymbol{0}$, the necessary and sufficient condition for $\boldsymbol{x}^t\boldsymbol{A}\boldsymbol{x}>0$ is that $\boldsymbol{B}$ is a positive definite. In addition, from [Theorem 5.6; (i) ⇔ (v)], the fact that $\boldsymbol{B}$ is a positive definite is equivalent to the matrix equation of any k-order leading principal minor matrix $\boldsymbol{B}_{K_kK_k}$ taken from the top left corner of $\boldsymbol{B}$. However, $K_k=\{1,\cdots,k\}$ and $k\in\{1,\cdots,n-m\}$. Hereinafter, for simplification, $\boldsymbol{B}_{K_kK_k}$ will be simply be denoted $\boldsymbol{B}_k$. Using (5.36) to calculate $\boldsymbol{B}_k$,

$$\boldsymbol{B}_k=\left(\boldsymbol{C}_2^{(k)}\right)^t(\boldsymbol{C}_1^{-1})^t\boldsymbol{A}_{11}\boldsymbol{C}_1^{-1}\boldsymbol{C}_2^{(k)}-\left(\boldsymbol{C}_2^{(k)}\right)^t(\boldsymbol{C}_1^{-1})\boldsymbol{A}_{12}^{(k)}-\boldsymbol{A}_{21}^{(k)}\boldsymbol{C}_1^{-1}\boldsymbol{C}_2^{(k)}+\boldsymbol{A}_{22}^{(k)}. \tag{5.37}$$

Hence, the following matrix is considered to explore the sign of the matrix equation of $\boldsymbol{B}_k$. However, $\boldsymbol{D}_k$ is a matrix stipulated by.

$$\begin{aligned}&\begin{bmatrix}\boldsymbol{I}_m & \boldsymbol{0} & \boldsymbol{0}\\ \boldsymbol{0} & \boldsymbol{I}_m & \boldsymbol{0}\\ \boldsymbol{0} & -\left(\boldsymbol{C}_2^{(k)}\right)^t(\boldsymbol{C}_1^{-1})^t & \boldsymbol{I}_k\end{bmatrix}\boldsymbol{D}_k\begin{bmatrix}\boldsymbol{I}_m & \boldsymbol{0} & \boldsymbol{0}\\ \boldsymbol{0} & \boldsymbol{I}_m & -\boldsymbol{C}_1^{-1}\boldsymbol{C}_2^{(k)}\\ \boldsymbol{0} & \boldsymbol{0} & \boldsymbol{I}_k\end{bmatrix}\\ &=\begin{bmatrix}\boldsymbol{0} & \boldsymbol{C}_1 & \boldsymbol{0}\\ \boldsymbol{C}_1^t & \boldsymbol{A}_{11} & -\boldsymbol{A}_{11}\boldsymbol{C}_1^{-1}\boldsymbol{C}_2^{(k)}+\boldsymbol{A}_{12}^{(k)}\\ \boldsymbol{0} & -\left(\boldsymbol{C}_2^{(k)}\right)^t(\boldsymbol{C}_1^{-1})^t\boldsymbol{A}_{11}+\boldsymbol{A}_{21}^{(k)} & \boldsymbol{B}_k\end{bmatrix}.\end{aligned} \tag{5.38}$$

Here if the final matrix on the left-hand side of (5.38) is defined as $\boldsymbol{G}_k$, $det(\boldsymbol{G}_k)=det(\boldsymbol{G}_k^t)=1$. Hence, the matrix equations on both sides of (5.38) can be rearranged as

$$\begin{aligned}det\left(\boldsymbol{G}_k^t\boldsymbol{D}_k\boldsymbol{G}_k\right)&=det(\boldsymbol{D}_k)\\ &=det\begin{bmatrix}\boldsymbol{0} & \boldsymbol{C}_1 & \boldsymbol{0}\\ \boldsymbol{C}_1^t & \boldsymbol{A}_{11} & -\boldsymbol{A}_{11}\boldsymbol{C}_1^{-1}\boldsymbol{C}_2^{(k)}+\boldsymbol{A}_{12}^{(k)}\\ \boldsymbol{0} & -\left(\boldsymbol{C}_2^{(k)}\right)^t(\boldsymbol{C}_1^{-1})^t\boldsymbol{A}_{11}+\boldsymbol{A}_{21}^{(k)} & \boldsymbol{B}_k\end{bmatrix}\end{aligned} \tag{5.39}$$

so by letting $\boldsymbol{Q}=-\boldsymbol{A}_{11}\boldsymbol{C}_1^{-1}\boldsymbol{C}_2^{(k)}+\boldsymbol{A}_{12}^{(k)}$, from characteristics of matrix equations

$$\begin{aligned}
det(\boldsymbol{D}_k) &= det\begin{bmatrix} \boldsymbol{0} & \boldsymbol{C}_1 & \boldsymbol{0} \\ \boldsymbol{C}_1^t & \boldsymbol{A}_{11} & \boldsymbol{Q} \\ \boldsymbol{0} & \boldsymbol{Q}^t & \boldsymbol{B}_k \end{bmatrix} \\
&= det\begin{bmatrix} \boldsymbol{C}_1 & \boldsymbol{C}_1 & \boldsymbol{0} \\ \boldsymbol{C}_1^t + \boldsymbol{A}_{11} & \boldsymbol{A}_{11} & \boldsymbol{Q} \\ \boldsymbol{Q}^t & \boldsymbol{Q}^t & \boldsymbol{B}_k \end{bmatrix} \\
&= det\begin{bmatrix} \boldsymbol{C}_1 & \boldsymbol{0} & \boldsymbol{0} \\ \boldsymbol{C}_1^t + \boldsymbol{A}_{11} & -\boldsymbol{C}_1^t & \boldsymbol{Q} \\ \boldsymbol{Q}^t & \boldsymbol{0} & \boldsymbol{B}_k \end{bmatrix} \\
&= det(\boldsymbol{C}_1)det\begin{bmatrix} -\boldsymbol{C}_1^t & \boldsymbol{Q} \\ \boldsymbol{0} & \boldsymbol{B}_k \end{bmatrix} \\
&= (-1)^m[det(\boldsymbol{C}_1)]^2 det(\boldsymbol{B}_k).
\end{aligned} \tag{5.40}$$

Therefore if (5.40) is multiplied by $(-1)^m$ and $[det(\boldsymbol{C}_1)]^2 > 0$ is noted, $sgn[(-1)^m \, det\,(\boldsymbol{D}_k)] = sgn\,[det(\boldsymbol{B}_k)]$ is established. Therefore, when for any $k \in \{1, \cdots, n-m\}$, $(-1)^m \, det\,(\boldsymbol{D}_k) > 0$, and only when this holds, $\boldsymbol{x}^t\boldsymbol{A}\boldsymbol{x} > 0$ for any $\boldsymbol{x} \neq \boldsymbol{0}$ which is $\boldsymbol{C}\boldsymbol{x} = \boldsymbol{0}$.

(ii) It can be shown using exactly the same approach as (i). QED

The next corollary can be obtained from this theorem.

[Corollary 5.10] The assumptions and notations are taken to be the same as in [Theorem 5.10]. Here, for all $\boldsymbol{x} \neq \boldsymbol{0}$ that satisfies $\boldsymbol{C}\boldsymbol{x} = \boldsymbol{0}$,

(i) The necessary and sufficient condition to hold $\boldsymbol{x}^t\boldsymbol{A}\boldsymbol{x} < 0$ is $(-1)^{m+k} \, det\,(\boldsymbol{D}_k) > 0$ $(k = 1, \ldots, n - m)$.

(ii) The necessary condition to hold $\boldsymbol{x}^t\boldsymbol{A}\boldsymbol{x} \leqq 0$ is $(-1)^{m+k} \, det\,(\boldsymbol{D}_k) \geqq 0$ $(k = 1, \ldots, n - m)$.

(Proof) Only (i) will be shown.

(i) For any $\boldsymbol{x} \neq \boldsymbol{0}$ that becomes $\boldsymbol{C}\boldsymbol{x} = \boldsymbol{0}$, $\boldsymbol{x}^t\boldsymbol{A}\boldsymbol{x} < 0$ and $\boldsymbol{x}^t(-\boldsymbol{A})\boldsymbol{x} > 0$ are mutually equivalent. Hence, the establishment of (5.3) must be shown in respect of $\boldsymbol{x}^t(-\boldsymbol{A})\boldsymbol{x}$. If $\boldsymbol{A}_{11}$, $\boldsymbol{A}_{12}^{(k)}$, $\boldsymbol{A}_{21}^{(k)}$ and $\boldsymbol{A}_{22}^{(k)}$ in [Theorem 5.10] is replaced by $-\boldsymbol{A}_{11}$, $-\boldsymbol{A}_{12}^{(k)}$, $-\boldsymbol{A}_{21}^{(k)}$, $-\boldsymbol{A}_{22}^{(k)}$ respectively,

$$
det\begin{bmatrix} \mathbf{0} & \boldsymbol{C}_1 & \boldsymbol{C}_2^{(k)} \\ \boldsymbol{C}_1^t & -\boldsymbol{A}_{11} & -\boldsymbol{A}_{12}^{(k)} \\ \left(\boldsymbol{C}_2^{(k)}\right)^t & -\boldsymbol{A}_{21}^{(k)} & -\boldsymbol{A}_{22}^{(k)} \end{bmatrix} = (-1)^{m+k}\begin{bmatrix} \mathbf{0} & \boldsymbol{C}_1 & \boldsymbol{C}_2^{(k)} \\ -\boldsymbol{C}_1^t & \boldsymbol{A}_{11} & \boldsymbol{A}_{12}^{(k)} \\ -\left(\boldsymbol{C}_2^{(k)}\right)^t & \boldsymbol{A}_{21}^{(k)} & \boldsymbol{A}_{22}^{(k)} \end{bmatrix}
$$

$$
= (-1)^{2m+k} det(\boldsymbol{D}_k).
$$

Therefore, the optimality needs to be shown under

$$
(-1)^m det\begin{bmatrix} \mathbf{0} & \boldsymbol{C}_1 & \boldsymbol{C}_2^{(k)} \\ \boldsymbol{C}_1^t & -\boldsymbol{A}_{11} & -\boldsymbol{A}_{12}^{(k)} \\ \left(\boldsymbol{C}_2^{(k)}\right)^t & -\boldsymbol{A}_{21}^{(k)} & -\boldsymbol{A}_{22}^{(k)} \end{bmatrix} = (-1)^{m+k} det(\boldsymbol{D}_k).
$$

Thus, the conclusion is evident. QED

Characteristics of a matrix that slightly generalizes the matrix $\boldsymbol{D}_k$ in [Theorem 5.10] are shown by the following lemma.

[Lemma 5.5] Let $\boldsymbol{A}$, $\boldsymbol{B}$, and $\boldsymbol{C}$ be n-order square matrix, m-order square matrix, and $m \times n$ matrix, respectively, and determine $(m + n)$-order square matrix $\boldsymbol{G}$ as

$$
\boldsymbol{G} = \begin{bmatrix} \boldsymbol{B} & \boldsymbol{C} \\ \boldsymbol{C}^t & \boldsymbol{A} \end{bmatrix}.
$$

If $\boldsymbol{G}$ is nonsingular,

$$
rank(\boldsymbol{A}) \geqq n - rank(\boldsymbol{C}) \text{ and } rank(\boldsymbol{B}) \geqq m - rank(\boldsymbol{C}).
$$

(Proof) $rank(\boldsymbol{A}) \geqq n - rank(\boldsymbol{C})$ will be shown. If the conclusion is contradicted and let $rank(\boldsymbol{A}) < n - rank(\boldsymbol{C})$, $rank(\boldsymbol{A}) < n$ and $p \equiv rank(\boldsymbol{C}) < n - rank(\boldsymbol{A}) \equiv q$. Therefore if the base of $N_A = \{\boldsymbol{x} | \boldsymbol{Ax} = \mathbf{0}\}$ is expressed as $\{\boldsymbol{\xi}^1, \cdots, \boldsymbol{\xi}^q\}$, from direct calculations,

$$
\boldsymbol{G}\begin{bmatrix} \mathbf{0} \\ \boldsymbol{\xi}^j \end{bmatrix} = \begin{bmatrix} \boldsymbol{C\xi}^j \\ \boldsymbol{A\xi}^j \end{bmatrix} = \begin{bmatrix} \boldsymbol{C\xi}^j \\ \mathbf{0} \end{bmatrix}, \quad (j = 1, \cdots, q). \tag{5.41}
$$

From the assumed nonsingularity of $\boldsymbol{G}$ and $\boldsymbol{\xi}^j \neq \mathbf{0}$, for any $j \in \{1, \cdots, q\}$, $\boldsymbol{C\xi}^j \neq \mathbf{0}$. Hence each $\boldsymbol{C\xi}^j$ can be represented as a linear combination of p linearly independent row vectors of $\boldsymbol{C}$ and assumption implies $p < q$. Hence $\boldsymbol{C\xi}^1$, $\cdots$, $\boldsymbol{C\xi}^q$is linearly dependent. Thus, there exists $\boldsymbol{y} \neq \mathbf{0}$ that satisfies $[\boldsymbol{C\xi}^1, \cdots, \boldsymbol{C\xi}^q]\boldsymbol{y} = \mathbf{0}$. Here if (5.41) is considered,

$$G\begin{bmatrix} \mathbf{0} & \cdots & \mathbf{0} \\ \xi^1 & \cdots & \xi^q \end{bmatrix} y = \begin{bmatrix} C\xi^1 & \cdots & C\xi^q \\ A\xi^1 & \cdots & A\xi^q \end{bmatrix} y = \begin{bmatrix} [C\xi^1, \cdots, C\xi^q]y \\ \mathbf{0} \end{bmatrix} = \mathbf{0}. \quad (5.42)$$

However, from $y \neq \mathbf{0}$ and linear independence of $\xi^1, \cdots, \xi^q$, $[\xi^1, \cdots, \xi^q]y \neq \mathbf{0}$. Therefore,

$$\begin{bmatrix} \mathbf{0} & \cdots & \mathbf{0} \\ \xi^1 & \cdots & \xi^q \end{bmatrix} y = \begin{bmatrix} \mathbf{0} \\ [\xi^1 & \cdots & \xi^q]y \end{bmatrix} \neq \mathbf{0}.$$

Therefore (5.42) is inconsistent with the nonsingularity of G.

$rank(B) \geqq m -\ rank\ (C)$ can be shown in the same way. QED

(Notes on [Lemma 5.5]): Okuguchi (1977, p. 194, Theorem 5.1), given B which is a m-order square matrix, c a m-order column vector, α a scalar, show that if $G = \begin{bmatrix} B & c \\ c^t & \alpha \end{bmatrix}$ is nonsingular, $rank(B) \geqq m - 1$ holds. Thus, this theorem could be said to be a slightly expanded version of Okuguchi's Theorem 5.1.

[Lemma 5.6] Let A be a n-order symmetric matrix, C a $m \times n$ matrix and assume that $rank(C) = m < n$. For all $x \neq \mathbf{0}$ that satisfy $Cx = \mathbf{0}$, $x^t A x > (<) 0$, $D = \begin{bmatrix} \mathbf{0} & C \\ C^t & A \end{bmatrix}$ is nonsingular.

(Proof) If D is taken to be a singular matrix, for some $\eta^t = (y^t \quad z^t)$, $D\eta = \mathbf{0}$ and $\eta \neq \mathbf{0}$. Therefore,

$$Cz = \mathbf{0}, \quad (5.43)$$

$$C^t y + Az = \mathbf{0}. \quad (5.44)$$

Assuming $z = \mathbf{0}$, from (5.44), $C^t y = \mathbf{0}$. From the assumption, $m = rank\,(C) = rank\,(C^t)$. Therefore, all of the rows of C^t are linearly independent thus, $y = \mathbf{0}$ is obtained but this opposes $\eta \neq \mathbf{0}$. Accordingly, $z \in \{\xi \neq \mathbf{0} | C\xi = \mathbf{0}\}$. Hence, if (5.44) is multiplied by z^t on the left,

$$z^t C^t y + z^t A z = 0. \quad (5.45)$$

From (5.43), $z^t C^t = (Cz)^t = \mathbf{0}$, hence (5.45) means $z^t A z = 0$, but this opposes the assumption. Therefore, D is nonsingular. QED

Let us use these lemmas to derive useful theorems relating to the characteristics of the substitution matrix.

[Theorem 5.11] Let $\boldsymbol{A}$ be a n-order symmetric matrix, $\boldsymbol{C}$ a $m \times n$ matrix and assume that $rank(\boldsymbol{C}) = m < n$. Furthermore, if for all $\boldsymbol{x} \neq \boldsymbol{0}$ that satisfy $\boldsymbol{Cx} = \boldsymbol{0}, \boldsymbol{x}^t\boldsymbol{Ax} > (<)0$,

$$\boldsymbol{D} = \begin{bmatrix} \boldsymbol{0} & \boldsymbol{C} \\ \boldsymbol{C}^t & \boldsymbol{A} \end{bmatrix} \text{ and } \boldsymbol{D}^{-1} = \begin{bmatrix} \boldsymbol{D}_{11} & \boldsymbol{D}_{12} \\ \boldsymbol{D}_{21} & \boldsymbol{D}_{22} \end{bmatrix} \text{ are defined,}$$

the following hold.

(i) $\boldsymbol{D}_{12} = (\boldsymbol{D}_{21})^t$ and $rank(\boldsymbol{D}_{12}) = rank\ (\boldsymbol{D}_{21}) = m$.
(ii) $\boldsymbol{D}_{22}$ is not a negative definite (nonnegative definite) and $rank(\boldsymbol{D}_{22}) = n - m$.

(Proof) From [Lemma 5.6], $\boldsymbol{D}$ is nonsingular so it is almost obvious that $\boldsymbol{D}^{-1}$ is stipulated.

(i) From the definition of $\boldsymbol{D}$, $\boldsymbol{D} = \boldsymbol{D}^t$. Hence, $(\boldsymbol{D}^{-1})^t = \boldsymbol{D}^{-1}$. Therefore, $\boldsymbol{D}_{12} = (\boldsymbol{D}_{21})^t$ and $rank(\boldsymbol{D}_{12}) = rank[(\boldsymbol{D}_{21})^t] = rank\ (\boldsymbol{D}_{21})$. If $\boldsymbol{DD}^{-1} = \boldsymbol{I}_{m+n}$ is divided,

$$\boldsymbol{CD}_{21} = \boldsymbol{I}_m \tag{5.46}$$

$$\boldsymbol{CD}_{22} = \boldsymbol{0} \tag{5.47}$$

$$\boldsymbol{C}^t\boldsymbol{D}_{11} + \boldsymbol{AD}_{21} = \boldsymbol{0} \tag{5.48}$$

and

$$\boldsymbol{C}^t\boldsymbol{D}_{12} + \boldsymbol{AD}_{22} = \boldsymbol{I}_n. \tag{5.49}$$

Hence

$$\begin{aligned} m = \mathrm{rank}(\boldsymbol{I}_m) &= rank(\boldsymbol{CD}_{21}) \\ &\leqq rank(\boldsymbol{D}_{21}) \\ &\leqq m. \end{aligned}$$

Therefore, $rank(\boldsymbol{D}_{12}) = rank\ (\boldsymbol{D}_{21}) = m$.
(ii) Applying [Lemma 5.5] to $\boldsymbol{D}^{-1}$ gives

$$rank(\boldsymbol{D}_{22}) \geqq n - rank(\boldsymbol{D}_{12}) = n - m. \tag{5.50}$$

In addition, from (5.47), $\boldsymbol{D}_{22}$'s each column $\in \boldsymbol{N}_{\boldsymbol{C}} \equiv \{\boldsymbol{x} | \boldsymbol{Cx} = \boldsymbol{0}\}$. Assumption implies $rank(\boldsymbol{C}) = m$ hence $dim(\boldsymbol{N}_{\boldsymbol{C}}) = n - m$. Therefore $\boldsymbol{D}_{22}$ does not contain linearly independent column vectors exceeding $(n - m)$. Thus,

$$rank(\boldsymbol{D}_{22}) \leqq n - m. \tag{5.51}$$

Accordingly, from (5.50) and (5.51), $rank(\boldsymbol{D}_{22}) \leqq n - m$ is obtained. Next, if (5.49) is multiplied on the left by $\boldsymbol{D}_{22}$, $\boldsymbol{D}_{22}\boldsymbol{C}^t\boldsymbol{D}_{12} + \boldsymbol{D}_{22}\boldsymbol{AD}_{22} = \boldsymbol{D}_{22}$ is obtained but from (5.47), $D_{22}AD_{22} = D_{22}$. Furthermore, for any $x \neq 0$, and from (5.47), $C(D_{22}x) = (CD_{22})x = 0$. Therefore, $D_{22}x \in N_C$. Therefore for any $x \neq 0$,

$$\begin{aligned} x^t D_{22} A D_{22} x &= x^t D_{22} x \\ &\geqq (\leqq) 0. \end{aligned}$$

QED

5.4 Stability of Matrices

This section discusses the theorems relating to the stability of matrices using arguments shown in Sects. 5.2 and 5.3. In addition, the matrices and the concept of stability of matrices used in this section can be defined as follows.

⟨**Definition 5.12**⟩ Let A be a n-order square matrix.

⟨D5.12-1⟩ A diagonal matrix A whose diagonal elements are positive (negative) is called a **positive diagonal matrix** (hereinafter abbreviated to **p.d.m.**) (**negative diagonal matrix (n.d.m.)**). In this section, diagonal matrices A which are not a unit matrix are represented as $\widehat{A}$.
⟨D5.12-2⟩ If each of the non-diagnoal elements of A is nonnegative, A is called a **Metzler matrix**, **Metzlerian** or a **gross substitution matrix**.
⟨D5.12-3⟩ If there exist nonempty subsets J and K of $N = \{i = 1, \cdots, n\}$, and $J \cup K = N$, $J \cap K = \phi$, A_{JJ}, A_{KK} is a Metzler matrix and $A_{JK} \leqq [0]$ as well as $A_{KJ} \leqq [0]$ hold, A is called a **Morishima matrix.**[17]
⟨D5.12-4⟩ If the real part of all the eigenvalues of A is negative, A is said to be **stable**.
⟨D5.12-5⟩ If $\widehat{D}A$ is stable in respect of a p.d.m $\widehat{D}$, A is said to be ***D*-stable**.
⟨D5.12-6⟩ If any leading principal minor matrix of A is D-stable, A is said to be **totally stable**.

[Lemma 5.7] Let A be an n-order real positive matrix and there exists a p.d.m. $\widehat{W}$ such that $A\widehat{W}$ is a negative semidefinite, A and A^{-1} are stable.

(Proof) Firstly, we will add a few words about the singularity of A. If under the assumption A is taken to be a singular matrix, there exists $y \neq 0$ that satisfies $y^t A = 0$. Thus, we obtain $0 = y^t \left(A\widehat{W}\right) y < 0$, however, this is self-contradicting. Hence, A is certainly regular.

[17] See Morishima (1964), Morishima and Fujimoto (1974), Kemp and Kimura (1978). If you are interested in Morishima matrix, please refer to Tarr (1976), Maybee and Maybee (1983) and Bassett et al. (1967).

Next, let $\boldsymbol{J}$ be any subset of $\boldsymbol{N}$ and let $\widehat{\boldsymbol{D}}$ be a #$\boldsymbol{J}$-th order p.d.m. Moreover, let any eigenvalues of $\widehat{\boldsymbol{D}}\boldsymbol{A}_{JJ}\left(\widehat{\boldsymbol{D}}(\boldsymbol{A}^{-1})_{JJ}\right)$ be $\lambda(\mu)$ and $\boldsymbol{x}(\boldsymbol{y})$ be eigenvectors corresponding to $\lambda(\mu)$. Here, λ is an eigenvalue of $\left[\left(\widehat{\boldsymbol{W}}_J\right)^{-1}\widehat{\boldsymbol{D}}\right]\left(\boldsymbol{A}_{JJ}\widehat{\boldsymbol{W}}_J\right)$. This is because

$$\left[(\hat{\boldsymbol{W}}_J)^{-1}\hat{\boldsymbol{D}}\right](\boldsymbol{A}_{JJ}\hat{\boldsymbol{W}}_J)\left[(\hat{\boldsymbol{W}}_J)^{-1}\boldsymbol{x}\right]=\lambda\left[(\hat{\boldsymbol{W}}_J)^{-1}\boldsymbol{x}\right] \text{ and } (\hat{\boldsymbol{W}}_J)^{-1}\boldsymbol{x}\neq\boldsymbol{0}.$$

On the other hand, from the definitions of μ and $\boldsymbol{y}$,

$$\left(\widehat{\boldsymbol{D}}\widehat{\boldsymbol{W}}_J\right)\left[\left(\widehat{\boldsymbol{W}}_J\right)^{-1}(\boldsymbol{A}^{-1})_{JJ}\right]\boldsymbol{y}=\mu\boldsymbol{y}.$$

Therefore μ becomes one of the eigenvalues of $\left(\widehat{\boldsymbol{W}}_J\right)^{-1}(\boldsymbol{A}^{-1})_{JJ}$. Furthermore, $\left(\widehat{\boldsymbol{W}}_J\right)^{-1}\widehat{\boldsymbol{D}}$ and $\widehat{\boldsymbol{D}}\widehat{\boldsymbol{W}}_J$ are p.d.m. so these are both positive definites. From the assumption, it is evident that both $\boldsymbol{A}_{JJ}\widehat{\boldsymbol{W}}_J=\left(\boldsymbol{A}\widehat{\boldsymbol{W}}\right)_{JJ}$ and $\left(\widehat{\boldsymbol{W}}_J\right)^{-1}(\boldsymbol{A}^{-1})_{JJ}=\left[(\boldsymbol{A}\widehat{\boldsymbol{W}})^{-1}\right]_{JJ}$ are quasi-negative definite.

When an n-order real square matrix $\boldsymbol{F}$ is a quasi-negative definite, if any leading principal minor matrix of $\boldsymbol{F}$ is defined to be $\boldsymbol{C}$, $\boldsymbol{C}$ is again a quasi-negative definite. Hence if it is shown that $\boldsymbol{C}$ is D-stable, the proof is complete. Thus, if $\widehat{\boldsymbol{M}}$ is taken to be a p.d.m. of the same order as $\boldsymbol{C}$, $\widehat{\boldsymbol{M}}$ is a positive definite hence from [Theorem 5.7], there exists a nonsingular $\boldsymbol{B}$ which is of the same order as $\widehat{\boldsymbol{M}}$ which is $\boldsymbol{B}\boldsymbol{B}^t=\widehat{\boldsymbol{M}}$. Therefore

$$\begin{aligned}\lambda\boldsymbol{I}-\widehat{\boldsymbol{M}}\boldsymbol{C}&=\lambda\boldsymbol{I}-\boldsymbol{B}\boldsymbol{B}^t\boldsymbol{C}\\&=\boldsymbol{B}\left(\lambda\boldsymbol{B}^{-1}-\boldsymbol{B}^t\boldsymbol{C}\right)\\&=\boldsymbol{B}(\lambda\boldsymbol{I}-\boldsymbol{B}^t\boldsymbol{C}\boldsymbol{B})\boldsymbol{B}^{-1}.\end{aligned}$$

Hence,

$$det\left(\lambda\boldsymbol{I}-\widehat{\boldsymbol{M}}\boldsymbol{C}\right)=det(\lambda\boldsymbol{I}-\boldsymbol{B}^t\boldsymbol{C}\boldsymbol{B}).$$

In other words, the eigenvalues of $\widehat{\boldsymbol{M}}\boldsymbol{C}$ are the same as the eigenvalues of $\boldsymbol{B}^t\boldsymbol{C}\boldsymbol{B}$. In addition, $\boldsymbol{B}$ is nonsingular so for any $\boldsymbol{x}\neq\boldsymbol{0}$, $\boldsymbol{B}\boldsymbol{x}\neq\boldsymbol{0}$. Thus,

$$\boldsymbol{x}^t\boldsymbol{B}^t\boldsymbol{C}\boldsymbol{B}\boldsymbol{x}=(\boldsymbol{B}\boldsymbol{x})^t\boldsymbol{C}(\boldsymbol{B}\boldsymbol{x})<0$$

and $\boldsymbol{B}^t\boldsymbol{C}\boldsymbol{B}$ becomes a negative definite. Therefore,

$$Re(\lambda) \leqq \frac{1}{2}(\boldsymbol{B}'\boldsymbol{C}\boldsymbol{B} + \boldsymbol{B}'\boldsymbol{C}\boldsymbol{B})'\text{s maximum eigenvalue}$$
$$< 0.$$

QED

The following theorem can be derived from this lemma.

[Theorem 5.12] When a n-order real square matrix $\boldsymbol{A}$ has an n.d.d., the following holds.

(i) There exists a p.d.m. $\widehat{\boldsymbol{W}}$ such that $\boldsymbol{A}\widehat{\boldsymbol{W}}$ is a quasi-negative definite.
(ii) $\boldsymbol{A}$ and $\boldsymbol{A}^{-1}$ are stable.[18]

(Proof)
(i) The assumption means that there exists a $\boldsymbol{x} > \boldsymbol{0}$ which is $[\boldsymbol{B}(\boldsymbol{A})]\boldsymbol{x} > \boldsymbol{0}$. Moreover, the definition indicates that $[-\boldsymbol{B}(\boldsymbol{A})]$ is a Metzler matrix. Accordingly, based on [Theorem 4.7], each column of $[-\boldsymbol{B}(\boldsymbol{A})]^{-1}$ is semipositive. Accordingly, there exists a $\boldsymbol{y} > \boldsymbol{0}$ which satisfies $[\boldsymbol{B}(\boldsymbol{A})]'\boldsymbol{y} > \boldsymbol{0}$. Here if we define $\widehat{\boldsymbol{W}} = diag(x_1/y_1, \ldots, x_n/y_n)$ (however, $x_i(y_i)$ is the ith element of $\boldsymbol{x}(\boldsymbol{y})$), $\widehat{\boldsymbol{W}}$ is p.d.m. and

$$\left\{\widehat{\boldsymbol{W}}[\boldsymbol{B}(\boldsymbol{A})]' + \boldsymbol{B}(\boldsymbol{A})\widehat{\boldsymbol{W}}\right\}\boldsymbol{y} > \boldsymbol{0}. \tag{5.52}$$

Furthermore, from the following relationship

$$\boldsymbol{B}\left(\widehat{\boldsymbol{W}}\boldsymbol{A}' + \boldsymbol{A}\widehat{\boldsymbol{W}}\right) \geqq \left\{\widehat{\boldsymbol{W}}[\boldsymbol{B}(\boldsymbol{A})]' + [\boldsymbol{B}(\boldsymbol{A})]\widehat{\boldsymbol{W}}\right\} \tag{5.53}$$

holds. Therefore from (5.52), (5.53) and $\boldsymbol{y} > \boldsymbol{0}$, $\boldsymbol{B}\left[\widehat{\boldsymbol{W}}\boldsymbol{A}' + \boldsymbol{A}\widehat{\boldsymbol{W}}\right]\boldsymbol{y} > \boldsymbol{0}$. Moreover, from the assumption, each of the main diagonal elements of $\left(\widehat{\boldsymbol{W}}\boldsymbol{A}' + \boldsymbol{A}\widehat{\boldsymbol{W}}\right)$ are negative. Hence $\left(\widehat{\boldsymbol{W}}\boldsymbol{A}' + \boldsymbol{A}\widehat{\boldsymbol{W}}\right)$ has an n.d.d. Next, if we assume that some eigenvalue λ_0 of $\left(\widehat{\boldsymbol{W}}\boldsymbol{A}' + \boldsymbol{A}\widehat{\boldsymbol{W}}\right)$ has a nonnegative real part,

$$\begin{aligned} |\lambda_0 - g_{ii}| &\geqq Re(\lambda_0 - g_{ii}) \\ &= Re(\lambda_0) - g_{ii} \\ &= Re(\lambda_0) + |g_{ii}| > 0. \end{aligned} \tag{5.54}$$

However, g_{ij} represents the (i, j)th element of $\left(\widehat{\boldsymbol{W}}\boldsymbol{A}' + \boldsymbol{A}\widehat{\boldsymbol{W}}\right)$. Therefore, for $\boldsymbol{d} = (d_1, \cdots, d_n)' > \boldsymbol{0}$,

[18] (ii) is an expansion of Mckenzie (1960, p.49, Theorem 2).

$$\begin{aligned}|\lambda_0 - g_{ii}|d_i &\geqq [Re(\lambda_0) + |g_{ii}|]d_i \\ &= Re(\lambda_0)d_i + |a_{ii}|d_i \\ &\geqq 0 + |g_{ii}|d_i \\ &> \sum_{j\neq 1}^{j=1^n} |g_{ij}|d_j, (i = 1, \cdots, n).\end{aligned}$$

Hence $\left[\lambda_0 \boldsymbol{I}_n - \left(\widehat{\boldsymbol{W}}\boldsymbol{A}^t + \boldsymbol{A}\widehat{\boldsymbol{W}}\right)\right]$ is nonsingular and contradicts the definition of λ_0. Therefore for any eigenvalue λ of $\left(\widehat{\boldsymbol{W}}\boldsymbol{A}^t + \boldsymbol{A}\widehat{\boldsymbol{W}}\right)$, $Re(\lambda) < 0$. Furthermore, $\left(\widehat{\boldsymbol{W}}\boldsymbol{A}^t + \boldsymbol{A}\widehat{\boldsymbol{W}}\right)$ is a symmetric matrix hence $Re(\lambda) = \lambda$. Therefore from [Theorem 5.6], $\left(\widehat{\boldsymbol{W}}\boldsymbol{A}^t + \boldsymbol{A}\widehat{\boldsymbol{W}}\right)$ is a negative definite. Therefore $\boldsymbol{A}\widehat{\boldsymbol{W}}$ is a quasi-negative definite.
(ii) It is evident from this Theorem's (i) and [Lemma 5.7]. QED

The following corollary is obtained as direct induction of this theorem.

[Corollary 5.12] If matrix $\boldsymbol{A}$ in [Theorem 5.12] is a Metzler matrix or a Morishima matrix, the following (i)–(iv) are mutually equivalent.

(i) $\boldsymbol{A}$ has an n.d.d.
(ii) There exists a p.d.m. $\widehat{\boldsymbol{W}}$ Such that $\boldsymbol{A}\widehat{\boldsymbol{W}}$ is a quasi-negative definite.
(iii) $\boldsymbol{A}$ and $\boldsymbol{A}^{-1}$ are stable.
(iv) $\boldsymbol{A}$ is stable.

(Proof) [Theorem 5.12] has already shown that (i) ⇒ (ii) ⇒ (iii). (iii) ⇒ (iv) is self-evident hence only (iv) ⇒ (i) will be shown.

(iv) ⇒ (i) Let $\boldsymbol{A} = [a_{ij}]$ be a Metzler matrix and λ_A and λ be the Frobenius root of $\boldsymbol{A}$ and an eigenvalue of $\boldsymbol{A}$, respectively. From the definition of λ_A, λ_A is a real number and is stable, thus, $\lambda_A = Re(\lambda_A) < 0$. Additionally, the first-order leading principal minor matrix $a_{ii}(i = 1, \cdots, n)$ of $\boldsymbol{A}$ are the eigenvalues of a_{ii}, thus, combining $\lambda_A < 0$ and [Theorem 4.5; (ii)], we obtain for any $i \in \boldsymbol{N}$, $a_{ii} \leqq \lambda_A < 0$. In addition, from [Theorem 4.5; (iv)], if $\lambda_A < 0$, $(0 \cdot \boldsymbol{I}_n - \boldsymbol{A})^{-1} = (-\boldsymbol{A})^{-1} \geq [\boldsymbol{0}]$. Thus, there exists $\boldsymbol{d} = (-\boldsymbol{A})^{-1}\boldsymbol{b} > \boldsymbol{0}$ for any $\boldsymbol{b} > \boldsymbol{0}$ and $(-\boldsymbol{A})\boldsymbol{d} = (-\boldsymbol{A})(-\boldsymbol{A})^{-1}\boldsymbol{b} = \boldsymbol{b} > \boldsymbol{0}$ holds.
Next, let $\boldsymbol{A} = \begin{bmatrix}\boldsymbol{A}_{JJ} & \boldsymbol{A}_{JK} \\ \boldsymbol{A}_{KJ} & \boldsymbol{A}_{KK}\end{bmatrix}$ be a Morishima matrix. However, let $\boldsymbol{J}$ and K be as per ⟨D5.12-3⟩. Here if we set $\boldsymbol{H} = \begin{bmatrix}\boldsymbol{I}_{\#J} & \boldsymbol{0} \\ \boldsymbol{0} & -\boldsymbol{I}_{\#K}\end{bmatrix}$, $\boldsymbol{M} = \boldsymbol{HAH}$ becomes a Metzler matrix and $\boldsymbol{H} = \boldsymbol{H}^{-1}$. Hence

$$\begin{aligned}det(\lambda\boldsymbol{I}_n - \boldsymbol{M}) &= det[\boldsymbol{H}(\lambda\boldsymbol{I}_n - \boldsymbol{HMH})\boldsymbol{H}] \\ &= det(\lambda\boldsymbol{I}_n - \boldsymbol{HMH}) \\ &= det(\lambda\boldsymbol{I}_n - \boldsymbol{A}).\end{aligned}$$

Accordingly, the eigenvalues of the Morishima matrix matches the eigenvalues of some Metzler matrix ($\boldsymbol{HAH}$). In addition, in respect of the Morishima matrix $\boldsymbol{A}, \boldsymbol{B}(\boldsymbol{A}) = \boldsymbol{B}(\boldsymbol{HAH})$. This concludes the proof. QED

References

Bassett, L., Habibagahi, H., & Quirk, J. (1967). Qualitative economics and Morishima matrices. *Econometrica, 35*(2), 221–233.

Hadamard, J. (1903). *Leçons sur la propagation des ondes et les équations de l'hydrodynamique, par Jacques Hadamard*. Librairie Scientifique A. Hermann. ISBN-13: 978-2876472990. Reprinted by Nabu Press, 2010. ISBN-13: 978-1149097908.

Henderson, J., & Quandt, R. (1980). Microeconomic Theory: A mathematical approach. (2nd revised ed.). McGraw-Hill. ISBN-13: 978-007066321.

Kemp, M. C., & Kimura, Y. (1978). *Introduction to mathematical economics*. Springer-Verlag. ISBN-13: 978-1461262800.

Kojima, T. (1914). On a theorem of Hadamard's and its Application. *Tohoku Mathematical Journal, 5*, 54–60.

Maybee, J. S., & Maybee, S. J. (1983). An algorithm for identifying Morishima and anti-Morishima matrices and balanced digraphs. *Mathematical Social Sciences, 6*(1), 99–103.

Mckenzie, L. (1960). Matrices with dominant diagonals and economic theory. In K. J. Arrow, S. Karlin, & P. Suppes (Eds.), *Mathematical methods in the social science* (pp. 47–62).

Morishima, M. (1964). *Equilibrium, stability, and growth—A multi-sectoral analysis*. Oxford University Press.

Morishima, M., & Fujimoto, T. (1974). The Frobenius theorem, its Solow-Samuelson extension and Kuhn-Tucker theorem. *Journal of Mathematical Economics, 1*, 199–205.

Nakayama, K. (1993a). Jaku yutaikaku gyoretsu no tokuisei ni tuite—Hadamard no teiri no kakucho- [On the singularity of weakly dominant diagonal matrix—Extension of Hadamard's theorem]. In M. Tawada & H. Kondo (Eds.), *Modern economic theory and its application* (pp. 52–65). Chuo Keizaisha. (in Japanese).

Nakayama, K. (1993b). Dominant diagonal wo motsu gyoretsu no teigi ni tsuite [On the definition of a dominant diagonal matrix]. *Chukyo Economic Review, 4*, 163–177. (in Japanese).

Okuguchi, K. (1977). Keizai bunseki no sugaki kiso [Mathematics basics of economic analysis]. McGraw-Hill. (in Japanese).

Tarr, D. G. (1976). Distributed lags, Morishima matrices, and the stability of economic models. *Econometrica, 44*(3), 597–600.

Tarr, D. G. (1977). A note on the dominant diagonal matrix and its extensions. *Economic Studies Quarterly, 21*, 305–313.

Taussky, O. (1949). A recurring theorem on determinants. *American Mathematical Monthly, 56*, 672–676.

Uekawa, Y. (1971). Generalization of the Stolper-Samuelson theorem. *Econometrica, 39*, 197–217.

Morgenstern, O. (1954). Experiment and large-scale computation in economics. In *Economic activity analysis*. Wiley & Sons.

Chapter 6
Application to Economics

6.1 Introduction

The eigenvalue and nonlinear programming methods in the title of this publication are important tools used in a wide range of fields. They are not only important concepts in linear algebra but also in other scientific disciplines, such as in the case of matrix eigenvalues and eigenvectors, which are considered essential concepts in statistics, economics, physics, statistics, computer science, and mathematics as well. They apply to such fields as meteorology, psychology, and information science. Further, similarly to the nonlinear programming problem, they are usable in various fields beyond engineering and economics. Problems they have recently found application to include optimal design selection, optimal trajectory calculation, portfolio optimization, and model calibration.

Some simple examples of cases in which eigenvalues and nonlinear programming are used in economics are given below.

6.2 Eigenvalues in Matrices and Economics

6.2.1 *What Is the Eigenvalue Problem?*

For an nth-degree square matrix $\boldsymbol{A}$, the eigenvalues of matrix $\boldsymbol{A}$ are the scalar λ that satisfy the following relationship, and vector $\boldsymbol{x}$ is the eigenvector corresponding to that eigenvalue.

$$\boldsymbol{A}\boldsymbol{x} = \lambda\boldsymbol{x} \tag{6.1}$$

© Springer Nature Singapore Pte Ltd. 2024

K. Nakayama, *Eigenvalue Problem and Nonlinear Programming Problem*,
New Frontiers in Regional Science: Asian Perspectives 70,
https://doi.org/10.1007/978-981-99-2943-6_6

However, as the relationship that vector $\boldsymbol{x}$ is a zero vector is self-evident, we assume that vector $\boldsymbol{x}$ is not zero. If we change the shape of (6.1), it becomes (6.2), and for this to have a non-zero solution, it is necessary that this not be an inverse of the coefficient matrix (as, for vector x, an inverse would be a zero solution).

$$[\boldsymbol{A} - \lambda \boldsymbol{I}]\boldsymbol{x} = \boldsymbol{0} \tag{6.2}$$

Put another way, the determinant of the coefficient matrix must be zero. The conditions for (6.2) having a non-zero solution are as shown in (6.3). Incidentally, this equation is known as an eigenequation.

$$det(\boldsymbol{A} - \lambda \boldsymbol{I}) = 0 \tag{6.3}$$

Matrix $\boldsymbol{A}$ is an $n \times n$ square matrix, so (6.3) has n solutions (eigenvalues) (including duplicates).

The Eigenvalue Problem is the problem of finding eigenvalues and eigenvectors in relation to a given square matrix.

6.2.2 Eigenvalue Display for Determinant

The nth order square matrix $\boldsymbol{A}$ determinant det$\boldsymbol{A}$ is expressed as the product of n eigenvalues.

$$det\boldsymbol{A} = \lambda_1 \lambda_1 \cdots \lambda_n \tag{6.4}$$

While not in an economic sense, if the square matrix is a 2-degree square matrix, this represents the parallelogram formed using the two-column vectors of matrix $\boldsymbol{A}$. In the same way (if it is a 3-degree square matrix), this expresses a parallelepiped made up of three-column vectors of matrix $\boldsymbol{A}$. Put another way, the area and volume of the shape made from the column vectors constituting the matrix are the product of the eigenvectors.

6.2.3 Power of the Matrix

In economics, the value of the power of a matrix may be important; therefore, we introduce the concept of diagonalization of matrices. Let us suppose that $\boldsymbol{\Lambda}$ is a diagonal matrix of eigenvalues and $\boldsymbol{X}$, is a matrix of corresponding eigenvectors.

$$\Lambda = \begin{bmatrix} \lambda_1 & & \mathbf{0} \\ & \ddots & \\ \mathbf{0} & & \lambda_n \end{bmatrix} \quad (6.5)$$

$$X = [\boldsymbol{x}_1, \ \cdots, \ \boldsymbol{x}_n] \quad (6.6)$$

At this time, column $\boldsymbol{A}$ is diagonalized as follows.

$$X^{-1}AX = \Lambda \quad (6.7)$$

The powers of both sides (for example, the mth power) can be expressed as follows using this relationship:

$$X^{-1}AX \cdots X^{-1}AX = X^{-1}A^mX = \Lambda^m \quad (6.8)$$

It can also be rewritten as follows:

$$A^m = X\Lambda^mX^{-1} \quad (6.9)$$

What (6.9) means is that if the maximum eigenvalue is 1, if $\boldsymbol{A}^m$ is a power degree larger, then it will converge on some value, and if the maximum eigenvalue is less than 1, $\boldsymbol{A}^m$ will converge on the zero columns. Conversely, if the maximum eigenvalue is greater than 1 and it is a power degree m larger, it follows that $\boldsymbol{A}^m$ diverges.

6.2.4 Application to Industry-Related Input-Output Analysis

The Input-Output model is expressed as follows. We call matrix $\boldsymbol{A}$ the input coefficient matrix.

$$\boldsymbol{x} = \boldsymbol{A}\boldsymbol{x} + \boldsymbol{b} \quad (6.10)$$

This is rewritten as follows in this equation.

$$\boldsymbol{x} = \boldsymbol{b} + \boldsymbol{A}\boldsymbol{b} + \boldsymbol{A}^2\boldsymbol{b} + \boldsymbol{A}^3\boldsymbol{b} + \cdots \quad (6.11)$$

The production volume vector $\boldsymbol{x}$ does not diverge for any final demand vector $\boldsymbol{b}$. In other words, to make production possible, it is necessary and sufficient that the maximum eigenvalue of input coefficient matrix $\boldsymbol{A}$ is smaller than 1.

In other words, if the maximum eigenvalue of input coefficient matrix $\boldsymbol{A}$ is smaller than 1, the following equation will produce a non-negative solution $\boldsymbol{x}$ for any final demand $\boldsymbol{b}$. This shows a special case for Frobenius's theorem.

$$x = [I - A]^{-1} b \tag{6.12}$$

6.2.5 Application to Population Distribution Analysis

We refer to matrices where the matrix elements are non-negative and the vertical total is 1 as stochastic matrices, transition matrices, or Markov matrices. Let $\boldsymbol{A}$ be a stochastic matrix. We know that the maximum eigenvalue of stochastic matrix $\boldsymbol{A}$ is 1, and all others are non-negative values smaller than 1.

Here, vector $\boldsymbol{x}$ is the population, matrix $\boldsymbol{A}$ expresses the probability that a population will move from a certain region to another region, and vector $\boldsymbol{b}$ expresses "births minus deaths" in each region.

$$x_t = Ax_{t-1} + b \tag{6.13}$$

If we consider population transitions such as (6.13), if vector $\boldsymbol{b}$ is zero, this shows us that the total population is a constant value.

6.3 Linear Programming, Nonlinear Programming, and Economics

6.3.1 What Is Linear Programming?

The linear programming problem is an optimization problem where the objective function and constraints are all linear (maximization problem, minimization problem), and the methodology for solving this is called the linear programming method. If we take the maximization problem as an example, the linear programming problem is expressed as the following equation:

$$max\, \boldsymbol{cx} \tag{6.14}$$

$$\text{subject to}\, \boldsymbol{A}\, \boldsymbol{x} \leqq \boldsymbol{b} \tag{6.15}$$

Here, let $\boldsymbol{A}$ be the $m \times n$ matrix, $\boldsymbol{x}$ be the nth order longitudinal vector, $\boldsymbol{b}$ be the mth order longitudinal vector, and $\boldsymbol{c}$ is the nth order transverse vector.

To repeat, the objective functions and constraints are all expressed linearly in this way.

To go back a bit, linear programming problems and linear programming are said to have come to the fore when G. Dantzig proposed the simplex method in 1947 while working in the Applied Mathematics Division of the U.S. Department of Defense. This research also considered how to conserve resources, as might be expected of research conducted by the Department of Defense. The development of the simplex method finally allowed the practical solving of linear programming problems.

After that, this method became commonly used in economics and business studies. In 1975, the "Study concerning the optimal allocation of scarce resources" won the Nobel Prize for economics, but this prize was awarded to T. Koopmans and L. Kantorovich. There are various theories why Dantzig was displaced as the father of this methodology, but it seems one reason was that the research was not sufficiently applied to economics.

6.3.2 Application of the Linear Programming Problem

The linear programming problem is one of how to allocate scarce resources, so it has a wide field of applications.

For example, if there are sufficient funds, it is necessary to consider what kind of food supply would be the cheapest option that still enables the population's intake of various nutrients to keep them above the line necessary for maintaining health (this is called the homemaker's problem but is in terms of the country and households). This problem also applies to menu planning (optimization of nutritional balance and cost of school meals).

One well-known application of this to economics is the problem of determining production quantity in Input-Output analysis. We can look at this as the minimum amount of domestic production required when considering the quantity of final demand, something used in the recovery effort after the second world war and for calculating the production quantity required to achieve a certain level of economic growth given the current state of technology. Incidentally, W. Leontief, the inventor of the Input-Output analysis concept, received the Nobel Prize for economics in 1973.

Additionally, the linear programming method is also known to apply to the transportation problem. The transportation problem is the problem of working out how much to transport from which production site to which consumption site in a way that will minimize the total transportation cost when multiple production sites and consumption sites exist, we know the production and consumption quantities of each, as well as the transportation cost per unit from the production site to the consumption site. The reason for which logistics in Japan are so efficient is that this kind of thinking has been certainly incorporated by corporate management since long ago.

6.3.3 What Is the Nonlinear Programming Problem?

The nonlinear programming problem is an optimization problem where either the objective functions or constraints are nonlinear. The Kuhn–Tucker theorem, published as a paper by Kuhn and Tucker in 1951, is considered where the nonlinear programming problem starts. The history of the nonlinear programming method is considered approximately the same length as the linear programming method and is promoted by researchers with backgrounds in mathematics and economics. Compared to the linear programming method, problems related to the nonlinear programming method are extremely diverse. We can also think of linear programming as a special case of nonlinear programming.

The most well-known problem handled by microeconomics is the optimized behavior of consumers. Household finances have a certain nonlinear utility function. The explanatory variables in the utility function are the various consumable goods. However, there are also budgetary constraints on household finances. If the price is fixed, the budgetary constraints are linear. The problem of optimizing consumption when utility is maximized on the basis of budgetary constraints exemplifies a case where the constraints are linear, but objective functions are nonlinear.

Moreover, in an economy, multiple goods are produced, but if we consider the situation in which the production limits of the elements produced are reduced on a plane with output as the axis, the frontier of the set that can be produced is a convex curve from the source. In this situation, if we consider the problem of maximizing GDP, which is measured in given international prices, the constraints are nonlinear, but the objective functions are limited.

6.3.4 Optimization Problem and Eigenvalues

A positive-definite matrix refers to a matrix in which the elements of any vector multiplied from the right have a positive value.

Imagine a vessel like a wine glass without a foot. If this is an objective function, the part that originally had a foot (base of the glass) is the minimum value of the objective function. As for the minimum value, the first-order derivative is zero since the value does not change there. So, how about the second-order derivative? Since deviating from the minimum value means that whatever direction you go in will be positive, the second-order derivative will be positive.

A matrix of second-order partial differential coefficients is referred to as a Hessian matrix, and as this is also a situation in which "whatever direction you go in, it will be positive", the Hessian must be a positive-definite matrix. However, it is practically impossible to multiply every vector and find out that the answers are positive. Therefore, it is also a convenient method for investigating whether it is a positive-definite matrix or not. All eigenvalues being positive is a necessary and sufficient condition of the matrix being a positive-definite matrix. Details are not provided here, but matrix diagonalization and the optimization problem are also connected to the matrix eigenvalues.

6.4 Books Recommended to Read

Eigenvalues and the nonlinear programming method should not be considered on their own, and it is no exaggeration to say that economics has developed based on the existence of these two tools. These tools have made a significant contribution to the field of nonlinear dynamics. The reason for this is that the nature of eigenvalues is deeply connected to determining steady-state stability, economic growth, business cycles, and so on. Additionally, in recent years, we have seen numerous reference documents entitled "Nonlinear Eigenvalue Problem", and this may suggest the increasing importance of eigenvalues and nonlinear programming in different fields.

Finally, I would like to introduce you to some well-known works that have taken up the topics of the Nonlinear Programming Problem, the Eigenvalue Problem, and other works in English found useful by the author when writing problem-handling forms.

First, I recommend textbooks in economic mathematics such as Takayama (1993, 2010), Kemp and Kimura (1978) and Sydsaeter et al. (2022) to fill in the necessary knowledge for nonlinear programming and eigenvalue problems.

Among texts looking at the convex set and convex programming problem, Nikaido (1968) and Rockafellar (1970) is the clear standout. Horn and Johnson (1985) excel at commentating on linear algebra and, in particular, the matrices needed to define eigenvalues. Moreover, Ortega and Rheinbokdt (1970) summarize the linear algebra and calculus commonly used in optimization theory. On optimization, Intriligator (1971) and Jahn (1996) are also useful texts. The optimization conditions for Kuhn–Tucker are discussed in most texts related to nonlinear optimization, but a detailed commentary on constraint qualifications is given in Whittle (1971) and Bazaraa and Shetty (1976). Mangasarian (1969) is also reputed as an introductory text for the nonlinear programming method. This book comprehensively deals with the basic theory required for the nonlinear programming method. It explains in detail the basic theory of convex sets and convex functions as well as optimality and duality. Additionally, linear matrix inequality and the alternative theorem are systematically summarized, and this book can be considered a classic of nonlinear programming. Another work that should be treated as a classic of nonlinear programming is Avriel (1976), regarded as a model for later texts. Further, in nonlinear dynamics, Satty and Bram (1964) and Hardley (1964) are rated highly. Also, refer to Luenberger (1979) for the relationship between dynamics and eigenvalues. If you are interested in the relationship between nonlinearity and economic dynamics, see Lorenz (1993) for details.

All of the books mentioned here are from quite some time ago, but they are masterpieces that cannot be overlooked when discussing nonlinear programming problems and eigenvalue problems. I would like to add that there is still a lot of research in this area, and they all cite these as references.

References

Avriel, M. (1976). *Nonlinear programming: Analysis and methods*. Prentice-Hall.

Bazaraa, M. S., & Shetty, C. M. (1976). *Foundation of optimization*. Springer-Verlag.

Hardley, G. (1964). *Nonlinear and dynamic programming*. Addison-Wesley Publishing Company.

Horn, R. A., & Johnson, C. R. (1985). *Matrix analysis*. Cambridge University Press.

Intriligator, M. D. (1971). *Mathematical optimization and economic theory*. Prentice Hall.

Jahn, J. (1996). *Introduction to the theory of nonlinear optimization*. Springer. ISBN-13: 978-3662032718.

Kemp, M. C., & Kimura, Y. (1978). *Introduction to mathematical economics*. Springer-Verlag. ISBN-13: 978-1461262800.

Lorenz, H. W. (1993). *Nonlinear dynamical economics and chaostic motion* (2nd ed.). Springer-Verlag. ISBN-13: 978-0387568812.

Luenberger, D. G. (1979). *Introduction to dynamic systems*. John Wiley & Sons. ISBN-13: 978-0471025948.

Mangasarian, O. L. (1969). *Nonlinear programming*. McGraw-Hill. ISBN-13: 978-0898713412.

Nikaido, H. (1968). *Convex structures and economic theory* (Mathematics in science and engineering: A series of monographs and textbooks) (Vol. 51). Academic Press. ISBN-13: 978-0125194501.

Ortega, J. M., & Rheinboldt, W. C. (1970). *Iterative solution of nonlinear equations in several variables*. Academic Press. ISBN-13: 978012-5285506.

Rockafellar, R. T. (1970). *Convex analysis*. Princeton University Press. ISBN-13: 978-0691015866.

Satty, T. L., & Bram, J. (1964). *Nonlinear mathematics*. McGraw-Hill. Reprinted by Dover Publications, 2010. ISBN-13: 978-0486642338.

Sydsaeter, K., Hammond, P., & Storm, A. (2022). *Essential mathematics for economics analysis* (6th ed.). Pearson. ISBN-13: 978-1292-359298.

Takayama, A. (1993). *Analytical methods in economics*. University of Michigan Press. ISBN-13: 978-0472081356.

Takayama, A. (2010). *Mathematical economics*. (2nd revised ed.). Cambridge University Press. ISBN-13: 978-0521314985.

Whittle, P. (1971). *Optimization under constraints: Theory and practice of nonlinear programming*. John Wiley & Sons. ISBN-13: 978-0471941309.

Appendix

⟨**Definition 1**⟩ When there exists a real-valued function $d(\boldsymbol{x},\ \boldsymbol{y})$ defined on $\boldsymbol{X}$ which satisfy the following three conditions for any element $\boldsymbol{x},\ \boldsymbol{y}$ of a set $\boldsymbol{X}$:

(i) $d(\boldsymbol{x},\ \boldsymbol{y}) \geqq 0$; $d(\boldsymbol{x},\ \boldsymbol{y}) = 0 \Leftrightarrow \boldsymbol{x} = \boldsymbol{y}$
(ii) $d(\boldsymbol{x},\ \boldsymbol{y}) = d(\boldsymbol{y},\ \boldsymbol{x})$ (symmetry)
(iii) for any $z \in \boldsymbol{X}$, $d(\boldsymbol{x},\ \boldsymbol{y}) \leqq d(\boldsymbol{x},\ z) + d(z,\ \boldsymbol{y})$ (triangular inequality)

$(\boldsymbol{X},\ d)$ or simply $\boldsymbol{X}$ is referred to as a **metric space** and $d(\boldsymbol{x},\ \boldsymbol{y})$ is called a **metric function** of $\boldsymbol{x}$ and $\boldsymbol{y}$ or the **distance**. Geometrically, each element of $\boldsymbol{X}$ is called a **point** and a subset of $\boldsymbol{X}$ is called a point set.

⟨**Definition 2**⟩ Let $\boldsymbol{X}$, $\boldsymbol{Y}$ and $\boldsymbol{Z}$ be sets.

(i) A rule f which maps one element $\boldsymbol{y}$ to each element $\boldsymbol{x}$ of $\boldsymbol{X}$ is called a **mapping** from $\boldsymbol{X}$ to $\boldsymbol{Y}$ and is denoted as $f : \boldsymbol{X} \rightarrow \boldsymbol{Y}$. Here, $\boldsymbol{X}$ is called the **domain** of f and $\boldsymbol{Y}$ the **range** of f. Furthermore, the mapping may also be referred to as a **function** or a **transformation**.
(ii) When $\boldsymbol{A} \subseteq \boldsymbol{X}$, $\boldsymbol{B} \subseteq \boldsymbol{Y}$ and $f : \boldsymbol{X} \rightarrow \boldsymbol{Y}$, $f(\boldsymbol{A}) = \{f(\boldsymbol{x}) | \boldsymbol{x} \in \boldsymbol{A}\}(f^{-1}(\boldsymbol{B}) = \{\boldsymbol{x} \in \boldsymbol{X} | f(\boldsymbol{x}) \in \boldsymbol{B}\})$ is called the image of $\boldsymbol{A}$ due to f (inverse image of $\boldsymbol{B}$ due to f).
(iii) When $f : \boldsymbol{X} \rightarrow \boldsymbol{Y}$ is $f(\boldsymbol{X}) = \boldsymbol{Y}$, f is called an **onto-mapping** or a **surjection** from $\boldsymbol{X}$ to $\boldsymbol{Y}$. For different $\boldsymbol{x}^1,\ \boldsymbol{x}^2 \in \boldsymbol{X}$, if $f(\boldsymbol{x}^1) \neq f(\boldsymbol{x}^2)$, f is called a **one-to-one mapping** or an **injection**.
(iv) Let $f : \boldsymbol{X} \rightarrow \boldsymbol{Y}$ and $g : \boldsymbol{Y} \rightarrow \boldsymbol{Z}$. For $\boldsymbol{x} \in \boldsymbol{X}$, the mapping $h : \boldsymbol{X} \rightarrow \boldsymbol{Z}$ defined by $h(\boldsymbol{x}) = g(f(\boldsymbol{x}))$ Is called a **composite** of f and g and is represented as $h = g \circ f$.
(v) When f is a one-to-one mapping from $\boldsymbol{X}$ to $\boldsymbol{Y}$, in respect of $f : \boldsymbol{X} \rightarrow \boldsymbol{Y}$, $f^{-1} : \boldsymbol{Y} \rightarrow \boldsymbol{X}$ is called an **inverse mapping** of f.

⟨**Definition 3**⟩ Let $\tilde{\boldsymbol{N}}$ be a set of natural numbers and $f : \tilde{\boldsymbol{N}} \rightarrow \boldsymbol{X} \subseteq \boldsymbol{R}^m$.

(i) If the image $\equiv f(n)$ of $n \in \tilde{\boldsymbol{N}}$ by f is represented as $\boldsymbol{x}^n$, the whole $\equiv f(\tilde{\boldsymbol{N}})$ of $f(n)$ is written as $\{\boldsymbol{x}^n\}$ and $\{\boldsymbol{x}^n\}$ is called a **sequence of points** of $\boldsymbol{X}$.

© Springer Nature Singapore Pte Ltd. 2024

K. Nakayama, *Eigenvalue Problem and Nonlinear Programming Problem*,
New Frontiers in Regional Science: Asian Perspectives 70,
https://doi.org/10.1007/978-981-99-2943-6

(ii) When $\tilde{\tilde{N}} = \{n_1, n_2, \cdots\}$ is taken to be an infinite subset of $\tilde{N}$ that is $n_1 < n_2 < \cdots, f\left(\tilde{\tilde{N}}\right) = \{\boldsymbol{x}^{n_j}\}$ is called a **partial sequence** of $\{\boldsymbol{x}^n\}$.

(iii) Given a sequence of points $\{\boldsymbol{x}^n\}$ of $\boldsymbol{X}$, if there exists $\boldsymbol{x} \in \boldsymbol{R}^n$, $n_\varepsilon \in \tilde{N}$ is defined for any $\varepsilon > 0$, and $d(\boldsymbol{x}^n, \boldsymbol{x}) < \varepsilon$ holds for all n that are $n > n_\varepsilon$, $\{\boldsymbol{x}^n\}$ is said to **converge** to $\boldsymbol{x}$ and is expressed as $\lim_{n\to\infty} \boldsymbol{x}^n = \boldsymbol{x}$. Here, $\boldsymbol{x}$ is called the **limit** of $\{\boldsymbol{x}^n\}$.

(Notes on ⟨Definition 3⟩): When $f : \tilde{\boldsymbol{N}} \to \boldsymbol{X} \subseteq \boldsymbol{R}$, the sequence of points $\{x^n\}$ of $\boldsymbol{X}$ is called a **real number sequence** of $\boldsymbol{X}$ or simply a **number sequence**.

⟨Definition 4⟩ A real number sequence $\{x^n\}$, should it be $x^n \leqq x^{n+1} (x^n \geqq x^{n+1})$ for any natural number n, is said to be **monotonically increasing (monotonically decreasing)** or **monotonically non-decreasing (monotonically non-increasing)** and a number sequence that is monotonically increasing or monotonically decreasing is abbreviated to the monotonic number sequence.

[Theorem 1] Characteristics relating to $\boldsymbol{R}$

(i) $\boldsymbol{R}$ is a **field**. In other words, if x, y, z is taken to be an element of $\boldsymbol{R}$, the product of x and y, $x + y$, and the product xy are defined to be elements of $\boldsymbol{R}$ and the following conditions (i-i)–(i-ix) hold.
(i-i) $x + y = y + x$, (Commutative law)
(i-ii) $(x + y) + z = x + (y + z)$, (Associative law)
(i-iii) For $x \in \boldsymbol{R}$, there exists $0 \in \boldsymbol{R}$ which becomes $x + 0 = x$,
(i-iv) For any $x \in \boldsymbol{R}$, there exists one element $-x$ of $\boldsymbol{R}$ that is $x + (-x) = 0$,
(i-v) $xy = yx$, (Commutative law)
(i-vi) $(xy)z = x(yz)$, (Associative law)
(i-vii) $(x + y)z = xz + yz$, (Distributive law)
(i-viii) For any $x \in \boldsymbol{R}$, there exists $1 \in \boldsymbol{R}$ which is $1x = x$.
(i-ix) There exists one element x^{-1} (or $1/x$) in $\boldsymbol{R}$ which is $xx^{-1} = 1$ for any $x(\neq 0) \in \boldsymbol{R}$.
(ii) For any $x, y, z \in \boldsymbol{R}$, one of $x > y$, $x = y$, $x < y$ always holds (total order) and if $x \leqq y$ and $y \leqq z$, $x \leqq z$ holds (transitive law).
(iii) For any positive $x, y \in \boldsymbol{R}$, there exists a natural number n such that $x < ny$. (axiom of Archimedes).
(iv) For any $x, y \in \boldsymbol{R}$ such that $x < y$, there exists $z \in \boldsymbol{R}$ that satisfies $x < z < y$ (density of real numbers).
(v) When $\boldsymbol{R}_1$ and $\boldsymbol{R}_2$ are subsets of $\boldsymbol{R}$ that satisfy the following conditions:
(v-i) $\boldsymbol{R}_1 \cap \boldsymbol{R}_2 = \phi$,
(v-ii) $\boldsymbol{R}_1 \cup \boldsymbol{R}_2 = \boldsymbol{R}$,
(v-iii) $\boldsymbol{R}_1 \neq \phi$, $\boldsymbol{R}_2 \neq \phi$
(v-iv) If $x \in \boldsymbol{R}_1$ and $y \in \boldsymbol{R}_2$, and they are subsets of $\boldsymbol{R}$ that satisfy $x < y$, $\boldsymbol{R}_1$ has the largest element or $\boldsymbol{R}_2$ has the smallest element (continuity of real numbers).

(Notes on [Theorem 1]):Above, we do not refer to a strict definition of real numbers (Dedekind's definition, Cantor's definition) .

⟨**Definition 5**⟩ Let $\boldsymbol{S} \subseteq \boldsymbol{R}$.

(i) If there exists $a \in \boldsymbol{R}$ and for any $x \in \boldsymbol{S}$, $a \geqq (\leqq) x$, $\boldsymbol{S}$ is said to be **bounded above** (**bounded below**) and a is called an **upper bound** (**lower bound**) for $\boldsymbol{S}$. When $\boldsymbol{S}$ is bound above and below, $\boldsymbol{S}$ is said to be **bound**.
(ii) The minimum element (maximum element) of all of the sets bounded above (bounded below) of $\boldsymbol{S}$ is called the **spremum**, **least upper bound** (**infimum**) or the **greatest lower bound** of $\boldsymbol{S}$ and is expressed as $sup\boldsymbol{S}(inf\boldsymbol{S})$.

[Theorem 2]

(i) If a subset $\boldsymbol{S}$ of $\boldsymbol{R}$ has an upper bound (lower bound),
(i-i) it is possible to produce a sequence $\{a^n\}$ of $\boldsymbol{S}$ with an upper limit (lower limit) and furthermore,
(i-ii) that converges to the upper limit (lower limit).
(ii) When a sequence of real numbers $\{a^n\}(\{b^n\})$ that converge to $a(b) \in \boldsymbol{R}$ satisfies $a^n \leqq b^n$ for any natural number n, $\lim_{n \to \infty} a^n = a \leqq b = \lim_{n \to \infty} b^n$.
(iii) Any sequence of real numbers in the bound subset $\boldsymbol{S}$ of $\boldsymbol{R}$ has a subsequence that converges.

⟨**Definition 6**⟩ Let $\boldsymbol{X}$ be a nonempty set, $\boldsymbol{F}$ a field, $\boldsymbol{x}$, $\boldsymbol{y}$, $\boldsymbol{z}$ any element of $\boldsymbol{X}$ and α, β any element of $\boldsymbol{F}$. For this $\boldsymbol{x}$, $\boldsymbol{y}$, suppose that an element $\boldsymbol{x} + \boldsymbol{y}$ is defined in $\boldsymbol{X}$ that is called their sum and an element $\alpha\boldsymbol{x}$ referred to as scalar multiple of $\boldsymbol{X}$ is defined. Here, if $\boldsymbol{X}$ satisfies the following conditions (i)–(viii) in respect of a sum and scalar multiple, $\boldsymbol{X}$ is called a **vector space** or a **linear space** on $\boldsymbol{F}$.

(i) $\boldsymbol{x} + \boldsymbol{y} = \boldsymbol{y} + \boldsymbol{x}$,
(ii) $(\boldsymbol{x} + \boldsymbol{y}) + \boldsymbol{z} = \boldsymbol{x} + (\boldsymbol{y} + \boldsymbol{z})$,
(iii) For any $\boldsymbol{x} \in \boldsymbol{X}$, there exists $\boldsymbol{0} \in \boldsymbol{X}$ that is $\boldsymbol{x} + \boldsymbol{0} = \boldsymbol{x}$,
(iv) For any $\boldsymbol{x} \in \boldsymbol{X}$, there exists a single element of $\boldsymbol{X}$ that is $\boldsymbol{x} + (-\boldsymbol{x}) = \boldsymbol{0}$, $-\boldsymbol{x}$.
(v) $\alpha(\boldsymbol{x} + \boldsymbol{y}) = \alpha\boldsymbol{x} + \alpha\boldsymbol{y}$,
(vi) $(\alpha\beta)\boldsymbol{x} = \alpha(\beta\boldsymbol{x})$,
(vii) $(\alpha + \beta)\boldsymbol{x} = \alpha\boldsymbol{x} + \beta\boldsymbol{x}$,
(viii) $1\boldsymbol{x} = \boldsymbol{x}$.

An element of $\boldsymbol{X}$ is called a **vector** or a **point** and an element of $\boldsymbol{F}$ is called a **scalar**. Particularly, when $\boldsymbol{F} = \boldsymbol{R}(\boldsymbol{C})$, a vector space on $\boldsymbol{F}$ is called a **real (complex) vector space**.

Next, let us explain the basic concept and facts (⟨Definition 7⟩, ⟨Definition 8⟩, [Theorem 3]) relating to the vector space $\boldsymbol{X}$ on $\boldsymbol{F}$.

⟨**Definition 7**⟩ When a nonempty subset $\boldsymbol{W}$ of the vector space $\boldsymbol{X}$ satisfies the following two conditions:

(i) for any $\boldsymbol{x}$, $\boldsymbol{y} \in \boldsymbol{W}$, $\boldsymbol{x} + \boldsymbol{y} \in \boldsymbol{W}$,
(ii) for any $\boldsymbol{x} \in \boldsymbol{W}$ and scalar α, $\alpha\boldsymbol{x} \in \boldsymbol{W}$

W is called a **linear subspace** of X or simply a **subspace**.

⟨**Definition 8**⟩ Let k be a positive integer and x^i and α_i $(i = 1, \cdots, k)$ a vector in the vector space X and a scalar, respectively.

(i) Vector $\sum_{i=1}^{k} \alpha_i x^i \in X$ is called a **linear combination** of $x^1, \cdots, x^k$.

(ii) If $\sum_{i=1}^{k} \alpha_i x^i = \mathbf{0}$ means $\alpha_1 = \cdots = \alpha_k = 0$, $x^1, \cdots, x^k$ is said to be **linearly independent**. and when $x^1, \cdots, x^k$ is not linearly independent, they are said to be **linearly dependent**.

(iii) If $x^1, \cdots, x^k$ is linearly independent and a $y \in X(W)$ can be expressed as a linear combination of $x^1, \cdots, x^k, x^1, \cdots, x^k$ is called the **basis** of $X(W)$. Moreover, here, k is called the **dimension** of $X(W)$ and is represented as $dim(X)$ or $dim(W)$.

[Theorem 3] When x^i $(i = 1, \cdots, q), y^j$ $(j = 1, \cdots, p)$ and y are elements of X, the following hold.

(i) If $x^1, \cdots, x^q$ can be represented as a linear combination of $y^1, \cdots, y^p$ and $p < q, x^1, \cdots, x^q$ is linearly dependent.

(ii) If $\{x^1, \cdots, x^q\}$ and $\{y^1, \cdots, y^p\}$ are taken to be any two bases of X, $q = p$.

(iii) If $x^1, \cdots, x^q$ is linearly independent and $x^1, \cdots, x^q, y$ is linearly dependent, y is represented as a single linear combination of $x^1, \cdots, x^q$.

(iv) The necessary and sufficient conditions for $x^1, \cdots, x^q$ to be a basis of X are that $x^1, \cdots, x^q$ is linearly independent and $(q + 1)$ vectors of X become linearly dependent.

⟨**Definition 9**⟩

(i) When there is a real number (complex number) $\langle x, y \rangle$ defined for an element x, y in a real (complex) vector space X and the following four conditions are satisfied,

(i-i) $\langle x, y \rangle \geqq 0$; $\langle x, x \rangle = 0 \Leftrightarrow x = \mathbf{0}$

(i-ii) $\langle x, y \rangle = \langle y, x \rangle$ $(\langle x, y \rangle = \langle y, \bar{x} \rangle)$

(i-iii) $\langle x + y, z \rangle = \langle x, z \rangle + \langle y, z \rangle$ for any $z \in X$

(i-iv) $\langle \alpha x, y \rangle = \alpha \langle x, y \rangle$ for any real number (complex real number) α,

$\langle x, y \rangle$ is called an **inner product** of x and y and can also be expressed as (x, y). Additionally, X in which an inner product is defined is called an **inner product space**.

Furthermore, when x, y, z is an element of a real vector space and α is any real number, for $\langle x, y \rangle$ which satisfy (i-i)–(i-iv), $\langle x, y + z \rangle = \langle y + z, x \rangle = \langle x, y \rangle + \langle x, z \rangle$ and $\langle x, \alpha y \rangle = \langle \alpha y, x \rangle = \alpha \langle x, y \rangle$ hold but this is called a **bilinearity** of inner product in real vector space.

(ii) When a mapping $\|\cdot\|$ from the vector space, X to $\boldsymbol{R}$ satisfies the following three conditions,

(ii-i) $\|x\| \geqq 0$; $\|x\| = 0 \Leftrightarrow x = \mathbf{0}$

(ii-ii) For a scalar α, $\|\alpha x\| = |\alpha| \|x\|$

(ii-iii) $\|x + y\| \leqq \|x\| + \|y\|$

$\|x\|$ is called the **norm** of x and X is called a **normed space**.

(Notes on ⟨Definition 1⟩ and ⟨Definition 9⟩): For any $\boldsymbol{x}$, $\boldsymbol{y}$ of $\boldsymbol{R}^n$, if the following are defined,

① $d(\boldsymbol{x}, \boldsymbol{y}) = \sqrt{\sum_{i=1}^{n}(x_i - y_i)^2}$,
② $\langle \boldsymbol{x}, \boldsymbol{y} \rangle = \sum_{i=1}^{n} x_i y_i$,
③ $\|\boldsymbol{x}\| = \sqrt{\langle \boldsymbol{x}, \boldsymbol{x} \rangle}$

①~③ satisfy the condition about the distance, condition about the inner product, and condition about the norm respectively. Each of these is an example of distance, inner product, and norm.

① is called a **Euclidean distance** but as $d(\boldsymbol{x}, \boldsymbol{y}) = \|\langle \boldsymbol{x} - \boldsymbol{y} \rangle, \langle \boldsymbol{x} - \boldsymbol{y} \rangle\|$ holds, ① is also referred to as a distance derived from the inner product of ②.

⟨Definition 10⟩ Let $\boldsymbol{R}(\boldsymbol{C})$ be all real numbers (complex numbers). The whole of the ordered set $(x_1, \cdots, x_n)$ of n elements $x_1, \cdots, x_n$ of $\boldsymbol{R}(\boldsymbol{C})$ is represented as $\boldsymbol{R}^n(\boldsymbol{C}^n)$.

If in $\boldsymbol{R}^n(\boldsymbol{C}^n)$, a sum is defined as $(x_1, \cdots, x_n) + (y_1, \cdots, y_n) = (x_1 + y_1, \cdots, x_n + y_n)$ and for $\alpha \in \boldsymbol{R}(\boldsymbol{C})$, a scalar multiple is defined as $\alpha(x_1, \cdots, x_n) = (\alpha x_1, \cdots, \alpha x_n)$, $\boldsymbol{R}^n(\boldsymbol{C}^n)$ becomes a vector space on $\boldsymbol{R}(\boldsymbol{C})$. $\boldsymbol{R}^n(\boldsymbol{C}^n)$ and its elements $\boldsymbol{x}$ are called ***n*-dimensional real (complex) vector space** and ***n*-dimensional real (complex) vector** or ***n*-tuple**.

In the following definitions ⟨Definition 11⟩ to ⟨Definition 13⟩, $\boldsymbol{Y}$ is a subset of the metric space $\boldsymbol{X}$ and $\boldsymbol{S}$ is a subset of $\boldsymbol{Y}$.

⟨Definition 11⟩ A subset $\boldsymbol{S}$ of $\boldsymbol{X}$, if there exists $\varepsilon > 0$ which is $\boldsymbol{B}(\boldsymbol{x}\ ;\ \varepsilon) \cap \boldsymbol{Y} \subseteq \boldsymbol{S}$ for some $\boldsymbol{x} \in \boldsymbol{S}$, is said to be open relative to $\boldsymbol{Y}$. Particularly, if $\boldsymbol{S}$ is open relative to $\boldsymbol{X}$, $\boldsymbol{S}$ is said to be open in $\boldsymbol{X}$.

⟨Definition 12⟩ $\boldsymbol{G}$ is a **family** of subsets of $\boldsymbol{X}$ and is represented as $\boldsymbol{G} = \{\boldsymbol{G}_i \in \boldsymbol{X},\ i \in \tilde{N}\}$. However, $\tilde{N}$ is a set of natural numbers.

(i) If $\cup_{i \in \tilde{N}} \boldsymbol{G}_i \supseteq \boldsymbol{S}$, $\boldsymbol{G}$ is said to be a **covering** of $\boldsymbol{S}$ or that $\boldsymbol{G}$ covers $\boldsymbol{S}$.
(ii) If $\cup_{i \in \tilde{N}} \boldsymbol{G}_i \supseteq \boldsymbol{S}$ and $\boldsymbol{G}$ are finite sets, $\boldsymbol{G}$ is called a **finite covering** of $\boldsymbol{S}$. Moreover, if $\cup_{i \in \tilde{N}} \boldsymbol{G}_i \supseteq \boldsymbol{S}$ and each $\boldsymbol{G}_i \subseteq \boldsymbol{G}$ are open sets, $\boldsymbol{G}$ is said to be $\boldsymbol{S}$'s **open covering**. Furthermore, when each $\boldsymbol{G}_i \subseteq \boldsymbol{G}$ is open in respect of $\boldsymbol{Y}$, $\boldsymbol{G}$ is said to be an open covering of $\boldsymbol{S}$ in respect of $\boldsymbol{Y}$.
(iii) Let $\tilde{\tilde{N}}$ be a subset of $\tilde{N}$ made up of a finite number of natural numbers. If $\cup_{i \in \tilde{\tilde{N}}} \boldsymbol{G}_i \supseteq \boldsymbol{S}$, $\{\boldsymbol{G}_i \in \boldsymbol{X},\ i \in \tilde{\tilde{N}}\}$ is said to be a **finite partial covering** of S. The finite open covering of $\boldsymbol{S}$ is a finite partial covering of an open covering of $\boldsymbol{S}$.

⟨Definition 13⟩

(i) If any open covering of $\boldsymbol{S}$ has a finite open covering, $\boldsymbol{S}$ is said to be **compact** (in respect of $\boldsymbol{X}$).
(ii) If any open covering in respect of $\boldsymbol{Y}$ of $\boldsymbol{S}$ has a bound open partial covering in respect of $\boldsymbol{Y}$, $\boldsymbol{S}$ is said to be compact in respect of $\boldsymbol{Y}$.

(iii) If each sequence of points of $\boldsymbol{S}$ has a subsequence of points that converge to a point in $\boldsymbol{S}$, $\boldsymbol{S}$ is said to be **sequentially compact**.

[Theorem 4] Let $\boldsymbol{X}$ be a metric space and $\boldsymbol{S} \subseteq \boldsymbol{Y} \subseteq \boldsymbol{X}$. $\boldsymbol{S}$ Being compact in respect of $\boldsymbol{Y}$ and $\boldsymbol{S}$ being compact in respect of $\boldsymbol{X}$ are equivalent to one another.

[Theorem 5] If $\boldsymbol{S}$ is a subset of metric space $\boldsymbol{X}$, the following holds.

(i) If $\boldsymbol{S}$ is sequentially compact, $\boldsymbol{S}$ is a bounded closed set.
(ii) $\boldsymbol{S}$ being compact and $\boldsymbol{S}$ being sequentially compact are equivalent to one another.

[Theorem 6] When $\boldsymbol{S}$ is a subset of $\boldsymbol{R}^n$, the following holds.

(i) If $\boldsymbol{S}$ is bounded, any sequence of $\boldsymbol{S}$ contains a converging subset.
(ii)
(ii-i) $\boldsymbol{S}$ is compact,
(ii-ii) $\boldsymbol{S}$ is sequentially compact,
(ii-iii) $\boldsymbol{S}$ is a bounded closed set,
these are mutually equivalent.

[Theorem 7] Separation Theorem

(i) Let $\boldsymbol{X}$ and $\boldsymbol{Y}$ be a nonempty convex set of $\boldsymbol{R}^n$ and $\boldsymbol{intX} \neq \phi$ and $\boldsymbol{X} \cap \boldsymbol{Y} = \phi$. In this case, there exist $\boldsymbol{p} \neq \boldsymbol{0}$ and real number α that satisfy the following two conditions:
(i-i) For any $\boldsymbol{x} \in \boldsymbol{X}$, $\langle \boldsymbol{p}, \boldsymbol{x} \rangle \geqq \alpha$
(i-ii) For $\boldsymbol{y} \in \boldsymbol{Y}$, $\langle \boldsymbol{p}, \boldsymbol{y} \rangle \leqq \alpha$.
(ii) If the condition in (i) is strengthened and $\boldsymbol{X}$ is a closed convex set, $\boldsymbol{Y}$ is compact and $\boldsymbol{X} \cap \boldsymbol{Y} = \phi$, there exists $\boldsymbol{p} \neq \boldsymbol{0}$ and real number α that establish (i-i) and (i-ii) with strict inequalities.

[Theorem 8] If $\boldsymbol{X}$ is a nonempty closed convex set of $\boldsymbol{R}^n$ and $\boldsymbol{a} \notin \boldsymbol{X}$,

$$\text{for any } \boldsymbol{x} \in \boldsymbol{X},$$

there exists $\boldsymbol{p} \neq \boldsymbol{0}$ and real number α such that $\langle \boldsymbol{p}, \boldsymbol{a} \rangle < \alpha < \langle \boldsymbol{p}, \boldsymbol{x} \rangle$ holds.

⟨Definition 14⟩ Let f be a real-valued function defined on the metric space $\boldsymbol{X}$.

(i) Let $\boldsymbol{x}^0 \in \boldsymbol{X}$. For any $\varepsilon > 0$, if there exists $\delta_\varepsilon > 0$ and $d(\boldsymbol{x}, \boldsymbol{x}^0) < \delta_\varepsilon$, when $f(\boldsymbol{x}) < f(\boldsymbol{x}^0) + \varepsilon(f(\boldsymbol{x}^0) - \varepsilon < f(\boldsymbol{x}))$ holds, f is said to be **upper semicontinuous** on $\boldsymbol{x}^0$ (**lower semicontinuous**). When f is upper semicontinuous (lower semicontinuous) on any $\boldsymbol{x}^0 \in \boldsymbol{X}$, f is said to be upper semicontinuous (lower semicontinuous) on $\boldsymbol{X}$.
(ii) If f is upper semicontinuous and lower semicontinuous $\boldsymbol{x}^0 \in \boldsymbol{X}$ (on $\boldsymbol{X}$) then f is said to be **continuous** on $\boldsymbol{x}^0$ (on $\boldsymbol{X}$). f being continuous on $\boldsymbol{X}$ can sometimes be referred to simply as f being continuous.

[Theorem 9] If the real-valued function defined on a compact set $\boldsymbol{S}$ is upper semicontinuous (lower semicontinuous) on $\boldsymbol{S}$, f takes the maximum value (minimum value) on $\boldsymbol{S}$.

(Proof) From the definitions, the fact that f is lower semicontinuous is the same as $-f$ being upper semicontinuous so it is sufficient just to prove the case when f is upper semicontinuous.

To show that $f(\boldsymbol{S})$ is bound from above, if we contradict the conclusion and suppose that for any $m > 0$, there exists $\boldsymbol{x}^m \in \boldsymbol{S}$ that is $f(\boldsymbol{x}^m) > m$, $\boldsymbol{x}^n \in \boldsymbol{S}$ is determined for any natural number n and becomes $f(\boldsymbol{x}^n) > n$. $\boldsymbol{S}$ is compact so the sequence of points $\{\boldsymbol{x}^n\}$ of $\boldsymbol{S}$ has a part $\{\boldsymbol{x}^{n_j}\}$ that converges to $\boldsymbol{x}^0 \in \boldsymbol{S}$. f is upper semicontinuous, thus, for any $\varepsilon > 0$, there exists $\boldsymbol{B}(\boldsymbol{x}^0\ ;\ d_\varepsilon)$ and if $d(\boldsymbol{x}^0,\ \boldsymbol{x}) < d_\varepsilon$, $f(\boldsymbol{x}) < f(\boldsymbol{x}^0) + \varepsilon$. Moreover, from the convergence of $\{\boldsymbol{x}^{n_j}\}$, there exists a number n_{d_ε} which becomes $d(\boldsymbol{x}^{n_j},\ \boldsymbol{x}^0) < d_\varepsilon$ for all n that are $n \geqq n_\varepsilon$. Here, for $n_j \geqq max\{n_{d_\varepsilon},\ f(\boldsymbol{x}^0) + \varepsilon\}$, $n_j < f(\boldsymbol{x}^{n_j}) < f(\boldsymbol{x}^0) + \varepsilon \leqq n_j$ holds but this is self-contradictory. From above, the set of real numbers $f(\boldsymbol{S})$ is bound above. Therefore let us express $\sup_{x \in \boldsymbol{S}} f(\boldsymbol{S})$exists ([Theorem 2; (i-i)]) but for simplification of notation, let us express this as $\overline{a}$. Furthermore, if the sequence of real numbers of $f(\boldsymbol{S})$ converging to $\overline{a}$ is taken to be $\{y^n\}$ ([Theorem 2; (i-ii)]), from each $y^n \in f(\boldsymbol{S})$, there exists a sequence of points $\{\boldsymbol{x}^n\}$ of $\boldsymbol{S}$ that is $f(\boldsymbol{x}^n) = y^n$ corresponding to this $\{y^n\}$ and from the compactness of the $\boldsymbol{S}$ assumed and $\{\boldsymbol{x}^n\}$ has a partial sequence $\{\boldsymbol{x}^{n_j}\}$ that converges to $\overline{\boldsymbol{x}} \in \boldsymbol{S}$. If to derive $\overline{a} \in f(\boldsymbol{S})$, we let $\overline{a} \notin f(\boldsymbol{S})$, $\overline{a} > f(\overline{\boldsymbol{x}})$. Here, if we select $\varepsilon = \frac{1}{4}(\overline{a} - f(\overline{\boldsymbol{x}}))$, $(\overline{a} - \varepsilon) - (f(\overline{\boldsymbol{x}}) + \varepsilon) = \frac{1}{2}(\overline{a} - f(\overline{\boldsymbol{x}})) > 0$. Therefore $\overline{a} - \varepsilon > f(\overline{\boldsymbol{x}}) + \varepsilon$. Additionally, from the upper semicontinuity of f, if for any $\varepsilon > 0$, $d(\boldsymbol{x},\ \overline{\boldsymbol{x}}) < d_\varepsilon$, there exists $d_\varepsilon > 0$ which satisfies $f(\boldsymbol{x}) < f(\overline{\boldsymbol{x}}) + \varepsilon$. In addition, from the convergence of $\{f(\boldsymbol{x}^n)\}$ and $\{\boldsymbol{x}^{n_j}\}$, for $n_j \geqq n_{d_\varepsilon}$, there exists n_{d_ε} which establish $d(\boldsymbol{x}^{n_j},\ \overline{\boldsymbol{x}}) < d_\varepsilon$, $|f(x^{n_j})$ and $-\overline{a}| = \overline{a} - f(\boldsymbol{x}^{n_j}) < \varepsilon$. Hence, when $n_j \geqq n_{d_\varepsilon}$, the following self-contradiction, $\overline{a} - \varepsilon < f(\boldsymbol{x}^{n_j}) < f(\overline{\boldsymbol{x}}) + \varepsilon < \overline{a} - \varepsilon$ is created. Thus, $\overline{a} = \sup_{x \in \boldsymbol{S}} f(\boldsymbol{S}) \in f(\boldsymbol{S})$, in other words, f achieves its maximum value on $\boldsymbol{S}$. QED

A continuous function is upper semi-continuous and lower semi-continuous hence the famous Weierstrass theorem is a series of this theorem.

[Theorem 10] Weierstrass Theorem The real-valued function f defined on a compact set $\boldsymbol{S}$ reaches its maximum value and its minimum value on $\boldsymbol{S}$. In addition, when $\boldsymbol{S}$ is a subset of $\boldsymbol{R}^n$, the compactness of $\boldsymbol{S}$ can be rewritten as $\boldsymbol{S}$ being a bounded closed set.

[Theorem 11] Let f be a real-valued function defined on a subset $\boldsymbol{S}$ of metric space $\boldsymbol{X}$. Here, if f is upper semi-continuous (lower semi-continuous), $\boldsymbol{S}_u = \{\boldsymbol{x} \in \boldsymbol{S} \mid f(\boldsymbol{x}) < \alpha\}$ ($\boldsymbol{S}_t = \{\boldsymbol{x} \in \boldsymbol{S} \mid f(\boldsymbol{x}) > \alpha\}$) is an open set. However, α is a given real number.

(Notes on [Theorem 11]):

① Upper semi-continuity (lower semi-continuity) of f means that $S_u^C\left(S_l^C\right)$ is a closed set.

② If f is continuous on $\boldsymbol{S}$, $\boldsymbol{S}_u$ and $S_l\left(\boldsymbol{S}_u^C \text{ and } \boldsymbol{S}_l^C\right)$ are open sets (closed sets).

⟨**Definition 15**⟩ **Let** $\boldsymbol{D}$ be an open set of $\boldsymbol{R}^n$ and $\boldsymbol{f}: \boldsymbol{D} \rightarrow \boldsymbol{R}^m$.

(i) f being **differentiable** on $\boldsymbol{x}^0 \in \boldsymbol{D}$ means that there exists some linear transformation $\boldsymbol{T}_{\boldsymbol{x}^0} : \boldsymbol{D} \subseteq \boldsymbol{R}^n \rightarrow \boldsymbol{R}^m$ and that

(i-i) $$\lim_{\boldsymbol{h}\rightarrow\boldsymbol{0}} \frac{\left\|\boldsymbol{f}\left(\boldsymbol{x}^0+\boldsymbol{h}\right)-\boldsymbol{f}\left(\boldsymbol{x}^0\right)-\boldsymbol{T}_{\boldsymbol{x}^0}(\boldsymbol{h})\right\|}{\|\boldsymbol{h}\|}=0$$

holds. If $\boldsymbol{\alpha}(\boldsymbol{x}^0,\boldsymbol{h}) = [\boldsymbol{f}(\boldsymbol{x}^0+\boldsymbol{h}) - \boldsymbol{f}(\boldsymbol{x}^0) - \boldsymbol{T}_{\boldsymbol{x}^0}(\boldsymbol{h})]/\|\boldsymbol{h}\|$ is defined, (i-i) is equivalent to $\lim\limits_{\boldsymbol{h}\rightarrow\boldsymbol{0}} \alpha(\boldsymbol{x}^0, \boldsymbol{h}) = \boldsymbol{0}$. Additionally, if a $m \times n$ matrix which has a one-to-one correspondence with a linear transformation $\boldsymbol{T}_{\boldsymbol{x}^0}$ is represented as $\boldsymbol{f}'(\boldsymbol{x}^0)$, for any $\boldsymbol{h}$, $\boldsymbol{T}_{\boldsymbol{x}^0}(\boldsymbol{h}) = \boldsymbol{f}'(\boldsymbol{x}^0)\boldsymbol{h}$, hence $\alpha(\boldsymbol{x}^0, \boldsymbol{h}) = [\boldsymbol{f}(\boldsymbol{x}^0+\boldsymbol{h}) - \boldsymbol{f}(\boldsymbol{x}^0) - \boldsymbol{f}'(\boldsymbol{x}^0)\boldsymbol{h}]/\|\boldsymbol{h}\|$. Thus, the differentiability of $\boldsymbol{f}$ on $\boldsymbol{x}^0$ can also be said to be the existence of some $m \times n$ matrix with the following conditions holding.

(i-ii) $\boldsymbol{f}(\boldsymbol{x}^0+\boldsymbol{h}) = \boldsymbol{f}(\boldsymbol{x}^0) + \boldsymbol{f}'(\boldsymbol{x}^0)\boldsymbol{h} + \boldsymbol{\alpha}(\boldsymbol{x}^0,\boldsymbol{h})\|\boldsymbol{h}\|$, where $\lim\limits_{\boldsymbol{h}\rightarrow\boldsymbol{0}} \alpha(\boldsymbol{x}^0, \boldsymbol{h}) = \boldsymbol{0}$. When $\boldsymbol{f}(\boldsymbol{x})$ is differentiable over some $\boldsymbol{x} \in \boldsymbol{D}$, $\boldsymbol{f}$ is said to be differentiable over $\boldsymbol{D}$.

(ii) Let $\boldsymbol{f}(\boldsymbol{x}) = (f_1(\boldsymbol{x}), \cdots, f_m(\boldsymbol{x}))^t$ and let each $f_i(\boldsymbol{x})(i = 1, \cdots, m)$ be a real-valued function on $\boldsymbol{D}(\subseteq\boldsymbol{R}^n)$. If we set $\boldsymbol{h} = t\boldsymbol{e}^j(j = 1, \cdots, n)$ in (i), the differentiability of $\boldsymbol{f}$ means the following.

$$\begin{aligned}\boldsymbol{0} &= \lim_{t\boldsymbol{e}^j\rightarrow 0} \frac{\left\|\boldsymbol{f}(\boldsymbol{x}^0+t\boldsymbol{e}^j)-\boldsymbol{f}(\boldsymbol{x}^0)-t\boldsymbol{f}'(\boldsymbol{x}^0)\boldsymbol{e}^j\right\|}{\|t\boldsymbol{e}^j\|}\\ &= \lim_{t\rightarrow 0}\left\|\frac{\boldsymbol{f}(\boldsymbol{x}^0+t\boldsymbol{e}^j)-\boldsymbol{f}(\boldsymbol{x}^0)-t\boldsymbol{f}'\left(\boldsymbol{x}^0\right)\boldsymbol{e}^j}{t}\right\|.\end{aligned}$$

However,

$$\begin{aligned}\lim_{t\rightarrow 0}\frac{\boldsymbol{f}(\boldsymbol{x}^0+t\boldsymbol{e}^j)-\boldsymbol{f}(\boldsymbol{x}^0)}{t} &= \boldsymbol{f}'\left(\boldsymbol{x}^0\right)\boldsymbol{e}^j\\ &= \left(f_{1j}\left(\boldsymbol{x}^0\right), \cdots, f_{mj}\left(\boldsymbol{x}^0\right)\right)^t, \quad (j=1, \cdots, n).\end{aligned}$$

Here, the ith element of $\boldsymbol{f}'(\boldsymbol{x}^0)\boldsymbol{e}^j$ can be expressed as

$$\lim_{t\rightarrow 0}\frac{f_i(\boldsymbol{x}^0+t\boldsymbol{e}^j)-f_i(\boldsymbol{x}^0)}{t} = f_{ij}\left(\boldsymbol{x}^0\right), \quad (i=1, \cdots, m)$$

This $f_{ij}(\boldsymbol{x}^0)$ is called a partial derivative of $f_i(\boldsymbol{x}^0)$ for x_j evaluated at $\boldsymbol{x}^0$ or a first-order partial derivative and is denoted $\left.\frac{\partial f_i}{\partial x_j}\right|_{\boldsymbol{x}^0}$. In addition, the $m \times n$ matrix $\boldsymbol{f}'(\boldsymbol{x}^0) = [f_{ij}(\boldsymbol{x}^0)]$

is called the Jacobian matrix of f in x^0 and when there exists $f'(x^0)$ on any $x \in D$, f is said to be a **partial derivative** or a **first-order partial derivative**. Furthermore, if f has partial derivatives on D and they are continuous, f is called a **continuously differentiable function** or a **function of class C^1**Function of class C1. Additionally, a function that is continuous over D is called a **function of class C^0**.

[Theorem 12] If D is an open set of R^n and $f: D \subseteq R^m$, if f is differentiable on $x^0 \in D$, f is continuous on x^0.

⟨Definition 16⟩ Let f be a real-valued function defined on an open set D of R^n.

(i) $f: D \subseteq R^n \to R$ being differentiable over $x^0 \in D$ means that there exists $f'(x^0) \equiv \nabla f(x^0)$ and the following holds.

(i-i) $f(x^0 + h) = f(x^0) + \nabla f(x^0)h + \alpha(x^0, h)\|h\|$, where $\lim_{h \to 0} \alpha(x^0, h) = 0$. Here $\nabla f(x^0) = \left(\left.\frac{\partial f}{\partial x_1}\right|_{x^0}, \cdots, \left.\frac{\partial f}{\partial x_n}\right|_{x^0} \right)$ is said to be a **gradient vector** of f in x^0.

(ii) $f: D \subseteq R^n \to R$ being **twice differentiable** over $x^0 \in D$ means

(ii-i) $f(x^0 + h) = f(x^0) + \nabla f(x^0)h + \frac{h'\nabla^2 f(x^0)h}{2} + \beta(x^0, h)\|h\|^2$, but that $\lim_{h \to 0} \beta(x^0, h) = 0$ holds. If f is twice differentiable on each point x on D, f is sent to be twice differentiable on D. Moreover, the n-order square matrix $f''(x^0) \equiv \nabla^2 f(x^0)$ is called a Hessian matrix on x^0 and its (i, j)th element is expressed as $\left[\nabla^2 f(x^0)\right]_{ij} = \left.\frac{\partial \left(\frac{\partial f}{\partial x_i}\right)}{\partial x_j}\right|_{x^0} = \left.\frac{\partial^2 f}{\partial x_i x_j}\right|_{x^0}$ $(i, j = 1, \cdots, n)$.

$f''(x^0)$ is also called a **second-order partial derivative coefficient** of f over x^0 and when there exists a second-order partial derivative coefficient of f for any $x \in D$, f is said to have a **second-order partial derivative**. Furthermore, if second-order partial derivative of f exists on D and if that is continuous, f is said to be a **twice continuously differentiable** function or a **C^2-class function**.

[Theorem 13] Let f be a real-valued function defined on an open set D of R^n. If f is (i) twice differentiable on $x^0 \in D$ or (ii) has a continuous second-order partial derivative, $\left.\frac{\partial^2 f}{\partial x_j x_i}\right|_{x^0} = \left.\frac{\partial^2 f}{\partial x_i x_j}\right|_{x^0}$ holds.

[Theorem 14] Mean value theorem for one-variable functions

When f is a differentiable real number function on the open interval (a, b) that is continuous on the closed interval $[a, b]$, the fact that $c \in (a, b)$ satisfying

$$f(b) - f(a) = f'(c)(b - a)$$

Exists is called the mean value theorem. Particularly, when $f(b) = f(a)$, it is called the Rolle's theorem.

[Theorem 15] Taylor's Theorem (2nd order) Let f be a real-valued twice differentiable function defined on the open convex set $\boldsymbol{D}$ of $\boldsymbol{R}^n$ and $\boldsymbol{x}^1$, $\boldsymbol{x}^2$ be any two different points from $\boldsymbol{D}$. Here, there exists $\lambda \in (0, 1)$ and

$$f(\boldsymbol{x}^2) - f(\boldsymbol{x}^1) = \nabla f(\boldsymbol{x}^1)(\boldsymbol{x}^2 - \boldsymbol{x}^1) + \frac{1}{2}(\boldsymbol{x}^2 - \boldsymbol{x}^1)^t \boldsymbol{H}(\boldsymbol{x}^\lambda)(\boldsymbol{x}^2 - \boldsymbol{x}^1)$$

holds. However, $\boldsymbol{H}(\boldsymbol{x}^\lambda)$ is a Hessian matrix of f evaluated by $\boldsymbol{x}^\lambda = (1 - \lambda)\boldsymbol{x}^1 + \lambda\boldsymbol{x}^2$.

[Theorem 16] Chain Rule for Derivatives Let $\boldsymbol{D}$ be an open set of $\boldsymbol{R}^n, f : \boldsymbol{D} \to \boldsymbol{R}^m$, $g : f(\boldsymbol{D}) \to \boldsymbol{R}^k$ and $F(\boldsymbol{x}) \equiv g(f(\boldsymbol{x}))$. If f is differentiable on $\boldsymbol{x}^0 \in \boldsymbol{D}$, and g is differentiable on $f(\boldsymbol{x}^0)$, $F(\boldsymbol{x})$ is differentiable on $\boldsymbol{x} = \boldsymbol{x}^0$ and $F'(\boldsymbol{x}^0) = g'(f(\boldsymbol{x}^0))f'(\boldsymbol{x}^0)$ is established.

Constrained optimization problems were examined in Chap. 3 but in optimization problems with no constraints, which are the simplest of nonlinear programming problems, the required and sufficient condition for a solution to be a local optimum can be given by the following theorem.

[Theorem 17] Let f be a real-valued function which is twice differentiable defined on an open set $\boldsymbol{D}$ of $\boldsymbol{R}^n$.

(i) The required condition for $f(\boldsymbol{x})$ to be maximum (minimum) at $\boldsymbol{x} = \boldsymbol{x}^0 \in \boldsymbol{D}$ is $\nabla f(\boldsymbol{x}^0) = \boldsymbol{0}$ and for any $\boldsymbol{y} \neq \boldsymbol{0}$ in $\boldsymbol{R}^n$, $\boldsymbol{y}^t\boldsymbol{H}_f(\boldsymbol{x}^0)\ \boldsymbol{y} \leqq (\geqq) 0$.
(ii) The sufficient condition for $f(\boldsymbol{x})$ to be at a maximum (minimum) at $\boldsymbol{x} = \boldsymbol{x}^0 \in \boldsymbol{D}$ is that, for a $\boldsymbol{y} \neq \boldsymbol{0}$ of $\nabla f(\boldsymbol{x}^0) = \boldsymbol{0}$ and $\boldsymbol{R}^n$, $\boldsymbol{y}^t\boldsymbol{H}_f(\boldsymbol{x}^0)\ \boldsymbol{y} < (>) 0$. However, $\boldsymbol{H}_f(\boldsymbol{x}^0)$ is the **Hessian matrix** of f evaluated at $\boldsymbol{x}^0$.

⟨**Definition 17**⟩ When there exists an n-order square matrix $\boldsymbol{X}$ that is $\boldsymbol{AX} = \boldsymbol{I}_n$, $\boldsymbol{A}$ is called a **nonsingular matrix**, $\boldsymbol{X}$ is called $\boldsymbol{A}$'s **inverse matrix** and is represented as $\boldsymbol{X} = \boldsymbol{A}^{-1}$. Any nonsingular matrix is called a **singular matrix**.

[Theorem 18]

(i) In respect of a n-order square matrix $\boldsymbol{A}$, (i-i)–(i-iv) are mutually equivalent.
(i-i) $\boldsymbol{A}$ is nonsingular.
(i-ii) All columns (rows) of $\boldsymbol{A}$ are linearly independent.
(i-iii) $det\boldsymbol{A} \neq 0$.
(ii) When a matrix $\boldsymbol{A}$ and $\boldsymbol{B}$ are square matrices which can be multiplied in this order,

$$(\boldsymbol{AB})^{-1} = \boldsymbol{B}^{-1}\boldsymbol{A}^{-1}, (\boldsymbol{A}^{-1})^{-1} = \boldsymbol{A} \text{ and } (\boldsymbol{A}^t)^{-1} = (\boldsymbol{A}^{-1})^t.$$

⟨**Definition 18**⟩ Let $\boldsymbol{A}$ be a $m \times n$ matrix.

(i) When on $\boldsymbol{A}$, $p(q)$ row (column) vectors are linearly independent and any $(p + 1)$ $((q + 1))$ row (column) vectors are linearly dependent, $p(q)$ is called $\boldsymbol{A}$'s **row (column) rank**.
(ii) When a r-order minor determinant of $\boldsymbol{A}$ contains something, which is not 0 and all values of $s(>r)$-order minor determinants are 0, r is called the **rank** of $\boldsymbol{A}$ and is expressed as rank $(\boldsymbol{A}) = r$.
(iii)
(iii-i) It is obvious that $\boldsymbol{Ax} = \boldsymbol{0}$ has a solution $\boldsymbol{x} = \boldsymbol{0}$, thus, this is called a **trivial solution**. Accordingly, a solution $\boldsymbol{x} \neq \boldsymbol{0}$, which satisfies (iii-i) is called a **non-trivial solution**. The set of solutions of (iii-i) is a linear subspace. The base of this linear subspace is called the **fundamental solution** of (iii-i).

[Theorem 19]

(i) If matrices $\boldsymbol{A}$ and $\boldsymbol{B}$ have the same form, rank$(\boldsymbol{A} + \boldsymbol{B}) \leqq$ rank $(\boldsymbol{A})$ + rank $(\boldsymbol{B})$.
(ii) If matrix $\boldsymbol{A}$ and $\boldsymbol{B}$ can be multiplied in this order, rank$(\boldsymbol{AB}) \leqq$ min {rank$(\boldsymbol{A})$, rank$(\boldsymbol{B})$}. Particularly, if $\boldsymbol{A}(\boldsymbol{B})$ is nonsingular, rank$(\boldsymbol{AB}) =$ rank $(\boldsymbol{B})$(rank$(\boldsymbol{A})$).
(iii) In respect of the matrix $\boldsymbol{A}$, the row rank of rank $(\mathbf{A}) = \boldsymbol{A} =$ column rank of $\boldsymbol{A}$ $=$ rank $(\mathbf{A}^t)$.
(iv) When $\boldsymbol{A}$ is a $m \times n$ matrix and $\boldsymbol{b}$ is a m-dimensional column vector,
(iv-i) The necessary and sufficient condition for $\boldsymbol{Ax} = \boldsymbol{b}$ to be solvable is that rank $(\mathbf{A}) =$ rank $(\boldsymbol{A} \vdots \boldsymbol{b})$ holds.
(iv-ii) If rank $(\mathbf{A}) = k$, $\boldsymbol{Ax} = \boldsymbol{0}$ has $(n - k)$ fundamental solutions.

⟨**Definition 19**⟩ For n-order square matrix $\boldsymbol{A}$,

(i) if there exists $\boldsymbol{x} \neq \boldsymbol{0}$ that satisfies $\boldsymbol{Ax} = \lambda\boldsymbol{x}$ and a scalar λ, λ is called $\boldsymbol{A}$'s **eigenvalue** or a **characteristic value**, and $\boldsymbol{x} \neq \boldsymbol{0}$ is called an **eigenvector** in respect of λ or simply an eigenvector.
(i) has a solution $\boldsymbol{x} \neq \boldsymbol{0}$ when
(ii) $det(\lambda\boldsymbol{I}_n - \boldsymbol{A}) = 0$ holds and it is limited to this case, and hence if $\phi_A(\lambda) = det\ (\lambda\boldsymbol{I}_n - \boldsymbol{A})$ is defined, eigenvalues of $\boldsymbol{A}$ can be obtained as the solutions of a n-order polynomial $\phi_A(\lambda) = 0$. $\phi_A(\lambda)$ and $\phi_A(\lambda) = 0$ are called **characteristic polynomials** and **characteristic equations** of $\boldsymbol{A}$.

[Theorem 20] The necessary and sufficient condition for the n-order complex square matrix $\boldsymbol{A}$ to be a normal matrix is for a n-order unitary matrix $\boldsymbol{P}$ that is $\boldsymbol{PAP}^* = \boldsymbol{\Lambda} =$ diag $(\lambda_1, \cdots, \lambda_n)$ to exist. However, $\lambda_i (i = 1, \ldots, n)$ is an eigenvalue of $\boldsymbol{A}$.

List of Symbols

Symbol	Name or meaning
$\boldsymbol{X}$, $\boldsymbol{Y}$ is a set.	
$a \in \boldsymbol{X}$	a is an element or a base of $\boldsymbol{X}$.
$a \notin \boldsymbol{X}$	a is not an element of $\boldsymbol{X}$.
$\boldsymbol{X} \subseteq \boldsymbol{Y}$	$\boldsymbol{X}$ is a subset of $\boldsymbol{Y}$, in other words, any $a \in \boldsymbol{X}$, $a \in \boldsymbol{Y}$.
$\boldsymbol{X} = \boldsymbol{Y}$	$\boldsymbol{X} \subseteq \boldsymbol{Y}$ and $\boldsymbol{X} \supseteq \boldsymbol{Y}$.
$\boldsymbol{X} \subset \boldsymbol{Y}$	$\boldsymbol{X}$ is a proper subset of $\boldsymbol{Y}$. $\boldsymbol{X} \subseteq \boldsymbol{Y}$ and $\boldsymbol{X} \neq \boldsymbol{Y}$.
$\boldsymbol{X} \cup \boldsymbol{Y} = \{a \vert a \in \boldsymbol{X}$ or $a \in \boldsymbol{Y}\}$	Union of $\boldsymbol{X}$ and $\boldsymbol{Y}$.
$\boldsymbol{X} \cap \boldsymbol{Y} = \{a \vert a \in \boldsymbol{X}$ and $a \in \boldsymbol{Y}\}$	Intersection of $\boldsymbol{X}$ and $\boldsymbol{Y}$.
$\# \boldsymbol{X}$	Number of elements belonging to $\boldsymbol{X}$.
$\boldsymbol{int\ X}$	Set of internal points of $\boldsymbol{X}$.
$\boldsymbol{ext\ X}$	Set of exterior points of $\boldsymbol{X}$.
$\boldsymbol{X}_L$	Set of limit points of $\boldsymbol{X}$.
∂X	Set of boundary points of $\boldsymbol{X}$.
$\overline{\boldsymbol{X}}$	Closure of $\boldsymbol{X}$.
When $\boldsymbol{X} \subseteq \boldsymbol{Y}$, $\boldsymbol{X}^C = \boldsymbol{Y} - \boldsymbol{X} = \boldsymbol{Y} \setminus \boldsymbol{X}$	Complementary set of $\boldsymbol{X}$ in respect of $\boldsymbol{Y}$.
ϕ	A set with no elements. Empty set.
$\boldsymbol{R}(\boldsymbol{C})$	All real numbers (complex numbers).
$\boldsymbol{R}^n(\boldsymbol{C}^n)$	All of the n-dimensional real (complex) vectors.
$\boldsymbol{x}, \boldsymbol{y}$ is a n-dimensional vector.	
$\boldsymbol{x} > \boldsymbol{y}$	$x_i > y_i$ $(i = 1, \ldots, n)$.
$\boldsymbol{x} \geqq \boldsymbol{y}$	$x_i \geqq y_i$ $(i = 1, \ldots, n)$.
$\boldsymbol{x} = \boldsymbol{y}$	$\boldsymbol{x} \geqq \boldsymbol{y}$ and $\boldsymbol{x} \leqq \boldsymbol{y}$.
$x \geq y$	$\boldsymbol{x} \geqq \boldsymbol{y}$ and $\boldsymbol{x} \neq \boldsymbol{y}$.
$\boldsymbol{0}$	Zero vector.
$\boldsymbol{e}^j = (0 \cdots 010 \cdots 0)^t$	n-order jth unit column vector with the jth element as 1 and the rest as 0.
$d(\boldsymbol{x}, \boldsymbol{y})$	Distance between $\boldsymbol{x}$ and $\boldsymbol{y}$.

© Springer Nature Singapore Pte Ltd. 2024

K. Nakayama, *Eigenvalue Problem and Nonlinear Programming Problem*,
New Frontiers in Regional Science: Asian Perspectives 70,
https://doi.org/10.1007/978-981-99-2943-6

Symbol	Name or meaning
$\langle \boldsymbol{x}, \boldsymbol{y} \rangle$ or $(\boldsymbol{x}, \boldsymbol{y})$	Inner product of $\boldsymbol{x}$ and $\boldsymbol{y}$.
$\|\boldsymbol{x}\|$	Norm of $\boldsymbol{x}$.
$\boldsymbol{R}^n_+ = \{\boldsymbol{x} \geqq \boldsymbol{0} \| \boldsymbol{x} \in \boldsymbol{R}^n\}$	Nonnegativity of $\boldsymbol{R}^n$.
$\boldsymbol{R}^n_{++} = \{\boldsymbol{x} > \boldsymbol{0} \| \boldsymbol{x} \in \boldsymbol{R}^n\}$	Positive quadrant of $\boldsymbol{R}^n$.
$\boldsymbol{N} = \{1, \ldots, n\}$	Number set.
$x_{\boldsymbol{J}}$	Partial vector of $\boldsymbol{x}$ composed of $x_j (j \in \boldsymbol{J} \subseteq \boldsymbol{N})$.
$f : \boldsymbol{X} \to \boldsymbol{Y}$($\mathbf{X}, \mathbf{Y}$ are sets)	Mapping from $\boldsymbol{X}$ to $\boldsymbol{Y}$.
$f(\boldsymbol{x})$ $(\boldsymbol{x} \in \boldsymbol{X})$	Image of $\boldsymbol{x} \in \boldsymbol{X}$ from f.
$\frac{\partial f(\boldsymbol{x})}{\partial x_i}$ or $\left.\frac{\partial f}{\partial x_i}\right\|_{\mathbf{x}}$ $(i = 1, \ldots, n)$	First-order partial derivative of $f(\boldsymbol{x})$ in respect of x_i evaluated at $\boldsymbol{x}$.
$f_{ij}(\boldsymbol{x}) = \frac{\partial \left(\frac{\partial f(x)}{\partial x_i} \right)}{\partial x_j}$ $(i, j = 1, \ldots, n)$	Second-order partial derivative of $f(\boldsymbol{x})$ evaluated at $\boldsymbol{x}$.
$\nabla f(\boldsymbol{x}) = \left(\frac{\partial f(\boldsymbol{x})}{\partial x_1}, \ldots, \frac{\partial f(\boldsymbol{x})}{\partial x_n} \right)$	Gradient vector of $f(\boldsymbol{x})$ evaluated at $\boldsymbol{x}$.
Let a, $b \in \boldsymbol{R}$.	
$[a, b] = \{x \| a \leqq x \leqq b\}$	Closed interval.
$(a, b) = \{x \| a < x < b\}$	Open interval.
$[a, b) = \{x \| a \leqq x < b\}$	
$(a, b] = \{x \| a < x \leqq b\}$	
$\|a\|$	Absolute value of a.
$\mathrm{sgn}(a)$	Sign of a.
$\alpha = a + ib \in \boldsymbol{C}$	Complex number. However, i is a complex unit that is $i^2 = -1$.
$Re\ (\alpha)$	Real part of α.
$Im(\alpha)$	Imaginary part of α.
$\overline{\alpha} = a - ib$	Conjugate complex number of α.
Let $\boldsymbol{A}$ be a matrix.	
$\boldsymbol{A} = [a_{ij}] = \begin{bmatrix} a_{11} & a_{12} & \cdots a_{1j} & \cdots & a_{1n} \\ a_{21} & a_{22} & \cdots a_{2j} & \cdots & a_{2n} \\ \vdots & \vdots & \vdots & & \vdots \\ a_{i1} & a_{i2} & \cdots a_{ij} & \cdots & a_{in} \\ \vdots & \vdots & \vdots & & \vdots \\ a_{m1} & a_{m2} & \cdots a_{mj} & \cdots & a_{mn} \end{bmatrix}$	$m \times n$ matrix.
$\boldsymbol{M} = \{1, \ldots, m\}$	Number sets.
a_{ij} $(i \in \boldsymbol{M}; j \in \boldsymbol{N})$	The (i, j)th element of $\boldsymbol{A}$.
$a_{ij} \in \boldsymbol{R}(\boldsymbol{C})$ $(i \in \boldsymbol{M}; j \in \boldsymbol{N})$	Real (complex) matrix.
$\boldsymbol{A}^t = [a_{ij}]$ $(A^* = [\overline{a}_{ji}])$	Transposition (conjugate transposition) matrix of the real (complex) matrix $\boldsymbol{A} = [a_{ij}]$.
$[0]$	Null matrix.
$rank\ (\boldsymbol{A})$	Order of $\boldsymbol{A}$.
$dim\ (\boldsymbol{A})$	Dimension of $\boldsymbol{A}$.
$m = n$ $\boldsymbol{A} = [a_{ij}]$ $(i, j = 1, \ldots, n)$	n-order square matrix $\boldsymbol{A}$.
a_{ii} $(i = 1, \ldots, n)$	Diagonal element of $\boldsymbol{A}$.
a_{ij} $(i \neq j; i, j = 1, \ldots, n)$	Nondiagonal element of $\boldsymbol{A}$.

Symbol	Name or meaning
$\boldsymbol{A}^{-1}$	Inverse matrix of $\boldsymbol{A}$.
$\boldsymbol{A} = diag\ (a_{11}, a_{22}, \ldots, a_{nn})$	n-order diagonal matrix $\boldsymbol{A}$ whose every nondiagonal elements are 0.
$\boldsymbol{I}_n = diag\ (1, 1, \ldots, 1)$	n-order unit matrix.
$\det(\boldsymbol{A})$	Determinant of $\boldsymbol{A}$.
$\boldsymbol{A}\binom{i}{j}$ $(i, j = 1, \ldots, n)$	Matrix with ith row and jth column removed from $\boldsymbol{A}$.
$\boldsymbol{A}_{ij} = (-1)^{i+j}\ \boldsymbol{x}\ (\boldsymbol{A}\binom{i}{j})$ $(i, j = 1, \ldots, n)$	Cofactor relating to a_{ij} of $\det(\boldsymbol{A})$.
$adj(\boldsymbol{A}) = [\boldsymbol{A}_{ji}]$ $(i, j = 1, \ldots, n)$	Matrix of cofactors of $\boldsymbol{A}$.
$\boldsymbol{A}_{\boldsymbol{LK}} = [a_{ij}]$ $(\boldsymbol{L} \subseteq \boldsymbol{M}; \boldsymbol{K} \subseteq \boldsymbol{N})$ When $\boldsymbol{L} = \boldsymbol{K}$ and $\sharp \boldsymbol{K} = k$,	a submatrix of $\boldsymbol{A}$ with $a_{ij}(i \in \boldsymbol{L}; j \subseteq \boldsymbol{K})$ as the (i, j) th element.
$\boldsymbol{A}_{\boldsymbol{KK}}$ $\det(\boldsymbol{A}_{\boldsymbol{KK}})$	k-order leading principal submatrix of $\boldsymbol{A}$. k-order leading principal minor matrix of $\boldsymbol{A}$. For simplification of notation, $\boldsymbol{A}_{\boldsymbol{KK}}$ may sometimes be expressed as $A(k)$ and $\boldsymbol{A}_{\boldsymbol{KK}}$ taken from the top left corner of $\boldsymbol{A}$ as $\boldsymbol{A}_{\boldsymbol{K}k\boldsymbol{K}k}(\boldsymbol{K}_k = \{1, \ldots, k\})$.

Index

© Springer Nature Singapore Pte Ltd. 2024

K. Nakayama, *Eigenvalue Problem and Nonlinear Programming Problem*, New Frontiers in Regional Science: Asian Perspectives 70,
https://doi.org/10.1007/978-981-99-2943-6

SPRINGER NATURE

GPSR Compliance

The European Union's (EU) General Product Safety Regulation (GPSR) is a set of rules that requires consumer products to be safe and our obligations to ensure this.

If you have any concerns about our products, you can contact us on ProductSafety@springernature.com

In case Publisher is established outside the EU, the EU authorized representative is:

Springer Nature Customer Service Center GmbH
Europaplatz 3
69115 Heidelberg, Germany

The manufacturer's authorised representative in the EU is Springer Nature Customer Service Centre GmbH, Europaplatz 3, 69115 Heidelberg, Germany. If you have any concerns regarding our products, please contact ProductSafety@springernature.com

Printed and bound by CPI Group (UK) Ltd, Croydon, CR0 4YY
01/10/2026
02225863-0001